Introduction to SPSS
Statistics in Psychology

For Version 19 and earlier

Fifth edition

Dennis Howitt Loughborough University
Duncan Cramer Loughborough University

Prentice Hall
is an imprint of

PEARSON

Harlow, England • London • New York • Boston • San Francisco • Toronto • Sydney • Singapore • Hong Kong
Tokyo • Seoul • Taipei • New Delhi • Cape Town • Madrid • Mexico City • Amsterdam • Munich • Paris • Milan

Pearson Education Limited
Edinburgh Gate
Harlow
Essex CM20 2JE
England

and Associated Companies throughout the world

Visit us on the World Wide Web at:
www.pearsoned.co.uk

First published 1997
Second edition published 2000
Second (revised) edition published 2002
Third edition published 2005
Fourth edition published 2008
Fifth edition published 2011

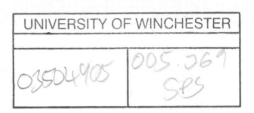
ISBN 978-0-273-73426-0

British Library Cataloguing-in-Publication Data
A catalogue record for this book is available from the British Library

Library of Congress Cataloging-in-Publication Data
Howitt, Dennis.
   Introduction to SPSS statistics in psychology : for version 19 and earlier / Dennis Howitt, Duncan Cramer. – 5th ed.
       p. cm.
   Rev. ed. of: Introduction to SPSS in psychology. 2008.
   ISBN 978-0-273-73426-0 (pbk.)
   1. Psychometrics.   2. SPSS for Windows.   3. Psychometrics–Computer programs.
I. Cramer, Duncan, 1948–   II. Howitt, Dennis. Introduction to SPSS in psychology.   III. Title.
   BF39.H718 2011
   150.1'5195–dc22

                                                                          2010053175

10   9   8   7   6   5   4   3   2   1
15   14   13   12   11

Typeset in 10/12pt Times by 35
Printed and bound by Rotolito Lombarda, Italy

Brief contents

THE FOLLOWING ADDITIONAL MATERIAL CAN BE FOUND ON THE WEBSITE
(www.pearsoned.co.uk/howitt)

Contents

Part 1 Introduction to SPSS Statistics   1

Part 3 Significance testing and basic inferential tests   125

Part 6 Advanced qualitative or nominal techniques   363

Part 8   Linear structural relationship (LISREL) analysis   469

THE FOLLOWING ADDITIONAL MATERIAL CAN BE FOUND ON THE WEBSITE
(www.pearsoned.co.uk/howitt)

Supporting resources

Visit www.pearsoned.co.uk/howitt to find valuable online resources

Companion Website for students
- Chapter overviews to introduce and give a feel for topics covered in the chapter
- Multiple choice questions to test your understanding
- Additional data sets with exercises for further practice and self testing
- A set of research scenarios and questions that enable you to check your understanding of when to use a particular statistical test or procedure
- An online glossary to explain key terms
- Interactive online flashcards that allow the reader to check definitions against the key terms during revision
- A guide to using Microsoft Excel to help you if doing statistical analyses using Excel
- A guide to statistical computations on the web for your reference
- Roadmaps to help you to select a test for analysis of data

For instructors
- PowerPoint slides that can be downloaded and used for presentations
- Testbank of question material

Also: The Companion Website provides the following features:
- Search tool to help locate specific items of content
- E-mail results and profile tools to send results of quizzes to instructors
- Online help and support to assist with website usage and troubleshooting

For more information please contact your local Pearson Education sales representative or visit **www.pearsoned.co.uk/howitt**

Guided tour

Background details

Outlines the background to the techniques discussed in each chapter to encourage a deeper understanding. This includes details of:

- What the technique is
- When you should use it
- When you should not use it
- The data required for the analysis
- Typical problems to be aware of

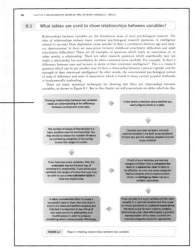

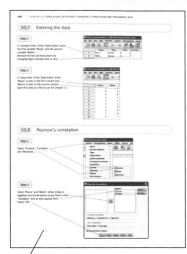

Step by step illustrations and screenshots of SPSS

This presents the stages of data entry, and data analysis visually to help you gain confidence in the processes and procedures of SPSS.

Interpreting the output

Offers a simple explanation of what the important parts of the output mean. SPSS statistical output is presented exactly as it appears on screen to help you become familiar with it.

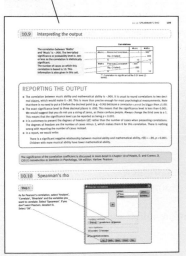

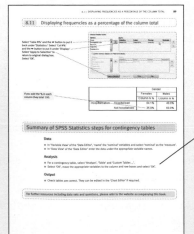

Summary of SPSS Statistics steps

Summarises the important steps so that you can easily see how to input, analyse and present data.

Homepage

Lists and links to the resources on the website.

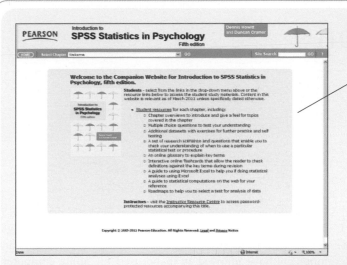

Multiple-choice questions

Allow you to check your knowledge and prepare for tests.

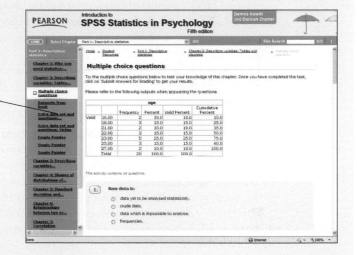

Data sets and exercises

Offer the opportunity to practise analyses with data sets.

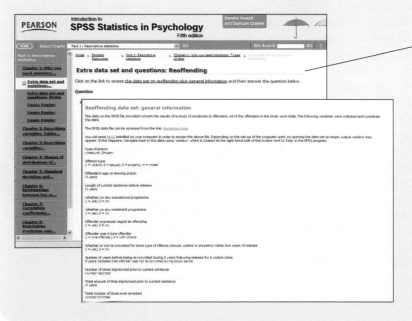

Introduction

Statistical Package for the Social Sciences (SPSS, an IBM company*) was initially developed in 1965 at Stanford University in California. Since then it has become the leading data analysis package in the field and available all over the world in universities and elsewhere. It dominates the field. Modern computing developments have allowed it to be used on home computers. Because of SPSS's popularity and universality, being able to use it is one of the most readily transferable of all research skills. Once learnt, SPSS can be used virtually anywhere in the world. SPSS is constantly being updated both in terms of the range of statistical techniques covered and the detail of the output.

This book is a stand-alone step-by-step approach to statistical analysis using SPSS for Windows and is applicable to Releases 10 to 19. It is suitable for students and researchers wishing to analyse psychological, sociological, criminological, health and similar data. SPSS does change with time but this book will also be helpful to those using earlier releases of SPSS as the changes which affect this book are generally hard to notice. Although the last three releases of SPSS have Statistics in the title and releases 17 and 18 were called PASW Statistics, we shall generally refer to all versions as SPSS unless we are speaking about particular versions, in which case we will give their release numbers. This is what is generally done by most users. The official name of the latest release at the time of publication is IBM® SPSS® Statistics 19 (Release 19.0.0).

This book updates the fourth edition of *Introduction to SPSS in Psychology* to cover recent changes in SPSS. Nevertheless, the structure provides the fastest possible access to computerised data analysis with SPSS even when using the most advanced techniques. Each statistical technique is carefully described step-by-step using screenshots of SPSS data analysis and output. The user's attention is focused directly on the screenshots, what each of them signifies, and why they are important. In other words, it is as close as is possible in a textbook to face-to-face individual instruction. Users with little or no previous computer skills will be able to quickly analyse quite complex data and appreciate what the output means.

The chapters have a common pattern. The computer steps (which keys to press) are given in exact sequence. However, this is not the end of any data analysis, and so there are also explanations of how to interpret and report the SPSS output. For this new edition we have added three new SPSS chapters on statistics techniques (simple mediational analysis, simultaneous multiple regression and interactions in multiple regression) and one on data transformation as well as including two further sections. The first is a section on structural equation modelling with the student version of LISREL (Linear Structural Relationships) which is freely downloadable from Scientific Software (http://www.ssicentral.com/lisrel/student.html). This method is becoming increasingly widely used and so some familiarity with it is useful. The last two chapters of this section have been put on the website. The second section introduces the topics of power analysis and meta-analysis also using software which is freely available on the Internet. The two chapters of this section have also been placed on the website.

---

* SPSS was acquired by IBM in October 2009.

The basic structure of the major chapters is as follows:

- A brief bulleted overview of the use of the procedure or statistical technique. This will be sufficient often to get a clear idea of where and when to use the techniques.

- A need-to-know account of what the technique is for and what needs to be known in preparation to doing the SPSS analysis. It also gives information about when the technique is used, when is should not be used, the data required for the analysis, and typical problems that we know from long experience that cause users difficulties.

- An illustrative example is given of the appropriate sorts of data for each statistical technique. These examples allow the user to work through our computations, and to gain confidence before moving on to their own data.

- Data entry for a particular statistical analysis is presented visually and explained in adjacent text.

- This is followed by a step-by-step, screenshot-by-screenshot description of how a particular statistical analysis is done using SPSS for Windows.

- The SPSS statistical output is included exactly as it appears on the monitor screen and in print-outs of the analysis. This is crucial – SPSS output can be confusing and unclear at first.

- The key features of the statistical output are highlighted on the output itself together with simple explanations of what the important parts of the output mean – SPSS output is infamous for its overinclusiveness.

- Suggestions are made on reporting the statistical findings in reports, theses and publications. These include samples of how to describe research findings and present tables clearly. The form followed is that recommended by the American Psychological Association (APA) which is also widely used by other publishers.

This book is based on the latest version of *SPSS Statistics for Windows* (that is, Release 19); but remains suitable for Releases 10 to 18 because of their similarity. Notes after this Introduction describe the main differences between these releases. Although SPSS is updated every year or so, usually there is little difficulty in adapting knowledge gained on the older versions to the new version.

Introduction to SPSS Statistics in Psychology is an excellent single source for data analysis. It is complete in itself and contains many features not available elsewhere. Unlike other SPSS books, it meets the needs of students and researchers at all levels. However, it is also part of a package of methodology books by the same authors designed to be comprehensive, authoritative and exhaustive. The three volumes in this series are closely tied to each other. The other two are:

- *Introduction to Statistics in Psychology* (2011) (5th edition) (Pearson Education: Harlow): This is a thorough introduction to statistics for all students. It consists of a basic introduction to key psychological statistics and, although maintaining its accessibility to students, it also covers many intermediate and advanced techniques in detail. It contains chapters on topics, such as meta-analysis, which are seldom covered in other statistics texts. Importantly, the structure of the statistics textbook is closely linked to this book. Thus, anyone following a chapter in the statistics book will, where appropriate, find an equivalent chapter in this book with details of how to do the analysis using SPSS. Similarly, anyone using this book will be able to find a detailed account of the technique in the statistics textbook.

- *Introduction to Research Methods in Psychology* (2011) (3rd edition) (Pearson Education: Harlow): This is a major textbook on research methods in psychology. It covers both quantitative and qualitative methods. There are major chapters on report writing, ethics in psychology and searching the literature. All aspects of experimental, field study, survey and questionnaire construction are covered, and guidance is given on qualitative data collection and analysis. There are numerous cross-references to this book and *Introduction to Statistics in Psychology*.

In other words, the three books offer a comprehensive introduction to conducting research in psychology. They may be used independently or in any combination.

Introduction to SPSS Statistics in Psychology can be used alongside virtually any statistics textbook to support a wide variety of statistics and practical courses. The range of statistical techniques covered is large and includes the simplest as well as the most important advanced statistical techniques. The variety of techniques described and the relative ease of using SPSS Statistics for Windows ensure that this guide can be used at introductory, intermediate and advanced levels of statistics teaching. The structure of the book is such that statistical procedures are described more or less in order of conceptual difficulty. Generally speaking, computing with SPSS is as easy for advanced statistical techniques as it is for simple ones.

Chapter 2 is essential reading, as it explains data entry and basic computer operating. However, the subsequent chapters can be used on a stand-alone basis if desired. Users with insufficient time to work chapter-by-chapter through the guide should find enough detail in the relevant chapters to complete an SPSS analysis successfully. Table 1.4, at the end of Chapter 1, states which chapter is most appropriate for which purpose, thereby enabling the reader to move directly to that part of the book.

Those who work steadily through the book will profit by doing so. They will have a much better overview of SPSS computing procedures. For most readers, this is possible in a matter of hours, especially if they have prior knowledge of statistics.

SPSS has an extensive catalogue of statistical procedures – far more than could be included. We have selected those suitable for most purposes when the range of possibilities is likely to confuse the average reader. The quickness and ease of SPSS mean that more advanced users can explore the alternatives by using the menus and dialog boxes. Most users will find our coverage more than sufficient.

The data and statistical analyses carried out in this book correspond almost always to those in the authors' accompanying statistics text, *Introduction to Statistics in Psychology* (2011) (5th edition) (Pearson Education: Harlow). This book is referred to as *ISP*, followed by the corresponding chapter or table number.

Dennis Howitt
Duncan Cramer

Acknowledgements

Author acknowledgements

One of the enjoyable things about writing a book like this is the formidable team of people that Pearson Education provides to do all of the stuff that we do not have a clue about. This is the first full-colour version of this book. This was entirely the idea of Janey Webb, our Editor at Pearson. This was the stimulus to a number of new features in the book but we should not forget the tremendous influence that Janey has had in almost every respect. She is remarkably determined to get the best possible product onto the market. Thanks also to Catherine Morrissey, Assistant Editor who has now moved on to a new role. During the production of this edition, it has been good to be able to welcome Jane Lawes as Janey's new assistant. The look of the book is down to the remarkable design work of Kevin Ancient on the text and Nicola Woowat on the cover design.

We were lucky to have Anita Atkinson in charge of the production of the book since she can do things that simply can't be done – without showing any signs of the frustration that we must cause her. Thanks also to Caterina Pellegrino, the Production Controller responsible for keeping this complicated project on track. Ros Woodward was the copy editor who made the design work and pieced together what is one of the most complex manuscripts imaginable into the coherent book that you have in your hands while dealing with us with the utmost patience. The proof reader was Jonathan Price who provided a final safe pair of hands for which we are most grateful. It is easy to overlook the boffins who work on the Website. So we would like to thank Louise Hammond for this work. Also Daniel Rhind for his insightful contributions to the content of the web pages.

Finally, we are indebted to a number of academic colleagues for providing us with comments and suggestions for improving this new edition. They are not unsung heroes but:

Andy Bell, Manchester Metropolitan University

Dr William M. Brown, University of East London.

Dr Jane Hutchinson, University of Central Lancashire

Dr Stephen Jones, University of the West of England

Professor Alan Pickering, Goldsmiths, University of London

Professor Ronnie Wilson, University of Ulster

Dennis Howitt and Duncan Cramer

■ Publisher acknowledgements

We are grateful to the following for permission to reproduce copyright material:

Screenshots
Screenshots on pages 18 middle and bottom, page 19, page 20 bottom, page 21, page 22, page 23, page 24, page 25, page 26, page 27, page 28, page 29, page 36, page 37, page 38, page 44, page 45, page 46, page 47, page 48, page 49, page 50, page 51, page 52, page 53, page 59, page 60, page 69, page 71, page 77, page 78, page 79, page 85, page 86, page 87, page 88, page 89, page 95, page 96, page 97, page 98 top, page 99, page 100, page 108, page 109, page 110, page 111, page 117, page 118, page 119, page 120, page 121, page 130, page 131, page 138, page 139, page 146, page 147, page 159, page 160, page 161, page 165, page 166, page 170, page 171, page 172, page 178, page 179, page 181, page 182, page 183, page 188, page 189, page 191, page 192, page 200, page 201, page 206, page 207, page 213, page 214, page 215, page 224, page 225, page 228, page 229, page 234, page 235, page 241, page 242, page 243, page 250, page 251, page 252, page 253, page 261, page 262, page 270, page 271, page 272, page 282, page 283, page 290, page 291, page 292, page 301, page 302, page 303, page 304, page 305, page 312, page 313, page 314, page 324, page 325, page 335, page 336, page 337, page 344, page 345, page 346, page 354, page 355, page 356, page 357, page 358, page 359, page 360, page 369, page 370, page 379, page 380, page 381, page 394, page 395, page 396, page 405 bottom, page 406, page 407, page 408, page 411, page 412, page 413, page 418, page 419, page 420, page 421, page 422, page 425, page 426, page 427, page 428, page 429, page 433, page 434, page 435, page 436, page 440, page 441, page 442, page 443, page 446, page 447, page 448, page 449, page 451, page 452, page 453, page 457, page 461, page 462 bottom, page 463, page 464, page 465 middle and bottom, page 466 and page 467 reprinted courtesy of International Business Machines Corporation, © SPSS Inc., an IBM Company. SPSS was acquired by IBM in October 2009; Screenshots on page 560 and page 561 from G*Power 3, with permission from the authors; Screenshots on page 569, page 570 and page 571 from The Meta-Analysis Calculator, http://www.lyonsmorris.com/ma1/index.cfm with permission from Larry C. Lyons; Microsoft product screenshots and screenshot frames reprinted with permission from Microsoft Corporation.

LISREL 8.80 for Windows screenshots are copyright Scientific Software, Inc. (SSI). LISREL, BILOG, TESTFACT, MULTILOG and MULTIVARIANCE are registered trademarks of SSI.

In some instances we have been unable to trace the owners of copyright material, and we would appreciate any information that would enable us to do so.

Key differences between IBM SPSS Statistics 19 and earlier versions

PASW Statistics 18

There seem to be very few differences between 19 and 18 and between 18 and 17 for the procedures described in this book.

PASW Statistics 17

This version of SPSS was called PASW which stands for Predictive Analytic Software. Otherwise, there appear to be very few differences between 17 and 16 for the procedures described in this book. Data Reduction which includes Factor Analysis is now called Dimension Reduction. Right clicking on the keys in the 'Select If: If' box no longer gives a description of what the keys do.

SPSS 16

There seem to be very few differences between 16 and 15 for the procedures described in this book. Basic tables are no longer available in the Tables procedure so the tables in Chapter 8 are produced with Custom Tables. In the dialog boxes the OK, Paste, Reset, Cancel and Help options are on the bottom of the box rather than on the right hand side while the analyses options are on the right-hand side rather than at the bottom of the box.

SPSS 15

There also appear to be very few differences between 15 and 14. The 'Graph' menu in 14 displays all available options. The procedure for 'Chart Builder' is described in Chapter 4 of this book as it was not available in releases before 14. An alternative procedure is shown in Chapters 6, 9 and 11.

SPSS 14

Similarly there seem to be very few differences between 14 and 13. Release 13 does not have 'Chart Builder' which was introduced in 14. In 14 the Properties dialog box of the 'Chart Editor' has separate boxes for 'Text Style' and 'Text Layout'.

SPSS 13

The major differences between 13 and 12 are to 'Compute Variable …', 'Scatter/Dot …' and the 'Chart Editor'. Also the plots in the output of 13 are shaded.

In 12 the 'Compute Variable' dialog box has a single 'Functions' menu from which options can be chosen. 'Scatter/Dot …' is called 'Scatter …', the 'Scatter/Dot' dialog box is called 'Scatterplot' and there is no 'Dot' option.

In 12, to label the slices of a pie diagram and add the percentages of cases in each, double click anywhere in the 'Chart Editor', double click on the pie diagram (to open the 'Properties' dialog box), select 'Data Value Labels' (in the 'Properties' dialog box), select 'Count' in the 'Contents' box, select the red '✕' (to put 'Count' in the 'Available' box), select the variable name (e.g. 'Occupation'), select the curved upward arrow (to put 'Occupation' in the 'Contents' box), select 'Percent' and the curved upward arrow (to put 'Percent' in the 'Contents' box), select 'Apply' and then 'Close'.

To fit a regression line to a scatterplot, click on a dot in the chart of the 'Chart Editor' so that the circles in the plot become highlighted, select 'Chart', select 'Add Chart Element', select 'Fit Line at Total' (which opens the 'Properties' dialog box). Assuming that the 'Fit Line' tab is active, select 'Linear' (this is usually the default) and then 'Close'.

SPSS 12

The major differences between 12 and 11 also apply to 10. They are relatively few. In 11 and 10 variable names cannot begin with a capital letter and are restricted to eight characters. The 'Data' and 'Transform' options are not available in the 'Viewer' or 'Output' window. Some output, such as partial correlation and reliability, is not organised into tables. The 'Chart Editor' works differently. To fit a regression line to a scatterplot, double click anywhere in the scatterplot to open the 'Chart Editor', select 'Chart', select 'Options …' (which opens the 'Scatterplot Options' dialog box), select 'Total' under 'Fit Line' and then 'OK'.

PART 1

Introduction to SPSS Statistics

CHAPTER 1

A brief introduction to statistics

Overview

- Unfortunately, there is no gain without some effort in statistics. There are a number of statistical concepts which need to be understood in order to carry out a decent statistical analysis. Each of these is discussed and explained in this chapter.

- Key ideas covered in this chapter include score variables versus nominal (category) variables, unrelated versus related designs, descriptive versus inferential statistics, and significance testing. With a knowledge of each of these it is possible to quickly develop a working knowledge of statistical analysis using SPSS Statistics.*

- The appropriate statistical analysis very much depends on the nature of the research design employed.

- The chapter provides detailed advice on how to select a statistical technique for the analysis of psychological data.

1.1 Basic statistical concepts essential in SPSS Statistics analyses

The elements of statistics are quite simple. The problem is in putting the elements together. Nobody can become expert in statistical analysis overnight but, with a very small amount of knowledge, quite sophisticated analyses can be carried out by inexperienced researchers. Mathematical ability has very little role to play in data analysis. Much more important is that the researcher understands some of the basic principles of research design. There are close links between different research designs and what is appropriate in terms of statistical analysis methods. At the most basic level, there are two broad classes of research design – the comparative and the correlational designs – though these have any number of variants. The type of research design involved in the study lays down broadly the sort of statistical tests, etc. which are needed for the analysis of the data from that study. Of course, sometimes the personal preferences of the researcher play a part since, quite often, there are several ways of achieving much the same ends.

---

* SPSS was acquired by IBM in October 2009.

Before we can discuss research designs, there are two basic concepts we need to understand as they are part of the jargon of statistics and SPSS Statistics:

● *Variable*   A variable is any concept that can be measured and which varies. Variables are largely inventions of the researcher and they vary enormously from study to study. There are a few fairly standard variables, such as age and gender, that are very commonly measured. Typically, however, the variables used tend to be specific to particular topics of study. Variables are the ways in which psychologists attempt to measure the concepts that they use in their research – a variable, generally, cannot perfectly measure a concept and so is an approximation to the concept. For this reason, it is important to understand that data and theory do not always map closely one on the other.

● *Cases*   A case is simply a member of the sample. In psychology a case is usually a person (i.e. an individual participant in the research). Cases are very much SPSS Statistics jargon and it is a wider and more embracing term than the participants which psychologist talk about.

Variables normally appear in SPSS analyses as the columns of the data spreadsheet. Cases (normally) appear in SPSS analyses as the rows of the data spreadsheet. In other words, variables and cases can be set out in the form of a matrix – the size of which depends on the number of variables and cases involved.

1.2   Basic research designs: comparative versus correlational designs

■ Comparative designs

The basic comparative design compares the typical or average score of a group of participants with that of another group. This might involve comparing a group of men with a group of women or comparing an experimental group with a control group in an experimental study. This design is illustrated in Table 1.1. Usually, in this sort of design, the comparison is between the average score for one group and the average score in the other group. Usually what most people refer to as the average is called by statisticians the mean. So the design can be used to assess whether, say,

Table 1.1	A simple comparative design such as an experiment		
Participant (case)	**Group A (e.g. experimental group)**	**Participant (case)**	**Group B (e.g. control group)**
1	13	11	5
2	12	12	8
3	10	13	6
4	7	14	9
5	5	15	3
6	9	16	6
7	5	17	5
8	14	18	4
9	12		
10	16		
Mean =	10.30	Mean =	5.75

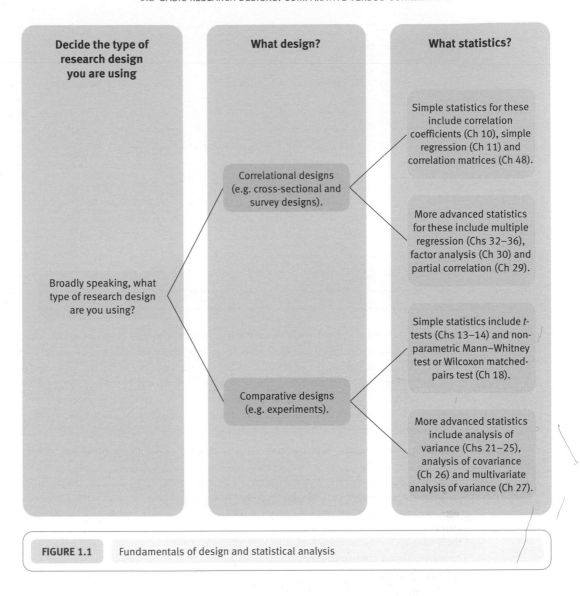

FIGURE 1.1   Fundamentals of design and statistical analysis

the average time taken by males getting ready for a first date is different from the average time taken by females.

This is the basic version of a whole range of statistical procedures which compare the average scores in different group in more complex research designs. The analysis of variance (ANOVA) involves a whole family of different research designs based on this basic principle. Look at Figure 1.1 for more information.

■ The correlational design

The basic correlational design is one in which the researcher measures several different things at the same time using a single group of participants. These things which are measured might be gender, age, IQ, extraversion and dogmatism. This basic correlational design is illustrated in Table 1.2.

The statistical analysis of this sort of design is usually based on the correlation coefficient or some other closely related statistical procedure based on the correlation coefficient. A correlation coefficient is a numerical index of the relationship between two measures. The data from a correlational design may be analysed using a variety of statistics as can be seen in Figure 1.1.

Correlational designs are sometimes called cross-sectional studies. They can be more complex, for example, when the researcher adds a time (temporal) dimension to the research design. There

Table 1.2	The basic correlational design				
Participant	**Gender**	**Age**	**IQ**	**Extraversion**	**Dogmatism**
1	Female	26	110	15	9
2	Male	31	130	19	6
3	Female	25	160	22	4
4	Female	22	110	34	8
5	Male	33	170	12	3
6	Female	28	140	17	7
7	Male	29	90	16	6
8	Male	34	130	22	5
9	Female	23	80	26	4
10	Male	27	70	11	2

are special statistics to deal with these more complex designs (e.g. causal modelling such as linear structural relationship, LISREL), of course, but these are essentially correlational in nature.

It would be misleading to pretend that the above covers every available statistical technique but a surprising range of statistics can be better understood if the underlying research design is clear to the researcher. Also remember that statistics is a mature discipline in its own right so it is unrealistic to assume that there are shortcuts to mastery of statistics in psychology. Getting basic concepts clear goes a long way towards this mastery, as does some experience.

1.3   The different types of variable in statistics

One's ability to use statistics in a practical context will be made much easier if some basic facts are learnt about the fundamental different types of variables in statistics. Different types of variable require different types of versions of statistical techniques for their analysis. So there are two basic questions that need to be asked:

- What types of variable have I got?

- What statistical tests analyse the variables in the way that I want?

Fortunately, there are just two main types of data so this is relatively straightforward. On the other hand there are many different statistical tests and techniques. Of course, the way to learn about each of these is to gain some experience trying each of them out by working through the chapters which follow in this book. Most of the chapters in this book cover just one statistical technique or test in each chapter. The important thing is that each chapter tells you exactly what sorts of data (variables) are appropriate for that test or technique – and then how to do the analysis using a computer.

■ Types of variable

For all practical purposes, variables can be classified as being of *two* types (see Figure 1.2):

- *Score variables* Some variables are scores. A score is when a numerical value is given to a variable for each case in the sample. This numerical value indicates the quantity or amount of the characteristic (variable) in question. So age is a score variable since the numerical value indicates an increasing amount of the variable age. One could also describe this variable as quantitative.

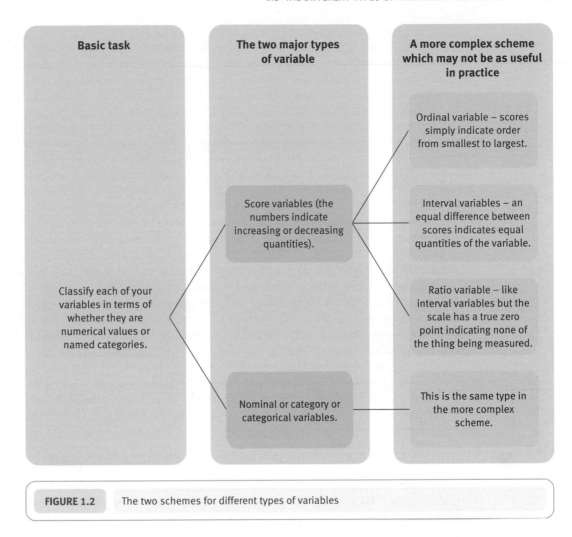

Basic task	The two major types of variable	A more complex scheme which may not be as useful in practice
Classify each of your variables in terms of whether they are numerical values or named categories.	Score variables (the numbers indicate increasing or decreasing quantities).	Ordinal variable – scores simply indicate order from smallest to largest.
		Interval variables – an equal difference between scores indicates equal quantities of the variable.
		Ratio variable – like interval variables but the scale has a true zero point indicating none of the thing being measured.
	Nominal or category or categorical variables.	This is the same type in the more complex scheme.

FIGURE 1.2   The two schemes for different types of variables

- *Nominal, category or categorical variables*   Some variables are measured by classifying cases into one of several named categories. These are also known as nominal, categorical or category variables. A better name for them might be qualitative variables because they measure the qualities of things rather than their quantities. For example, gender has two named categories – male and female. Nationality is another example: English, Welsh, Irish and Scottish are the nationalities of people of the United Kingdom. They have *no* numerical implications as such. To say that a person is Scottish is simply to put them into that named category. One could also describe this variable as qualitative. There is one risk of confusion – categories such as gender are usually entered into SPSS Statistics using different numbers to represent the different categories. For example, the variable 'gender' has two categories – males could be represented by the number 1 and females by the number 2 (or vice versa). The numbers used are arbitrary – it could be 1002 and 2005 if the researcher desired. It is vital not to confuse these numbers which merely represent different coding categories or qualities with scores. For this reason, it is important to label the different values of nominal variables in full in the SPSS data spreadsheet since the number codes, in themselves, mean nothing. This is easily done as is shown on pages 24–25.

The alternative traditional classification system

Sometimes variables are classified as nominal, ordinal, interval and ratio. This is mainly of conceptual interest and of little practical significance in selecting appropriate statistics. Generally speaking, we would advise that this system is ignored because it does not correspond with modern practice. *Nominal* is exactly the same as our classification of nominal (category) data and is important since a particular set of statistical techniques is called for to analyse category data. Of the other three, interval measurement is the most important. *Interval* measurement is where the steps on the scale of measurement are equal (just as centimetre steps on a rule are equal). Some psychologists are inclined to the view that this scale of measurement should reflect the underlying psychological variable being measured. Unfortunately, it is very difficult to identify whether a psychological measure has equal intervals but, nevertheless, it is a sort of holy grail to them. Others, ourselves included, take the view that so long as the numerical scale on which variables are measured has equal intervals (which is always the case except for nominal or category data, of course, from this perspective) then there is no problem since it is these numbers on which the statistical calculation is based and not some mystical underlying measurement scale. However, as a concession, we have mentioned equality of intervals as being desirable from time to time in the text though we (and no one else) can tell you how to establish this. *Ratio* measures have equal intervals and a zero point, which means one can calculate ratios and make statements such as one score is twice as big as another score. Unfortunately, yet again, it is impossible to identify any psychological variables which definitely are measured on a ratio measurement scale. Finally, *ordinal* data are data which do not have equal intervals so that scores only give the rank order of scores. Since the size of the intervals do not matter for ordinal data, then it is assumed that any psychological score data correspond to the ordinal measurement scale at a minimum. For this reason, some psychologists have advocated the use of non-parametric (distribution-free) statistics for the analysis of much psychological data. The problem is that these techniques are not so powerful or flexible as most statistics.

> You will find an extended discussion of ordinal, interval and ratio scales of measurement in Howitt, D. and Cramer, D. (2011). *Introduction to Research Methods in Psychology*, 3rd edition. Harlow: Pearson.

■ Importance of deciding the types of variable involved

It is essential to decide for each of your variables whether it is a nominal (category) variable or a score variable. Write a list of your variables and classify each of them if you are a beginner. Eventually you will be able to classify somewhat automatically and usually without much thought. The statistical techniques which are appropriate for score variables are generally inappropriate for nominal or category variables because they measure qualities. So, for example, it is appropriate to calculate the mean (numerical average) of any variable which is a score (e.g. average age). On the other hand, it is totally inappropriate to calculate the mean (average) for variables which consist of categories. It would be nonsense to say that the average nationality is 1.7 since nationality is not a score. The problem is that SPSS works with the numbers in the data spreadsheet and does not know whether they are scores or numerical codes for different categories. (Though SPSS does allow you to classify your variables as ordinal or nominal.)

1.4   Descriptive and inferential statistics compared

■ The difference between descriptive and inferential statistics

There are two main types of statistical techniques – *descriptive* and *inferential* statistics:

- Descriptive statistics chiefly describe the main features of individual variables: calculating the average age of a sample of people is an example of descriptive statistics. Counting the number of English people would be another example of descriptive statistics. If one variable is considered at a time this is known as univariate statistics. Bivariate statistics are used when the relationship between two variables is being described.

- Inferential statistics is a totally distinct aspect of statistics. It only addresses the question of whether one can rely on the findings based on a *sample* of cases rather than *all* cases. The use of samples is characteristic of nearly all modern research. The problem with samples is that some of them are not similar to the populations from which they are taken. The phrases 'statistically significant' and 'not statistically significant' simply indicate that any trends in the data can be accepted as substantial (i.e. statistically significant) or not substantial enough to rely on (i.e. not statistically significant). A statistically significant finding is one which is unlikely to be the result of chance factors determining the results in a particular sample.

Statistical significance is difficult to explain accurately in a few words. For a detailed discussion, see Box 1.1 in this chapter or Howitt, D. and Cramer, D. (2011). *Introduction to Statistics in Psychology*, 5th edition. Harlow: Pearson.

Box 1.1   Statistical Significance

A crucial fact about research is that it is invariably carried out on samples of cases rather than all possible cases. The reasons for this are obvious: economies of time and money. Sometimes it is very difficult to specify what the population is in words (for example, when one collects a sample of participants from a university restaurant). The realisation that research could be done on relatively small samples was the initial stimulus for many of the statistical techniques described in this book.

For many, statistics = tests of significance. This is a mistaken emphasis since, in terms of importance in any analysis, basic descriptive statistics are the key to understanding one's data. Statistical significance is about a very limited question – is it reasonably safe to generalise from my sample?

In order to do this, in statistics usually we take the information based on the sample(s) of data that we have collected and generalise to the population from which we might assume the sample is drawn. Sometimes one simply takes the sample characteristics and assumes that the population characteristics are the same. On other occasions the sample characteristics have to be slightly modified to obtain the best estimate of the population characteristics. Whichever applies, one then uses these estimated population characteristics to plot the distribution of the characteristics of random samples taken from the estimated population. The most important characteristic that is estimated from samples is the variation of scores in the data.

The distribution of these random samples forms a baseline against which the characteristic of our sample obtained in our research can be compared with what happens under conditions of randomness. If our actual sample characteristic is unlikely to have occurred as a consequence of randomness then we say that it is statistically

significant. All that we mean is that it is at the extremes of the distribution of random samples. If our sample is very typical of what happens by random sampling, then we say that our sample characteristic is not statistically significant.

Often in psychology this is put in terms of accepting the null hypothesis or rejecting the null hypothesis. The null hypothesis is basically that there is no relationship or no difference in our data. Usually the population is specified in terms of the null hypothesis. That is, in our population there is no correlation or in our population this is no difference.

Statistical significance is often set at the .05 or 5 per cent significance level. This is purely arbitrary and is not actually set in stone. Sometimes one would require a more stringent significance level (say .01 or 1 per cent) and in some circumstances one could relax the criterion a little. However, unless you are very experienced you perhaps ought to stick with the .05 or 5 per cent levels of statistical significance. These levels of significance simply mean that there is a one in 20 chance of getting a result as extreme as ours by random sampling from the estimated population.

One-tailed significance testing is used when the direction of the trend in the data is predicted on the basis of strong theory or consistent previous research. The prediction is made prior to collecting the data. Such conditions are rarely met in student research and it is recommended that you stick to two-tailed testing.

Finally, notice that the precision of this approach is affected by how representative the sample characteristics are of the population characteristics. One cannot know this, of course. This may help you understand that despite the mathematical sophistication of statistics, in fact it should be used as a guide rather than a seal of approval.

Every descriptive statistic has a corresponding inferential statistic. For example, the correlation coefficient is a descriptive statistic indicating the direction and the strength of the relationship between two variables. Associated with it is the inferential statistic – the significance of the correlation coefficient. The descriptive statistic is important for understanding the trends in the data – the inferential statistic simply deals with the reliance that can be placed on the finding.

1.5   Related versus unrelated designs

Researchers should be aware also of *two* different types of research design – that which uses *related measures* and that which uses *unrelated measures*. Related measures may also be called correlated measures or paired measures. Unrelated measures may also be called uncorrelated measures or unpaired measures. The terms are mostly used when the mean or averages of scores are being compared for two or more samples of data:

- Where the means of a single sample of individuals are compared on two (or more) measures of the same variable (e.g. taken at different points in time) then this is a related measures design.

- Where the means of two quite different samples of participants are compared on a variable, this is an unrelated design.

- Where two (or more) groups of participants have been carefully matched so that *sets* of participants in the two (or more) conditions are similar in some respects, then this is a related design too. In this case, members of each set are treated as if they were the same person. Normally, a researcher would know if the participants were matched in sets because it requires effort on the part of the researcher. For example, the researcher has to decide what characteristics to match sets on, then choose individuals for the sets on the basis of their similarity on these characteristics, and (often) has to allocate participants to the different samples (conditions) especially in experimental research.

The main point of using related designs is that variability due to sampling is reduced.

Almost without exception, the researcher will be using a variety of these techniques with the same data. Fortunately, once the data are entered, in many cases, data analysis may take just a minute or so. Note that in the related design there is always an equal number of cases (participants) and that each case contributes a score to more than one condition.

Table 1.3	The related and the unrelated research design	

The related design

Participant	Condition 1 (e.g. Time 1)	Condition 2 (e.g. Time 2)
Sarah	1	6
Callam	6	10
Dominic	5	11
Tracey	4	9
Kwame	7	12
Imogen	6	6
Claude	3	9
Means =	32/7 = 4.57	63/7 = 9.00

The unrelated design

Participant	Group 1	Participant	Group 2
1	6	7	9
2	4	8	12
3	8	9	14
4	5	10	9
5	2	11	7
6	3		
Mean =	28/6 = 4.67	Mean =	51/5 = 10.2

1.6 Quick summaries of statistical analyses

■ Succinct methods of reporting statistics

You will probably be aware of the short (succinct) methods used to report statistical findings in psychological research reports. We have, of course, used these in this book where we describe how to report research findings. The system is quite simple, though there is no universal standard and there will be some variation. Typical of the methods used would be the following, which follows the recommendations of the *Publication Manual of the American Psychological Association*:

> The hypothesis that students who just work and have no play was supported, $t(22) = 2.10$, $p < .05$.

What the stuff after the first comma says is that the statistical test used was the *t*-test, that the degrees of freedom are 22, and that the significance level is less than .05. In other words, our findings are statistically significant.

■ Confidence intervals versus point statistics

Traditionally psychological statistics use point statistics. These are where the characteristics of the data are defined by a single measure such as the average (mean) score. It is often recommended that confidence intervals should be used instead. Usually in statistics we are trying to estimate the characteristics of the population of scores from a sample of scores. Obviously samples tend to vary from the population characteristics so there is some uncertainty about what the population characteristic will be. Confidence intervals give the most likely range of values in the population and not simply a single value. In this way, the variability of the data is better represented. Chapter 15 discusses confidence intervals in more detail.

1.7 Which procedure or test to use

One common heartfelt plea is the demand to know how to choose appropriate statistical techniques for data. Over the years, writers of statistics textbooks have laboured to simplify the process of choosing. This is done largely by producing spreadsheets or tables which indicate what sorts of statistics are appropriate for different sorts of data. If you want that sort of approach then there are a number of websites which take you through the decision-making process:

This book explains what sort of design and data are needed for each of the statistical techniques discussed.

The basic task is to match the characteristics of your data and research design with a test or technique designed to deal with that situation.

Research designs may be broken down into two broad types.

It is also important to know whether each of your variables is a score or if it is a nominal category variable.

The appropriate statistical technique to use is partly determined by the type of design used. Each chapter of this book makes clear what type of design is involved for a particular statistical technique.

(a) comparative designs which compare group means and (b) correlational designs which investigate the relationships between different measures (variables).

Although psychologists generally collect data in the form of scores, sometimes they use nominal category data.

Once again, the choice of statistical techniques is dependent on this decision. Each chapter in this book makes it clear just what sorts of data should be used.

Of course, there may be more than one way of analysing any particular research study.

Descriptive methods are used to describe and summarise the data that the researcher has collected. They are very useful to help the researcher understand what is going on in their data.

Statistics also breaks down into (a) descriptive methods and (b) inferential methods.

There are some statistics which answer specific sorts of questions and others which address other sorts of research question. Study the following chapters to find out what is good for what.

Inferential statistics tend to take precedence in the eyes of novices though this is a misunderstanding.

Inferential statistics merely tell you whether your findings are unlikely to have occurred by chance if, in reality, there is no trend.

Inferential statistics do not tell you about any other aspect of the importance or significance of your finding.

Used wisely, this will help you carry out the task of data analysis thoughtfully and intelligently.

This book provides you with the important information about how, when and where to use a full range of statistical tests and techniques.

It is important to explore and think about your data and not simply apply statistical analyses without considering how they help you address your research question.

If you have time, you may find it helpful to work through the following chapters sequentially analysing the data we have provided using SPSS Statistics.

There is a lot to be learned about handling data using SPSS Statistics and the book will enable you to do this to a professional standard.

FIGURE 1.3   Steps towards a basic understanding of statistics

- http://www.socialresearchmethods.net/selstat/ssstart.htm
- http://www.graphpad.com/welcome.htm
- http://www.whichtest.info/

For basic statistics this is probably a useful approach. The difficulty increasingly is that research designs, even for student projects, are very varied and quite complex. Once psychology was almost a purely laboratory-based subject which concentrated on randomised experiments. Psychologists still use this sort of experimentation, but their methods have extended greatly, so extending the demands on their statistical knowledge. Therefore, there is a distinct limit to the extent to which a simple spreadsheet or flow diagram can help the researcher select appropriate statistical analyses.

One fundamental mistake that novice researchers make is to assume that data analysis is primarily driven by statistics. It is more accurate to regard statistics as being largely a tool which adds a little finesse to the basic task of research – to answer the researcher's research questions. Only the researcher can fully know what they want their research to achieve – what issues they want resolving through collecting and analysing research data. Unless the researcher clearly understands what they want the research to achieve, statistics can be of little help. Very often when approached for statistical advice we find that we have to clarify the objectives of the research first of all – and then try to unravel how the researcher thought that the data collected would help them. These are *not* statistical matters but issues to do with developing research ideas and planning appropriate data collection. So the first thing is to list the questions that the data were intended to answer. Too often sight of the purpose of the research is lost in the forest of the research practicalities. The following may help clarify the role of statistics in research:

- Much of the most important aspects of data analysis need little other than an understanding of averages and frequency counts. These are common in SPSS output. Many research questions may be answered simply by examining differences in means between samples or cross-tabulation tables or scattergrams. It is useful to ask oneself how one could answer the research questions just using such basic approaches. Too often, the complexities of statistical output become the focus of attention which can lead to confusion about how the data relate to the research question. It is not easy to focus on the research issues and avoid being drawn in unhelpful directions.

- Statistical analyses are actively constructed by the researcher. There is usually no single correct statistical analysis for any data but probably a range of equally acceptable alternatives. The researcher may need to make many decisions in the process of carrying out data analysis – some of these may have to be carefully justified but others are fairly arbitrary. The researcher is in charge of the data analysis – statistics is the researcher's tool. The analysis should not be a statistical tour de force, but led by the questions which necessitated data collection in the first place. There is no excuse – if you collected the data then you ought to know why you collected it.

The more research you read in its entirety the better you will understand how statistics can be used in a particular field of research. Very little research is carried out which is not related to other research. What are the typical statistical methods used by researchers in your chosen field? Knowing what techniques are generally used is often the best guide to what ought to be considered.

Figure 1.3 summarises some of the major steps in developing a basic understanding of statistics while Table 1.4 gives some insight into the styles of analysis which researchers may wish to apply to their data and what sections of this book describe these statistical techniques in detail.

For further resources including data sets and questions, please refer to the website accompanying this book.

Table 1.4	Major types of analysis and suggested SPSS procedures	
Type/purpose of analysis	**Suggested procedures**	**Chapter**
All types of study	Descriptive statistics, tables and diagrams	3–9
Assessing the relationship between two variables	Correlation coefficient	10
	Regression	11
Comparing two sets of scores for differences	Unrelated *t*-test	14
	F-ratio test	20
	Related *t*-test	13
	Unrelated ANOVA	21
	Related ANOVA	22
	Mann–Whitney	18
	Wilcoxon matched pairs	18
Comparing the means of three or more sets of scores	Unrelated ANOVA	21
	Related ANOVA	22
	Multiple comparisons	24
Computing groups for a multiple comparison of an interaction	Computing a new group variable	45
Comparing the means of two or more sets of scores (ANOVAs) while controlling for spurious variables influencing the data	ANCOVA	26
Complex experiments, etc. with *two* or more unrelated independent variables and *one* dependent variable	Two (or more)-way ANOVA	23
– if you have related *and* unrelated measures	Mixed-design ANOVA	25
– if other variables may be affecting scores on the dependent variable	Analysis of covariance	26
ANOVA designs with several conceptually relevant dependent variables	MANOVA	27
Eliminating third variables which may be affecting a correlation coefficient	Partial correlation	29
Finding predictors for a score variable	Simple regression	11
	Stepwise multiple regression	32
	Hierarchical multiple regression	33
	Simultaneous multiple regression	34
Testing interaction or moderator effects for continuous predictors of a score variable	Hierarchical multiple regression	35
Testing a simple path diagram without measurement error	Structural equation modelling	52
Testing a simple path diagram with measurement error	Structural equation modelling	53, 54*
Finding predictors for a category variable	Multinomial logistic regression	38
	Binomial logistic regression	39
Comparing frequency data for one or two unrelated variables	Chi-square	16
Comparing frequency data for two unrelated variables with some low expected frequencies	Fisher's test	16
Comparing frequency data for a related variable	McNemar's test	17
Comparing non-normally distributed data for three or more groups	Kruskal–Wallis	19
	Friedman	19

Type/purpose of analysis	Suggested procedures	Chapter
Comparing frequency data for three or more unrelated variables	Log-linear analysis	37
Analysing a questionnaire	Exploratory factor analysis	30
	Confirmatory factor analysis	51
	Alpha reliability	31
	Split-half reliability	31
	Recoding	42
	Computing a scale score	43, 44
Coding open-ended data using raters	Kappa coefficient	31
Determining sample size	Power analysis	55*
Averaging effect sizes	Meta-analysis	56*

*On the website.

CHAPTER 2

Basics of SPSS Statistics data entry and statistical analysis

Overview

- This chapter gives the basics of operating SPSS Statistics on a personal computer. It includes data entry as well as saving files. There will be small variations in how SPSS is accessed from location to location. The basics are fairly obvious and quickly learnt. By following our instructions, you will quickly become familiar with the essential steps in conducting a statistical analysis.

- The chapter provides detailed advice on how to select a statistical technique for the analysis of psychological data.

2.1 What is SPSS Statistics?

SPSS Releases 19, 18, 17, 16, 15, 14, 13 and 12 are commonly available on university and college computers. Individuals may still be using earlier versions such as Releases 11 and 10. It is by far the most widely used computer package for statistical analysis throughout the world. As such, learning to use SPSS is a transferable skill which is often a valuable asset in the job market. The program is used at all levels from students to specialist researchers, and in a great many academic fields and practical settings. One big advantage is that once the basics are learnt, SPSS is just as easy to use for simple analyses as for complex ones. The purpose of this *Introduction* is to enable beginners quickly to take advantage of the facilities of SPSS.

Most people nowadays are familiar with the basic operation of personal computers (PCs). The total novice though will not be at too much of a disadvantage since the elements of SPSS are quickly learnt. Users who have familiarity with, say, word processing will find much of this relevant to using SPSS – opening programs, opening files and saving files, for instance. Do not be afraid to experiment.

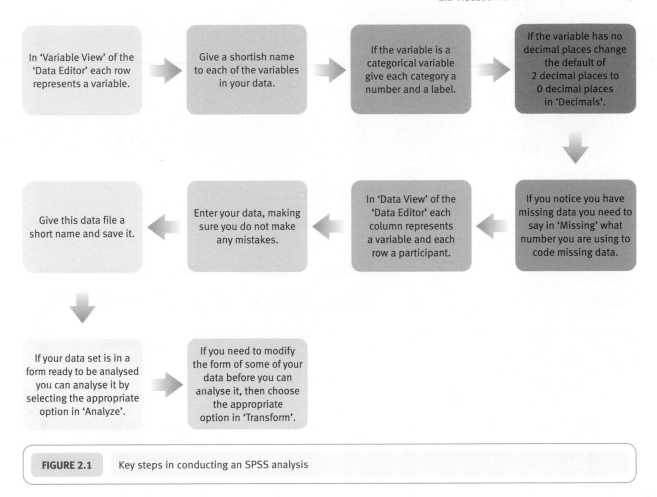

FIGURE 2.1   Key steps in conducting an SPSS analysis

Since SPSS is commonly used in universities and colleges, many users will require a user name and a password, which may be obtained from their institution. Documentation is often available on registration at the college or university. If not, such documentation is usually easily available.

SPSS may be found as a shortcut among the on-screen icons or it may be found in the list of programs on Windows. Each institution has its own idiosyncrasies as to where SPSS can be found. This book is based on Release 19 of SPSS. Although it is slightly different from earlier releases, the user of earlier releases will probably not notice the differences as they follow the instructions given in this book.

Figure 2.1 shows some of the keys steps in carrying out an SPSS analysis.

2.2   Accessing SPSS Statistics

SPSS Statistics for Windows is generally accessed using buttons and menus in conjunction with clicks of the mouse. Consequently the quickest way of learning is simply to follow our steps and screenshots on a computer. The following sequence of screenshots is annotated with instructions labelled Step 1, Step 2, etc.

Step 1

Double left click on the SPSS icon with the
mouse if it appears anywhere in the window.
Otherwise click the Start button to find the list
of programs, open the list of programs,
and click on SPSS.

Step 2

This dialog box appears after a few moments.
You could choose any of the options in the
window. However, it is best to close down
the superimposed menu by clicking on the
close-down button or 'Cancel' button. The
superimposed menu may not appear as it can
be permanently closed down.

2.3 Entering data

Step 1

The SPSS Data Editor can now be seen
unobstructed by the dialog box. The Data
Editor is a spreadsheet into which data are
entered. In the Data Editor the columns are
used to represent different variables, the
rows are the different cases (participants)
for which you have data.

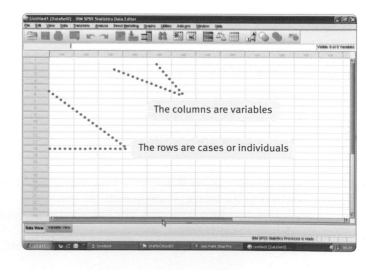

The columns are variables

The rows are cases or individuals

Step 2

To enter data into SPSS simply highlight one of the cells by clicking on that cell – SPSS may have one cell highlighted.

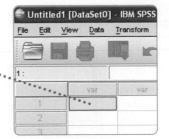

Step 3

Then type a number using the computer keyboard. On pressing return on the keyboard or selecting another cell with the mouse this number will be entered into the spreadsheet as shown here. The value 6.00 is the entry for the first row (first case) of the variable VAR00001.

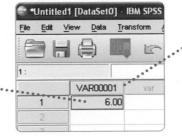

Notice that the variable has been given a standard name automatically. It can be changed – just click on variable name which bring up 'Variable View' and make the change (see Section 2.8).

Step 4

Correcting errors: simply highlight the cell where the error is using your mouse and type in the correction. On pressing return or moving to another cell, the correction will be entered.

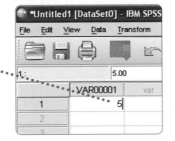

2.4   Moving within a window with the mouse

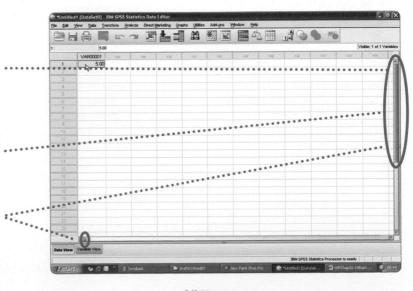

One can move a row or column at a time by clicking on the arrow-headed buttons near the vertical and horizontal scroll bars.

For major movements, drag the vertical and horizontal scroll bars to move around the page.

The relative position of the scroll bar indicates the relative position in the file.

2.5   Moving within a window using the keyboard keys with the mouse

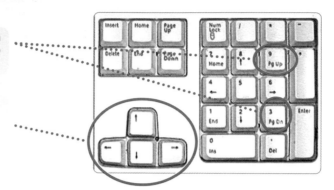

One can move one page up or down on the screen by pressing the Pg Up and Pg Dn keyboard keys.

The cursor keys on the keyboard move the cursor one space or character according to the direction of the arrow.

2.6   Saving data

Step 1

By selecting 'File' then 'Save As...' it is possible to save the data as a file.

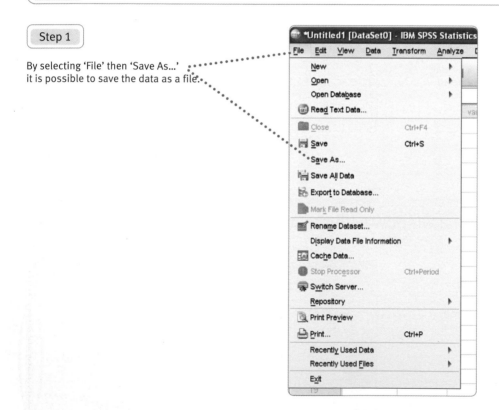

Step 2

The saved data file is automatically given the extension '.sav' by SPSS. A distinctive file name is helpful such as 'eg1' so that its contents are clear.

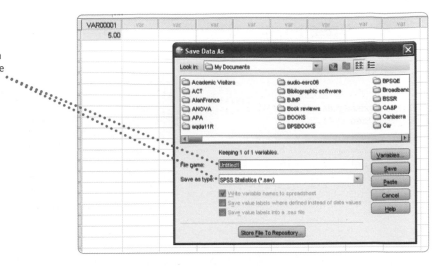

Step 3

To choose the place where the data file will be saved, indicate this place in the 'Look in:' box. Use the arrow to browse to the selected location.
Select 'Save' to save the file.

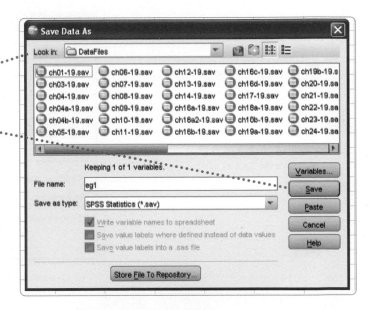

Step 4

The output window appears with the syntax commands for this procedure. Their use is described in Chapters 42–45.

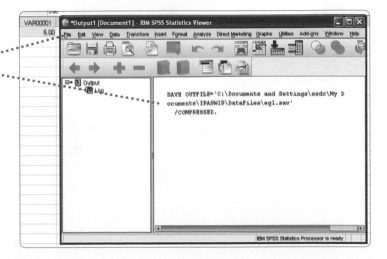

2.7   Opening up a data file

Step 1

To open an existing file, select 'File', 'Open' and 'Data...'.

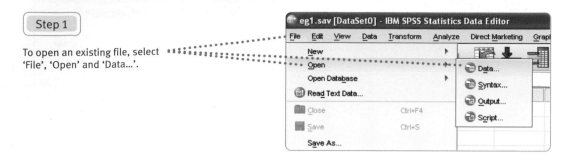

Step 2

Look in 'Look in:' if the file is not in the 'Open File' box (which it will be if you have just saved it), select the file name ('eg1') and then 'Open'.

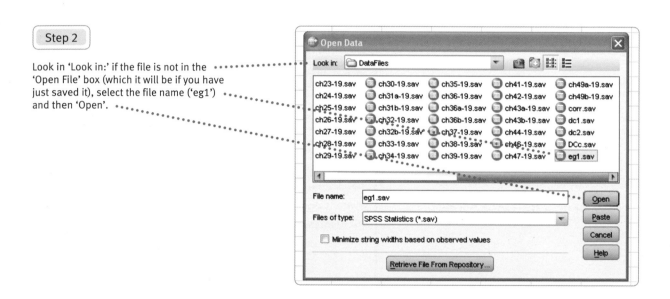

Step 3

To open a new file select 'File', 'New' and then 'Data'. This file can be saved as in Section 2.6.

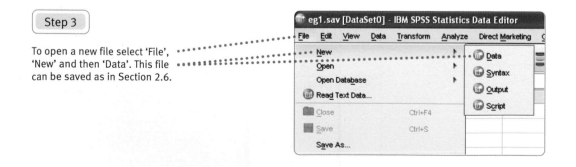

2.8 Using 'Variable View' to create and label variables

Step 1

Clicking on the 'Variable View' tab at the bottom changes the 'Data View' (the data spreadsheet) screen to one in which information about your variables may be entered conveniently.

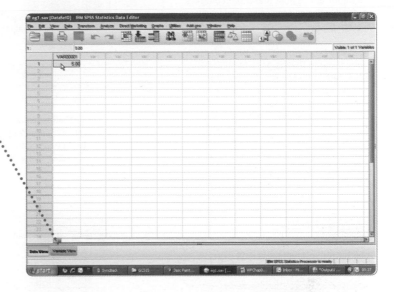

Step 2

This is the 'Variable View' spreadsheet. In this case one variable is already listed. We entered it in Section 2.3. But we can rename it and add extra variables quite simply by highlighting the appropriate cell and typing in a new or further variable names.

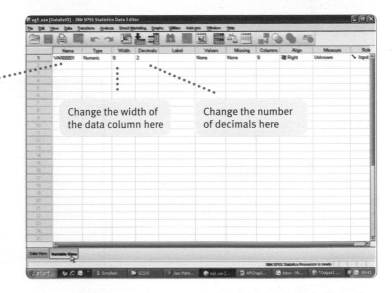

Change the width of the data column here

Change the number of decimals here

Step 3

There is no practical limit to the length of variable names in Release 12 and later. Earlier releases were limited to 8 characters. Highlight one of the cells under 'Name' and type in a distinct Variable name. The rest of the columns are given default values but may be changed. These renamed and newly defined variables will appear on the 'Data View' screen when this is selected. That is, new variables have been created and specified.

	Name	Type	Width	Decimals
1	Intelligence	Numeric	8	2
2	Age	Numeric	8	2
3	Gender	Numeric	8	2
4				

This is the number of decimals which will appear on screen – the calculation uses the actual decimal value which is larger than this.

Step 4

It is important to note that other columns are easily changed too.

'Label' allows one to give the variable a longer name than is possible using the Variable name column. Simply type in a longer name in the cells. This is no longer so important now that Variable names can be much longer in Release 12 and later.

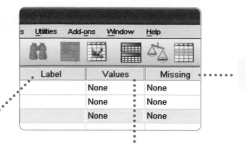

'Missing' values are dealt with in Chapter 41.

'Values' allows one to name the different categories or values of a nominal (category or categorical) variable such as gender. See Step 7 for details. It is recommended that 'Values' are given for all nominal variables.

Step 5

This 'button' appears. Click on it.

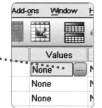

Step 6

This small window appears. Follow the next few steps. They show how male and female would be entered using the codes 1 for female and 2 for male.

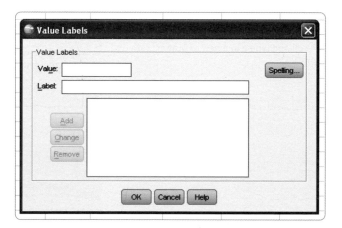

Step 7

Type in 1 next to 'Value:' and 'female' next to 'Label:'. Then click 'Add'.

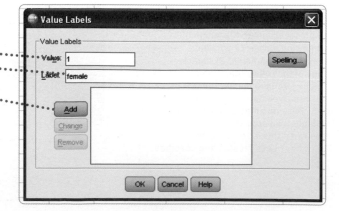

Step 8

This transfers the information into the large box.

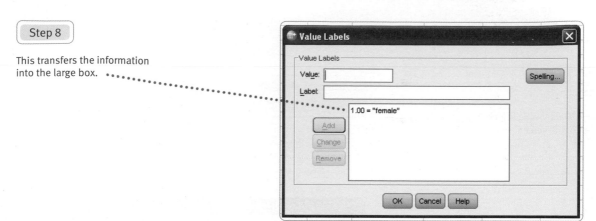

Step 9

Now type in 2 next to 'Value:' and 'male' next to 'Label:'. Then click 'Add'.

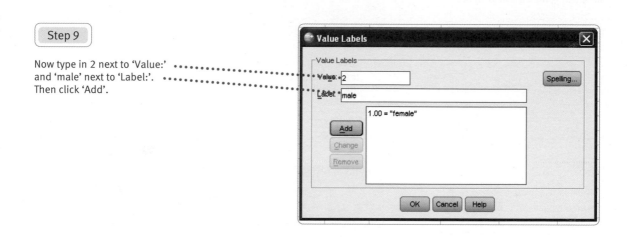

Step 10

This transfers the information into the large box.
Click OK to close Window.

Note: It is bad practice NOT to label Values in this way.

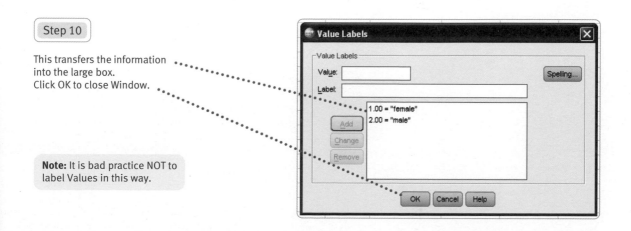

2.9 More on 'Data View'

Step 1

To return to 'Data View' click on this tab at the bottom left of the screen.

Step 2

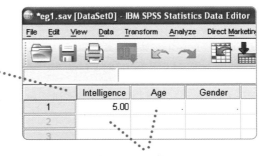

This is how 'Data View' looks now.
The data can be entered for all of the variables and cases. Remember that the value 5.00 was entered earlier along with the variable names. We can start entering the data in full now.

To enter data, simply highlight a cell, enter the number, then press return. You will be in the next cell down which will be highlighted.

Step 3

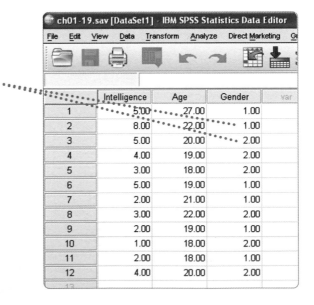

This shows how a typical data spreadsheet looks. Notice how the values for gender are coded as 1.00 and 2.00. It is possible to reveal their value labels instead. Click on 'View' on the task bar at the top.

	Intelligence	Age	Gender	var
1	5.00	27.00	1.00	
2	8.00	22.00	1.00	
3	5.00	20.00	2.00	
4	4.00	19.00	2.00	
5	3.00	18.00	2.00	
6	5.00	19.00	1.00	
7	2.00	21.00	1.00	
8	3.00	22.00	2.00	
9	2.00	19.00	1.00	
10	1.00	18.00	2.00	
11	2.00	18.00	1.00	
12	4.00	20.00	2.00	

Step 4

Then click on 'Value Labels'. Note you need to be in 'Data View' to do this.

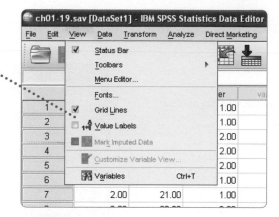

Step 5

Now the values are given as 'female' and 'male' – just as we coded them in Steps 5 to 10 in Section 2.8.

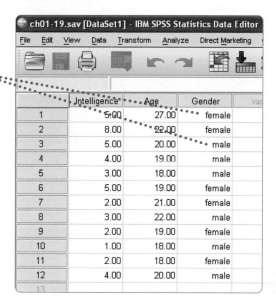

Select 'Data' to insert extra variables, extra cases, select cases and other data manipulations.

Select 'Window' to switch between the data spreadsheet and any output calculated on the data.

Step 6

There are many options available to you – including statistical analyses. Some of these options are shown here.

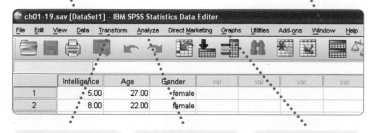

Select 'Transform' for a range of things that can be done with the data – such as recoding the values and computing combinations of variables.

Select 'Analyze' to access the full range of statistical calculations that SPSS calculates.

Select 'Graphs' for bar charts, scatterplots, and many other graphical representation methods.

2.10   A simple statistical calculation with SPSS

Step 1

To calculate the average (i.e. mean) age follow
the following steps:
Click 'Analyze'.
Select 'Descriptive Statistics'.
Select 'Descriptives...'.

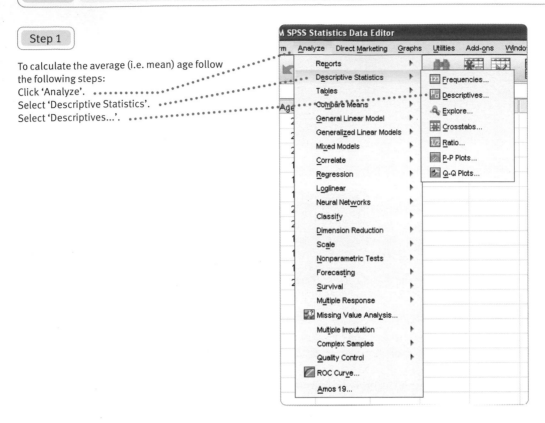

Step 2

This box appears. Highlight 'Intelligence' with
mouse.
Click on arrowed button to move 'Intelligence'
over into the 'Variable(s):' box.
Then click 'OK'.

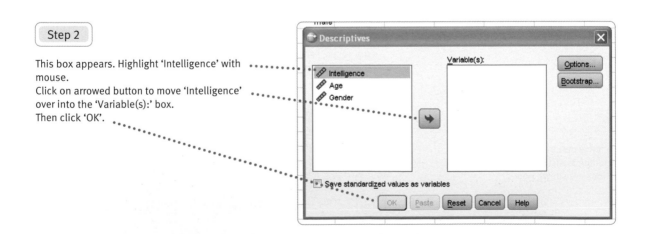

2.11   SPSS Statistics output

The 'Data Editor' window is replaced in view by the SPSS Output window.

Click on the 'Maximise' button to expand the window.
The first part of output is a list of commands that can be used to run this procedure. These kinds of command are discussed in Chapters 41–44.
The second part is a table of statistics. The average (mean) intelligence score is circled in for clarity here.

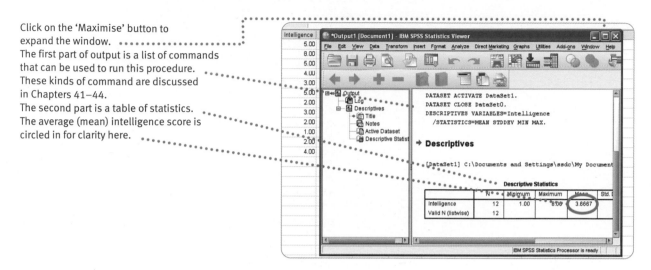

The good news is that anyone who can follow the above steps should have no difficulty in carrying out the vast majority of statistical analyses available on SPSS with the help of this book. It is worthwhile spending an hour or so simply practising with SPSS. You will find that this is the quickest way to learn.

More information about displaying variable names in dialog boxes is given in Chapter 43 (Section 43.3).

Summary of SPSS Statistics steps for a statistical analysis

Data

- In 'Variable View' of the 'Data Editor' 'name' the variable and 'label' each of its values if it's a 'nominal' or categorical 'measure'.
- In 'Data View' of the 'Data Editor', enter the values.
- Select 'File' and 'Save As ...'
- Name the data file, select the 'Save in:' file location and 'Save'.

Analysis

- Select 'Analyze' and then the type of analysis.

Output

- Output is presented in the 'Output' window.

For further resources including data sets and questions, please refer to the website accompanying this book.

PART 2

Descriptive statistics

CHAPTER 3

Describing variables

Tables

Overview

- Tables should quickly and effectively communicate important features of one's data. Complexity, for its own sake, is not a helpful characteristic of good tables.

- Clear and meaningful tables are crucial in statistical analysis and report writing. Virtually every analysis of data uses them in order to allow the distributions of variables to be examined. In this chapter we provide the basic computer techniques to allow the construction of tables to describe the distributions of individual variables presented one at a time.

- All tables should be clearly titled and labelled. Depending on the table in question, horizontal and vertical axes should be labelled, bars identified, scales marked and so forth. Great care should be taken with this. Generally speaking, SPSS tables require considerable work to make them optimally effective.

- Frequency tables merely count the number of times the different values of the variable appear in the data. A simple example would be a count of the number of males and the number of females in the research. Tables need to be relatively simple and this usually requires that the variable has only a small number of values or, if not, that a number of different values of the variable are grouped together.

3.1 What are tables?

Now this sounds like a really silly question because we all know what tables are, don't we? But simply because we know about these things does not mean that we understand their importance in a good statistical analysis or what makes a good table. The bottom line is that unless you very carefully study each of your variables using the methods described in this and the next few chapters then you risk failing to identify problems which may undermine the validity of your analysis.

Tables in data analysis are almost always summary tables of key aspects of the data since rarely are the actual data tabulated and reported in statistical analyses – usually because this would be very cumbersome. Tables then report important features of the data such as

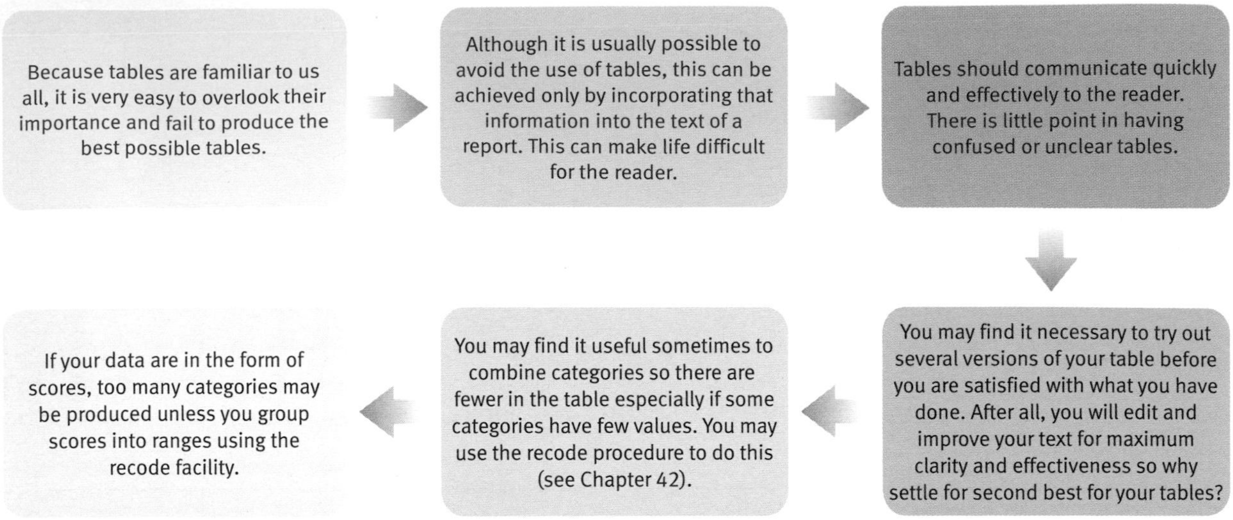

Because tables are familiar to us all, it is very easy to overlook their importance and fail to produce the best possible tables.

Although it is usually possible to avoid the use of tables, this can be achieved only by incorporating that information into the text of a report. This can make life difficult for the reader.

Tables should communicate quickly and effectively to the reader. There is little point in having confused or unclear tables.

If your data are in the form of scores, too many categories may be produced unless you group scores into ranges using the recode facility.

You may find it useful sometimes to combine categories so there are fewer in the table especially if some categories have few values. You may use the recode procedure to do this (see Chapter 42).

You may find it necessary to try out several versions of your table before you are satisfied with what you have done. After all, you will edit and improve your text for maximum clarity and effectiveness so why settle for second best for your tables?

FIGURE 3.1   Steps in understanding tables

(1) frequencies (counts) or the number of participants in the research having a particular characteristic such as the number of males and the number of females in the study and (2) the average score on a variable in different groups of participants such as the average (known as the mean in statistics) age of the male participants and the female participants in the research.

There is probably an infinite variety of possible tables. Table 3.1 is quite a simple example of a table. It gives a number of occupations together with the frequencies and percentage frequencies of participants in a study who have a particular type of job. Figure 3.1 outlines some basic steps in understanding tables.

3.2   When to use tables

Tables are a vital part of the analysis of data as well as being used to present data. The examination of basic tables which describe the data is an important aspect of understanding one's data set and familiarising oneself with it. There is nothing unsophisticated in using tables – quite the contrary, since it is poor statisticians who fail to make good use of them.

3.3   When not to use tables

Never use tables and diagrams as a substitute for reporting your research findings in words in the body of a report. To do so is not good because it leaves the reader with the job of interpreting the tables and diagrams, which is the task of the researcher.

3.4   Data requirements for tables

Different sorts of tables have different requirements in terms of the variables measured.

3.5   Problems in the use of tables

There are many problems in the use of tables:

- Remember to label clearly all of the important components of tables. Not to do so is counter-productive.

- Problems can occur because of rounding errors for percentages such that the total of the percentages for parts of the table do not add up to 100 per cent. Keep an eye open for this and make reference to it in your report if it risks confusing the reader.

- We would recommend that you *never* use tables directly copied and pasted from SPSS. For one thing, SPSS tables tend to contain more information than is appropriate to report. But the main reason is that often SPSS tables are simply confusing and could be made much better with a little work and thought.

You can find out more about tables in Chapter 2 of Howitt, D. and Cramer, D. (2011). *Introduction to Statistics in Psychology*, 5th edition. Harlow: Pearson.

3.6   The data to be analysed

SPSS is generally used to summarise raw data but it can use data which have already been summarised such as those shown in Table 3.1 (*ISP*, Table 2.1).

In other words, since the data in Table 3.1 are based on 80 people, the data would occupy 80 cells of one column in the 'Data Editor' window and each occupation would be coded with a separate number, so that nuns might be coded 1, nursery teachers 2 and so on. Thus, one would need 17 rows containing 1 to represent nuns, 3 rows containing 2 to represent nursery teachers and so on. However, it is possible to carry out certain analyses on summarised data provided that we appropriately weight the categories by the number or frequency of cases in them.

Table 3.1	Occupational status of participants in the research expressed as frequencies and percentage frequencies	
Occupation	**Frequency**	**Percentage frequency**
Nuns	17	21.25
Nursery teachers	3	3.75
Television presenters	23	28.75
Students	20	25.00
Other	17	21.25

3.7   Entering summarised categorical or frequency data by weighting

It seems better practice to define your variables in 'Variable View' of the 'Data Editor' before entering the data in 'Data View' because we can remove the decimal places where they are not necessary. So we always do this first. If you prefer to enter the data first then do so.

For this table we need two columns in 'Data View', one to say what the categories are and the other to give the frequencies for these categories. In 'Variable View' variables are presented as rows.

Step 1

Select 'Variable View' in 'Data Editor'.
Name the first two variables 'Occupation' and 'Freq'.
Remove the two decimal places.
Select the right side of the cell for the 'Values' of 'Occupation'.

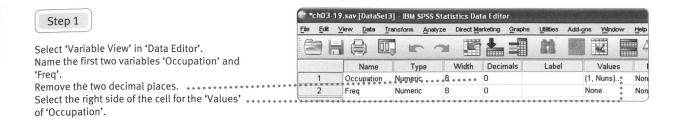

Step 2

Label the five 'Values' of 'Occupation' as shown and as described in Steps 5 to 10 in Section 2.8.
Select 'OK'.

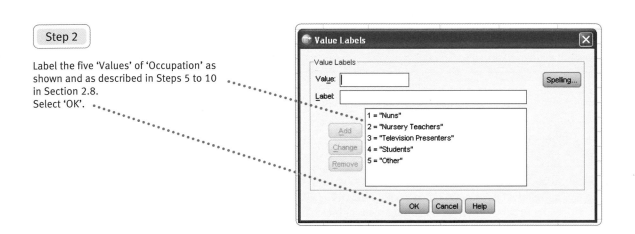

Step 3

Select 'Scale' for this variable and the downward arrow to give the drop-down menu.
Select 'Nominal'.

Step 4

Select 'Data View'. Enter the data as shown.

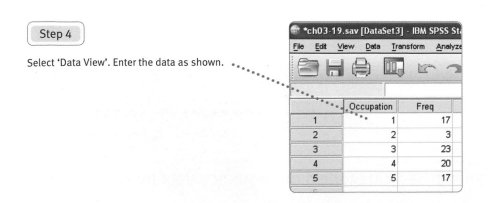

Step 5

Select 'Data' and 'Weight Cases...'.

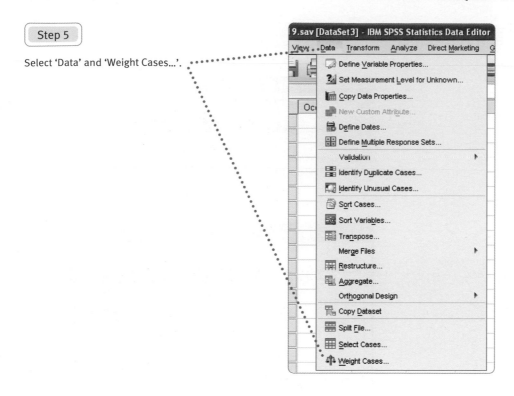

Step 6

Select 'Freq', 'Weight cases by' and the ➡ button to put it in the 'Frequency Variable:' box. Select 'OK'.

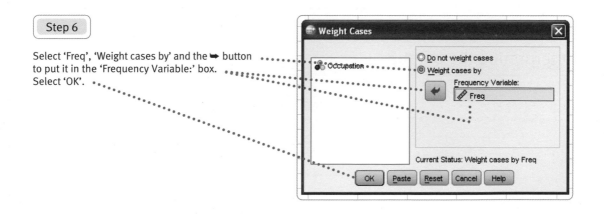

Step 7

The cases, which are the occupations, are now weighted by the frequencies as shown by the 'Weight On' message in the lower right corner of the 'Data Editor'.

3.8 Percentage frequencies

Step 1

Select 'Analyze', 'Descriptive Statistics' and 'Frequencies...'.

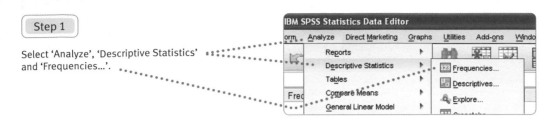

Step 6

Select 'Occupation' and the ➡ button to put 'Occupation' in the 'Variables(s):' box. Select 'OK'.

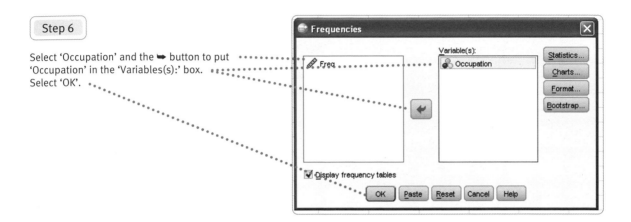

3.9 Interpreting the output

The third column gives the percentage frequency for each category, including any missing values of which there are none. So 17 is 21.3 per cent of a total of 80.

Occupation

		Frequency	Percent	Valid Percent	Cumulative Percent
Valid	Nuns	17	21.3	21.3	21.3
	Nursery Teachers	3	3.8	3.8	25.0
	Television Presenters	23	28.8	28.8	53.8
	Students	20	25.0	25.0	78.8
	Other	17	21.3	21.3	100.0
	Total	80	100.0	100.0	

The first column of this table gives the value of the five categories.

The fourth column gives the percentage frequency, excluding any missing values. As there are none, these percentages are the same as those in the second column.

The fifth column adds the percentages down the table so 25.0 of the cases are nuns or nursery teachers.

REPORTING THE OUTPUT

Only the category labels, the frequency and the percentage frequency need be reported, consequently you need to simplify this table if you are going to present it. If the occupation was missing for some of the cases you would need to decide whether you would present percentages including or excluding them. There is no need to present both sets of figures. Also omit the term 'Valid' in the first column as its meaning may only be familiar to SPSS users.

Summary of SPSS Statistics steps for frequency tables

Data

- In 'Variable View' of the 'Data Editor' 'name' the 'nominal' or categorical variable, 'label' each of its values and select 'nominal' as the 'measure'.
- Name the second variable.
- In 'Data View' of the 'Data Editor', enter the value of the category for each case and its frequency.
- Select 'Data', 'Weight Cases …', 'Weight cases by' and move the weighting variable to the 'Frequency Variable:' box.

Analysis

- Select 'Analyze', 'Descriptive Statistics' and 'Frequencies …'.
- Move the variables to be analyzed to the right-hand box.

Output

- The percentage of cases for each category is given for the whole sample in the 'Percent' column including any missing data, and in the 'Valid Percent' column excluding missing data. If there are no missing data, these percentages will be the same.

For further resources including data sets and questions, please refer to the website accompanying this book.

CHAPTER 4

Describing variables
Diagrams

Overview

- Diagrams, like tables, should quickly and effectively communicate important features of one's data. Complexity, for its own sake, is not a helpful characteristic of good diagrams.

- Clear and meaningful diagrams are crucial in statistical analysis and report writing. Virtually every analysis of data uses them in order to allow the distributions of variables to be examined. In this chapter we provide the basic computer techniques to allow the construction of diagrams to describe the distributions of individual variables presented one at a time.

- All diagrams should be clearly titled and labelled. Depending on the diagram in question, horizontal and vertical axes should be labelled, bars identified, scales marked and so forth. Great care should be taken with this. Generally speaking, SPSS Statistics diagrams require considerable work to make them optimally effective.

- Pie diagrams are effective and simple ways of presenting frequency counts. However, they are only useful when the variable being illustrated has a small number of different values. Pie diagrams are relatively uncommon in publications because they consume space, although they are good for conference presentations and lectures.

- A bar chart can be used in similar circumstances to the pie chart but can cope with a larger number of values of variables before becoming too cluttered. Frequencies of the different values of the variable are represented as physically separated bars of heights which vary according to the frequency in that category.

- A histogram looks similar to a bar chart but is used for numerical scores rather than categories. Thus the bars in a histogram are presented in size order of the scores that they represent. The bars in histograms are *not* separated by spaces. Often a histogram will need the ranges of scores covered by each bar to be changed in order to maximise the usefulness and clarity of the diagram. This can be done by recoding variables (Chapter 42) but SPSS also allows this to be done in the 'Chart Editor'. Producing charts like these is one of the harder tasks on SPSS.

4.1   What are diagrams?

This sounds like another really silly question because we all know what diagrams are, don't we? But simply because we know about these things does not mean that we understand their importance in a good statistical analysis or what makes a good diagram. The bottom line is that unless you very carefully study each of your variables using the methods described in this and the next few chapters then you risk failing to identify problems which may undermine the validity of your analysis.

Diagrams in statistics do much the same sort of job as tables but, of course, they are intended to have greater visual impact. So you are almost certain to be familiar with diagrams which illustrate frequencies of things in data. Pie charts represent the proportion of cases in each category of a variable using proportionate slices of a circle (the pie) (see the examples on pages 47–50). Very much the same sort of thing can be illustrated using bar charts in which each category of a variable is represented by a bar and the frequencies by the heights of the relevant bar (see the examples on pages 51–52). Neither is better than the other except a pie chart shows the proportions of each category compared with the total number of participants perhaps a little more quickly. Histograms are very similar in appearance to bar charts but are used when the categories are actually scores on a particular measure which are ordered from smallest to the largest. Examples of histograms are to be found on pages 53–54. Technically, a bar chart should have a gap between the bars but a histogram should not since it represents a dimension rather than a number of distinct categories.

Diagrams do much the same task as tables but usually in a less compact form so that less information can be put into the same physical space. They are much less common in professional publications than tables. Although it may be tempting to think that diagrams are preferable to tables, this is not the case. Complex data are often difficult to report in a simple diagram and possibly much easier to put into a table. Diagrams tend to be reserved for circumstances where the researcher is trying to communicate the findings of a study quickly and with impact. This is more likely to be the case with spoken presentations of research findings rather than written ones. So a diagram is often better as part of a lecture than a table so long as the diagram does not get too complicated.

Figure 4.1 shows some basic types of charts while Figure 4.2 emphasises the importance of nominal and score data when making charts.

Pie charts

- The familiar circular pie chart is primarily for use where you have nominal (category) data though it can be adapted for use with scores by making the segments refer to ranges of scores.
- It is a big mistake to have too many categories (segments) in a pie chart as the visual impact is lost.

Bar charts

- The bar chart (where the bars are separated) is used for nominal (category data).
- Like most forms of charts, care is needed to avoid making the bar chart too complex by, for example, having so many bars that it becomes difficult to read.

Histograms

- Histograms are superficially like bar charts but are used where the data are in the form of scores.
- The horizontal axis of a histogram indicates increasing values of the scores.

FIGURE 4.1   Some basic types of chart

> **Nominal category data**
> - Data which are just in the form of named categories (nominal/category data) can be used with pie charts and bar charts.

> **Scores**
> - Data which are in the form of scores is normally best presented in the form of a histogram.
> - However, if ranges of scores are used then pie diagrams can also be used.

FIGURE 4.2   The importance of nominal and score data when making charts

4.2   When to use diagrams

Diagrams can be an important part of the analysis of data as well as being used to present data. The examination of basic diagrams which describe the data is an important aspect of understanding one's data and familiarising oneself with it. There is nothing unsophisticated in using them – quite the contrary since it is poor statisticians who fail to make good use of them. We will look at a number of circumstances where diagrams can prevent us making basic mistakes.

4.3   When not to use diagrams

Never use diagrams as a substitute for reporting your research findings in words in the body of a report. To do so is not good because it leaves the reader with the job of interpreting the diagrams, which is the task of the researcher. In terms of practical reports and dissertations, you may well find that tables are generally more useful than diagrams because they can handle complexity better.

4.4   Data requirements for diagrams

Different kinds of diagrams have different requirements in terms of the variables measured. Bar charts and pie charts are largely used for nominal (category) data which consist of named categories. Histograms are used for score data in which the scores are ordered from the smallest values to the largest on the horizontal axis of the chart.

4.5   Problems in the use of diagrams

Problems in the use of diagrams are numerous:

- Remember to label clearly all of the important components of diagrams. Not to do so is counterproductive.

- Problems can occur because of rounding errors for percentages such that the total of the percentages for parts of the diagram do not add up to 100 per cent. Keep an eye open for this and make reference to it in your report if it risks confusing the reader.

- Technically, the bars of histograms should each have a width proportionate to the range of numbers covered by that bar. SPSS does not do this so the output can be misleading.

- You need to be very careful when a scale of frequencies on a bar chart or histogram does not start at zero. This is because if part of the scale is missing then the reader can get a very wrong impression of the relative frequencies as some bars can look several times taller than others whereas there is little difference in absolute terms. This is one of the 'mistakes' which has led to the suggestion that one can lie using statistics.

- We would recommend that you *never* use diagrams directly copied and pasted from SPSS without editing them first. SPSS diagrams are often simply confusing and could be made much better with a little work and thought. Many SPSS diagrams can be modified on SPSS to improve their impact and clarity. For example, there is no point in using coloured slices in a pie chart if it is to be photocopied in black and white. Such a chart may be very confusing because the shades of gray cannot be deciphered easily.

> You can find out more about tables and diagrams in Chapter 2 of Howitt, D. and Cramer, D. (2011). *Introduction to Statistics in Psychology*, 5th edition. Harlow: Pearson.

4.6 The data to be analysed

SPSS is generally used to summarise raw data but it can use data which have already been summarised such as those shown in Table 4.1 (*ISP*, Table 2.1). In other words, since the data in Table 4.1 are based on 80 people, the data would occupy 80 cells of one column in the 'Data Editor' window and each occupation would be coded with a separate number, so that nuns might be coded 1, nursery teachers 2 and so on. Thus, one would need 17 rows containing 1 to represent nuns, 3 rows containing 2 to represent nursery teachers and so on. However, it is possible to carry out certain analyses on summarised data provided that we appropriately weight the categories by the number or frequency of cases in them.

Table 4.1	Occupational status of participants in the research expressed as frequencies and percentage frequencies	
Occupation	**Frequency**	**Percentage frequency**
Nuns	17	21.25
Nursery teachers	3	3.75
Television presenters	23	28.75
Students	20	25.00
Other	17	21.25

4.7   Entering summarised categorical or frequency data by weighting

It seems better practice to define your variables in 'Variable View' of the 'Data Editor' before entering the data in 'Data View' because we can remove the decimal places where they are not necessary. So we always do this first. If you prefer to enter the data first then do so.

For this table we need two columns in 'Data View'. One is to say what the categories are. The other is to give the frequencies for these categories. In 'Variable View' variables are presented as rows. As we will be entering the data into columns we will refer to these rows as columns. If you have already still have the data from the previous chapter, simply use that data file.

Step 1

Select 'Variable View' in 'Data Editor'.
Name the first two variables
'Occupation' and 'Freq'.
Remove the two decimal places.
Select the right side of the cell for the
'Values' of 'Occupation'.

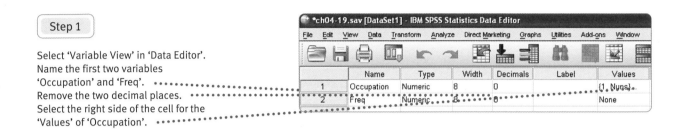

Step 2

Label the five 'Values' of 'Occupation' as shown
and as described in Steps 5 to 10 in Section 2.8.
Select 'OK'.

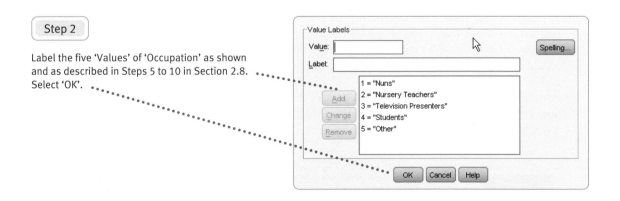

Step 3

Select 'Scale' for this variable and the
downward arrow to give the drop-down menu.
Select 'Nominal'.

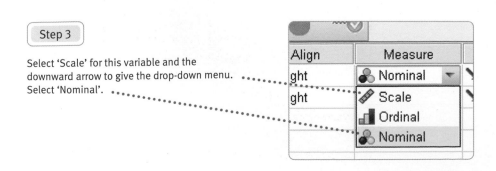

Step 4

Select 'Data View'. Enter the data as shown.

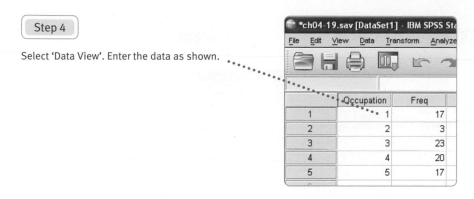

Step 5

Select 'Data' and 'Weight Cases...'.

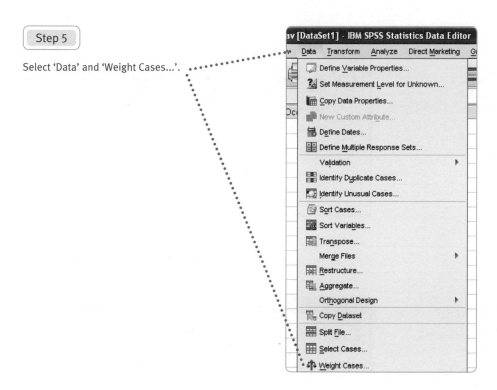

Step 6

Select 'Freq', 'Weight cases by' and the ➡ button to put it in the 'Frequency Variable:' box.
Select 'OK'.
The cases, which are the occupations, are now weighted by the frequencies as shown by the 'Weight On' message in the lower right corner of the 'Data Editor'.

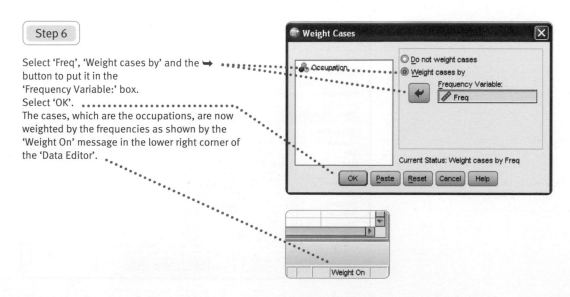

4.8   Pie diagram of category data

Step 1

Select 'Graphs' and 'Chart Builder...'.
An alternative way is to select 'Legacy Dialogs'.
This procedure is used in Chapters 6, 9 and 11.

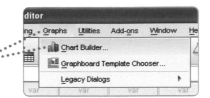

Step 2

If you have not defined 'Occupation' as a
nominal variable select 'Define Variable
Properties...'.
Otherwise select 'OK'.

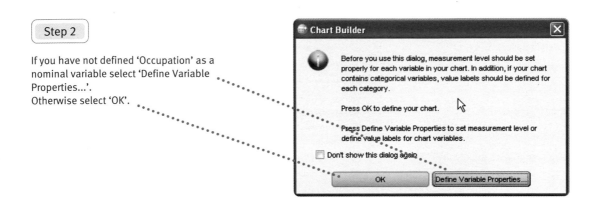

Step 3

Select 'Pie/Polar'. Drag the pie diagram into the
box above it.

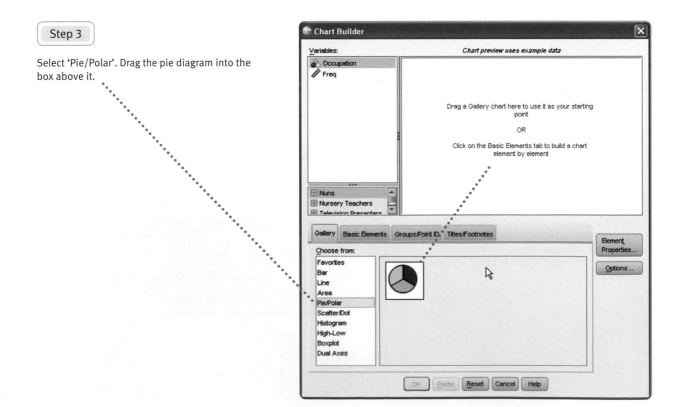

Step 4

Drag 'Occupation' into 'Slice by?'.
Select 'OK'.

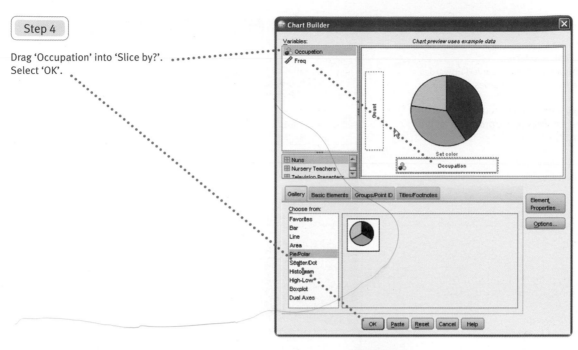

This is the way the pie diagram appears if the default options in SPSS have not been altered.

The slices are colour coded.

Features in this diagram may be altered with the 'Chart Editor'.

We will show how to label each slice (so that the reader does not have to refer to the colour code) and how to change the colours to monochrome patterns.

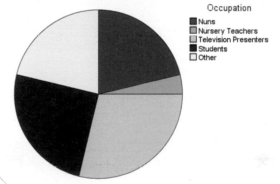

4.9 Adding labels to the pie diagram and removing the legend and label

Step 1

Double click anywhere in the rectangle containing the diagram to select the 'Chart Editor'.
Select 'Elements' and 'Show Data Labels'.

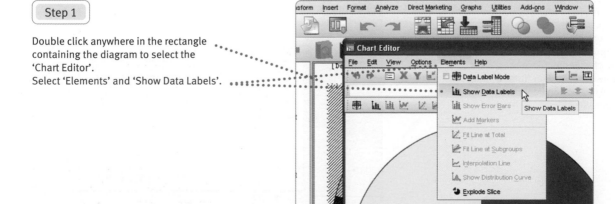

Step 2

Ensure 'Percent' is selected.
Select 'Occupation' and the curved upward green arrow to display the names of the occupations.
Select 'Apply' and 'Close' if you do not want to make further changes to the properties of the pie diagram.

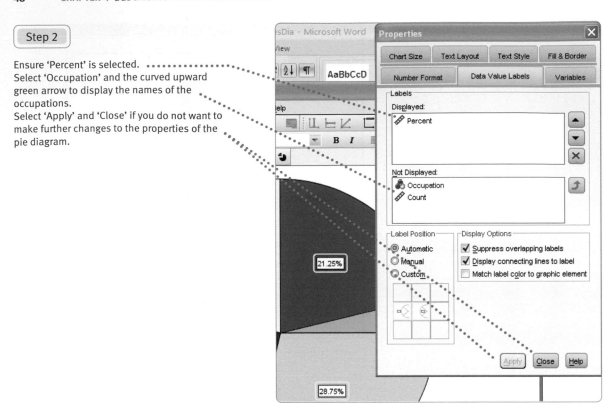

Step 3

Select 'Text Style' if you want to change aspects of the text such as making it bold and increasing its font size.
To make it bold, select the ▼ button next to 'Style' and 'Bold' from the dropdown menu.
To increase the font size to 14, select the ▼ button next to 'Size' and '14' from the dropdown menu.
NB Because of the small size of this slice, the font for Nursery Teachers will not be 14. To change this you need to click onto the box for this group.
If you do not want to make further changes, select 'Apply' and 'Close'
('Cancel' changes to 'Close').
To display full details for Nursery Teachers, select the text box and drag it to the left until all details are shown.

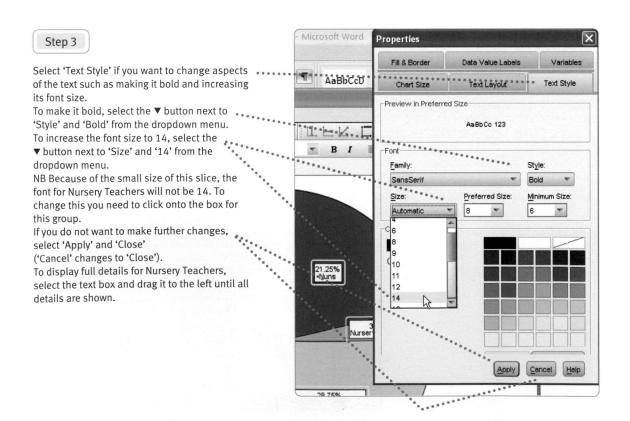

Step 4

To remove the legend, select 'Options' and
'Hide Legend'.

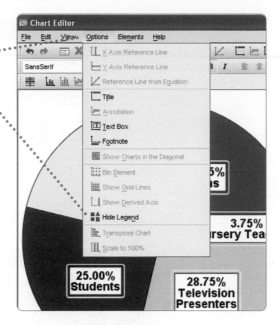

This is the pie diagram with the slices
named and the percentage frequency given.

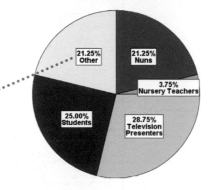

4.10 Changing the colour of a pie diagram slice to a black and white pattern

Step 1

In the 'Chart Editor', click on the slice you want
to change and then click again when the border
will have a yellow line around it.

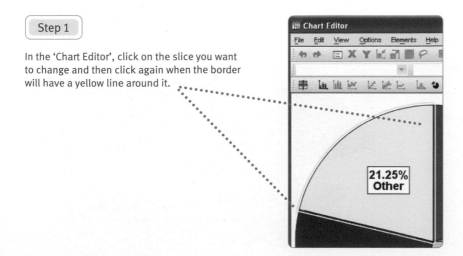

Step 2

Select 'Edit' and 'Properties'.

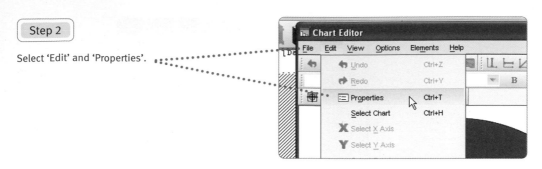

Step 3

Select 'Fill' and the colour of white.
To add a black border, select 'Border' and the colour of black.
To add a pattern, select the ▼ button beside 'Pattern' and the pattern you want.
Select 'Apply' and then 'Close'. Apply this procedure to the other four slices.

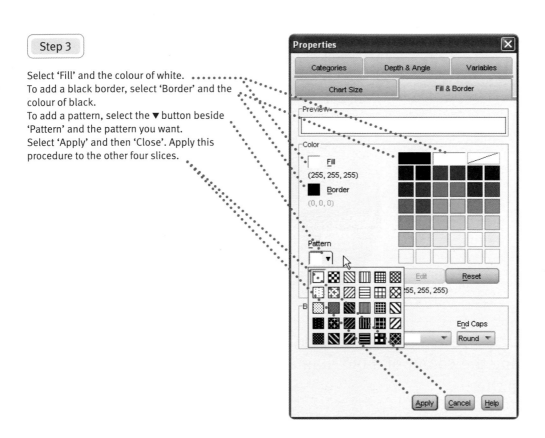

This is a pie diagram with black and white patterned slices.

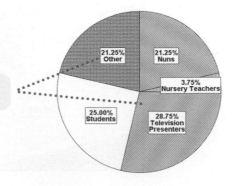

4.11   Bar chart of category data

Step 1

Follow the first two steps for the pie diagram, selecting 'Graphs' and 'Chart Builder...', and then 'OK' in the dialog box.
Select 'Bar' and then drag the 'Simple Bar' icon into the box above.

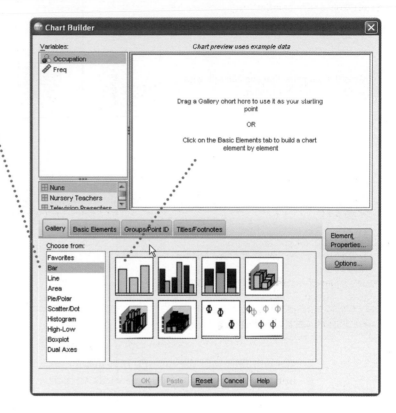

Step 2

Drag 'Occupation' into the 'X-axis?' box.
Select the 'Element Properties...' box
Select the ▼ button to reveal the drop-down menu.
Select 'Percentage (?)'.
Select 'Apply'.
Select 'OK'.

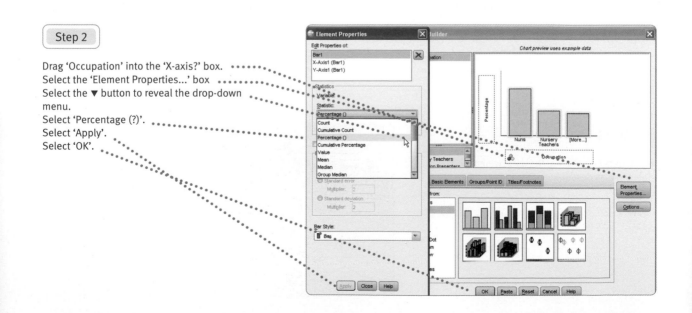

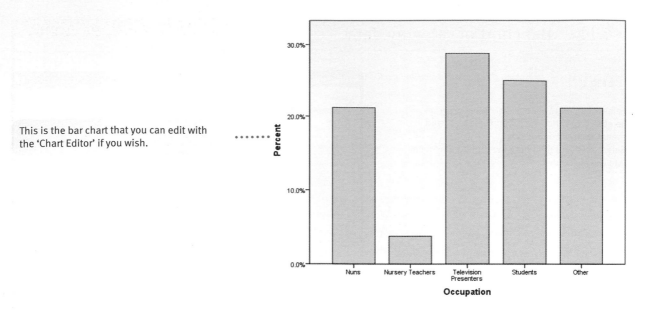

This is the bar chart that you can edit with the 'Chart Editor' if you wish.

4.12   Histograms

We will illustrate the making of a histogram with the data in Table 4.2 which shows the distribution of students' attitudes towards statistics. We have labelled this variable 'Response'.

Table 4.2	Distribution of students' replies to the statement 'Statistics is my favourite university subject'	
Response category	**Value**	**Frequency**
Strongly agree	1	17
Agree	2	14
Neither agree nor disagree	3	6
Disagree	4	2
Strongly disagree	5	1

Step 1

In the 'Data Editor' enter the data, weight and label them as described in Section 4.7.

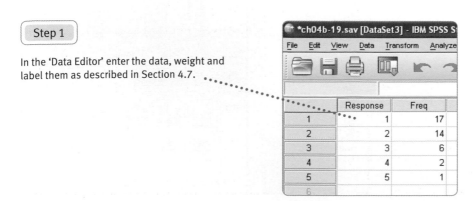

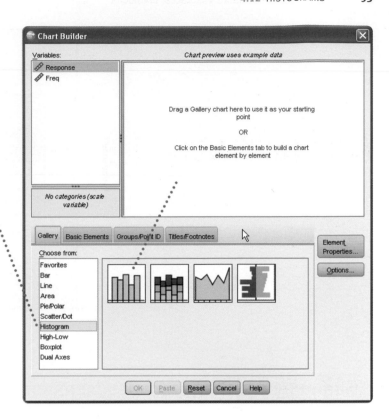

Step 2

Follow the first two steps for the pie diagram and bar chart, selecting
'Graphs' and 'Chart Builder...', and then
'OK' in the dialog box.
Select 'Histogram' and then drag the
'Simple Histogram' icon into the box above.

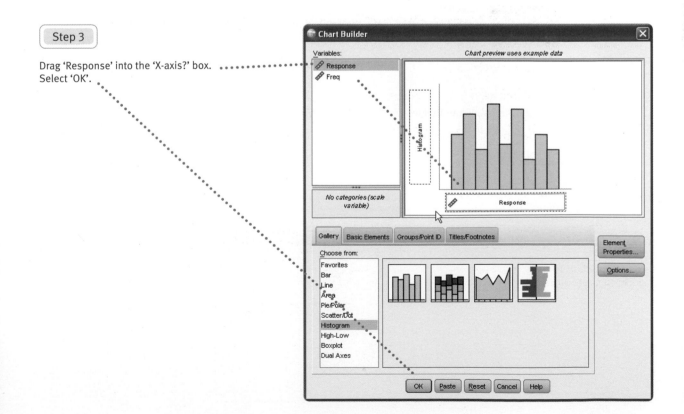

Step 3

Drag 'Response' into the 'X-axis?' box.
Select 'OK'.

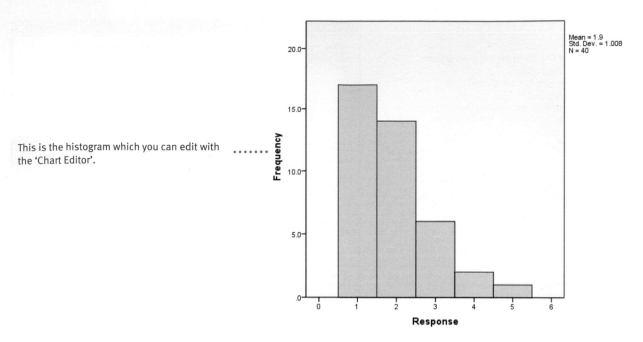

This is the histogram which you can edit with the 'Chart Editor'.

Summary of SPSS Statistics steps for charts

Data

- In 'Variable View' of the 'Data Editor' 'name' the 'nominal' or categorical variable, 'label' each of its values and select 'nominal' as the 'measure'.
- Name the second variable.
- In 'Data View' of the 'Data Editor', enter the value of the category for each case and its frequency.
- Select 'Data', 'Weight Cases ...', 'Weight cases by' and move the weighting variable to the 'Frequency Variable:' box.

Analysis

- Select 'Graphs', 'Chart Builder ...', 'OK' and move type of graph into the box above.
- Move variable to the 'X-Axis?' box.

Output

- Charts may be edited in the 'Chart Editor'.

For further resources including data sets and questions, please refer to the website accompanying this book.

CHAPTER 5

Describing variables numerically

Averages, variation and spread

Overview

- The computation of a number of statistics which summarise and describe the essential characteristics of each important variable in a study is explained. The techniques presented in this chapter involve individual variables taken one at a time. In other words they are single variable or univariate statistical techniques.

- Each technique generates a numerical index to describe an important characteristic of the data.

- With the exception of the mode which can be used for any type of data, all of the techniques are for data in the form of numerical scores.

- The mean is the everyday or numerical average of a set of scores. It is obtained by summing the scores and dividing by the number of scores.

- The mode is simply the most frequently occurring score. A set of scores can have more than one mode if two or more scores occur equally frequently. The mode is the value of the score occurring most frequently – it is *not* the frequency with which that score occurs.

- The median is the score in the middle of the distribution if the scores are ordered in size from the smallest to the largest. For various reasons, sometimes the median is an estimate of the score in the middle – for example, where the number of scores is equal, and so that there is no exact middle.

- The procedures described in this chapter can readily be modified to produce measures of variance, kurtosis and other descriptive statistics.

5.1   What are averages, variation and spread?

Most of us when we think of averages imagine what statisticians call the *mean*. This is the sum of a number of scores divided by the number of scores. So the mean of 5, 7 and 10 is 22/3 = 7.33. However, statisticians have other measures of the average or typical score or observation which are conceptually different from the mean. These include the median and mode. The *median* is obtained by ordering the scores from the smallest to the largest and the score which separates the first 50 per cent of cases from the second 50 per cent of cases is the median. If there is an even number of scores then there is no one score right in the middle and so a slight adjustment has to be made involving the mean of the score just below the middle and the score just above the middle. Sometimes even finer adjustments are made but SPSS will do the necessary calculation for you. The third measure of the average is called the *mode*. This is the most frequently occurring value in the data. This can be the most frequently occurring score if the data are score data or the most frequently occurring category if the data are nominal (category) data. The mode is the only measure of the average which can be used for nominal (category) data.

In statistics, rather than speak of an average it is more usual to refer to measures of *central tendency*. The mean, median and mode are all measure of central tendency. They are all measures of the typical score in the set of scores.

If a frequency distribution of scores is symmetrical (such as in the case of the normal distribution discussed in Chapter 6), then the value of the mean, the median and the mode will be identical – but only in those circumstances. Where the frequency distribution of scores is not symmetrical then the mean, the median and the mode will have different values. Figure 5.1 contains an example of an asymmetrical frequency distribution – though it is only slightly so. There is a total of 151 participants. The mean in this case is 5.47, the median is 6.00, and the mode is also 6.00. It is easy to see that the mode is 6.00 as this is the most frequently appearing score (i.e. it has the highest bar). Since there are 151 cases then the median is the 76th score from the left-hand side of the bar chart.

The spread of the scores is another characteristic which can be seen in Figure 5.1. Quite clearly the scores vary and the largest score is 10 and the smallest is 1. The difference between the largest score and the smallest score is known as the range, which in this case is 9. Sometimes the top and bottom quarters of scores are ignored to give the *interquartile range* which is less susceptible to the influence of unusual exceptionally large or small scores (outliers).

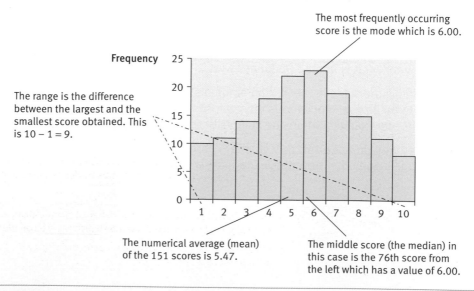

The most frequently occurring score is the mode which is 6.00.

The range is the difference between the largest and the smallest score obtained. This is 10 – 1 = 9.

The numerical average (mean) of the 151 scores is 5.47.

The middle score (the median) in this case is the 76th score from the left which has a value of 6.00.

| FIGURE 5.1 | A slightly asymmetrical frequency distribution |

Variance is the most important measure of the spread of scores. It is based on the squared deviation (difference) between each score and the mean of the set of scores in the data. In fact it is the average squared deviation from the mean. SPSS calculates this to be 6.06 for these data. (Actually SPSS calculates the variance estimate and not the variance, which is a little smaller than 6.06. This is no great problem as the variance estimate is generally more useful – it is an estimate of what the variance is in the population from which our sample came. The variance estimate is bigger than the variance since it is the sum of the squared deviations of each score from the mean score divided by the number of scores –1).

Figure 5.2 outlines some essential ideas about averages while Figure 5.3 highlights some measures related to variance.

Mean
- The mean is what most people call the average.
- It is simply the total of a set of scores divided by the number of scores.
- The mean is only the same value as the mode and median if the distribution is symmetrical.

Mode
- The mode is the most frequently occurring score or category.
- There can be more than one mode for a variable.
- Where there are two modes, then the distribution of scores is called bimodal.

Median
- The median is the middle score of the distribution ordered in terms of size.
- Sometimes it has to be estimated if there are even numbers of scores.
- It supplies different information from that of the mean and mode.

FIGURE 5.2   Essential ideas of averages

Variance
- This is a measure of the variability of scores.
- It is calculated by working out the average squared deviation of each score from the mean of the scores.

Standard deviation
- This is a sort of average of the deviation of each score from the mean score.
- It is simply the square root of the variance.

Variance estimate
- This is an estimate of the variance of the population based on the information from a sample.
- It is calculated in the same way as variance except rather than the average squared deviation from the mean, one divides by the number of scores minus 1.

Standard error
- This is essentially the standard deviation of sample means taken from the population unlike standard deviation which is based on the scores.
- It is simple to calculate the standard error: divide the standard deviation by the square root of the sample size.

FIGURE 5.3   Measures which are closely related to variance

5.2   When to use averages, variation and spread

These should be routinely (always) calculated when dealing with score variables. However, many of these can be calculated using different analysis procedures on SPSS and there may be more convenient methods than the ones in this chapter in some circumstances. For example, you will find that SPSS will calculate means as part of many tests of significance such as the *t*-tests or ANOVA though they may have to be requested.

5.3   When not to use averages, variation and spread

Apart from when using nominal (category) data, all measures of average, variation and spread discussed in this chapter are useful when exploring one's data.

5.4   Data requirements for averages, variation and spread

All of the techniques described in this chapter require a variable which is in the form of scores. The only exception is the mode, which can be calculated on any type of data.

5.5   Problems in the use of averages, variation and spread

The measures of central tendency (mean, median and mode) refer to quite different ideas and none of them is intrinsically better than the others. They each contain different information and each can be presented in any data analysis. There is, however, a tendency to ignore the median and mode, which is a mistake for anything other than symmetrical distributions of scores (for which the three different measures of central tendency give the same value). Be careful when considering the mode since a distribution has more than one mode if it has more than one peak. SPSS will not warn you of this but a glance at the frequency distribution for the scores will also alert you to this 'bimodal' or 'multimodal' characteristic of the distribution.

The range is sometimes given in error as being from 1 to 10, for example. This is not the range since the range is a single number such as 9. The value 1 is the *lower bound* of the range and the 10 is the *upper bound* of the range.

Initially when learning statistics, it is sometimes easy to understand a concept but difficult to see the importance or point of it. Variance is a good example of this since it is basically a very abstract notion (the average of the sum of squared differences between each score and the mean score). A particular value of variance is not intuitively meaningful since it does not refer to something that we regard as concrete such as, say, the mean score. Although variance is a fundamental statistical concept, it becomes meaningful only when we compare the variances of different variables or groups of participants. Nevertheless, it will rapidly become apparent that the concept of variance is involved in a great many different statistical techniques. Indeed, nearly all of the techniques in this book involve variance in some guise.

Be careful to note that variance as calculated by SPSS is really what is known technically as the variance estimate. It would be better to refer to it as the variance estimate but in this respect SPSS does not set a good example.

You can find out more about describing data numerically in Chapter 4 of Howitt, D. and Cramer, D. (2011). *Introduction to Statistics in Psychology*, 5th edition. Harlow: Pearson.

5.6   The data to be analysed

We will illustrate the computation of the mean, median and mode on the ages of university students – see Table 5.1 (*ISP*, Table 3.7).

Table 5.1	Ages of 12 students										
18	21	23	18	19	19	19	33	18	19	19	20

5.7   Entering the data

Step 1

In 'Variable View' of the 'Data Editor' name the first variable 'Age'.
Remove the two decimal places.

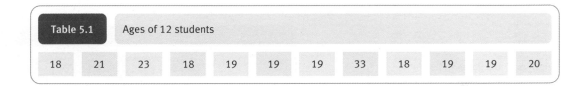

Step 2

In 'Data View' of the 'Data Editor' enter the ages in the first column.

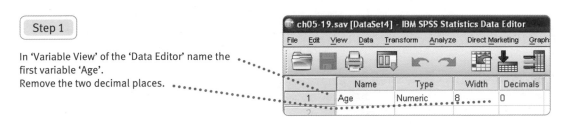

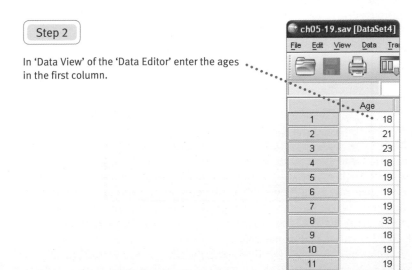

5.8 Mean, median, mode, standard deviation, variance and range

Step 1

Select 'Analyze', 'Descriptive Statistics' and 'Frequencies...'.

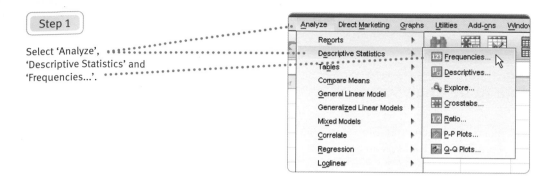

Step 2

Select 'Age' and the ➡ button to put it into the 'Variable(s):' box. De-select 'Display frequencies tables:'. Ignore the warning message. Select 'Statistics...'.

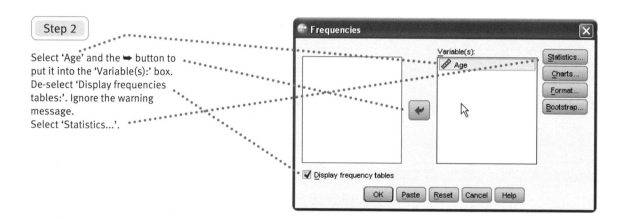

Step 3

Select 'Mean', 'Median', 'Mode', 'Std deviation', 'Variance' and 'Range'. Select 'Continue'. Select 'OK' (as in the above box).

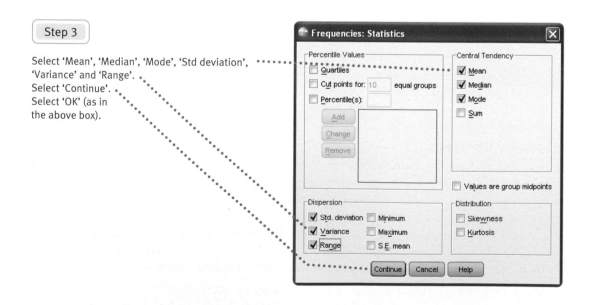

5.9   Interpreting the output

Statistics

Age

N	Valid	12
	Missing	0
Mean		20.50
Median		19.00
Mode		19
Std. Deviation		4.189
Variance		17.545
Range		15

There are 12 cases with valid data on which the analysis is based.

There is no (0) missing data. See Chapter 41 on 'missing values' for a discussion.

The mean age (mathematical average) = 20.5 years.

The median age (the age of the person halfway up a list of ages from smallest to largest) is 19.00.

The modal (most common) age is 19. Check the next table on the output. This gives frequencies for each value. A variable may have more than one mode.

The standard deviation is 4.189.

The variance is 17.545.

The range (the difference between the lowest and highest score) is 15.

5.10   Other features

You will see from the dialog box in Step 3 of Section 5.8 there are many additional statistical values which may be calculated. You should have little difficulty obtaining these by adapting the steps already described.

- *Percentiles* – indicate the cutoff points for percentages of scores. Thus the 90th percentile is the score which cuts off the bottom 90 per cent of scores in terms of size.

- *Quartiles* – values of a distribution which indicate the cutoff points for the lowest 25 per cent, lowest 50 per cent and lowest 75 per cent of scores.

- *Sum* – the total of the scores on a variable.

- *Skewness* – frequency distributions are not always symmetrical about the mean. Skewness is an index of the asymmetry or lop-sidedness of the distribution of scores on a variable. It takes a positive value if the values are skewed to the left and a negative value if they are skewed to the right.

- *Kurtosis* – an index of how much steeper or flatter the distribution of scores on the variable is compared with the normal distribution. It takes a '+' sign for steep frequency curves and a '–' sign for flat curves.

- *Standard deviation (estimate)* – this is a measure of the amount by which scores differ on average from the mean of the scores on a particular variable. Its method of calculation involves unusual ways of calculating the mean. In SPSS the standard deviation is calculated as an estimate of the population standard deviation. It is an index of the variability of scores around the mean of a variable. Some authors call this the sample standard deviation.

- *Variance (estimate)* – this is a measure of the amount by which scores on average vary around the mean of the scores on that variable. It is the square of the standard deviation and is obviously therefore closely related to it. It is also always an estimate of the population variance in SPSS. Some authors call this the sample variance. Like standard deviation, it is an index of the variability of scores around the mean of a variable but also has other uses in statistics. In particular, it is the standard unit of measurement in statistics.

- *Range* – the numerical difference between the largest and the smallest scores obtained on a variable. It is a single number.
- *Minimum (score)* – the value of the lowest score in the data for a particular variable.
- *Maximum (score)* – the value of the highest score in the data for a particular variable.
- *Standard error (SE mean)* – the average amount by which the means of samples drawn from a population differ from the population mean. It is calculated in an unusual way. Standard error can be used much like standard deviation and variance as an index of how much variability there is in the scores on a variable.

REPORTING THE OUTPUT

- The mean, median and mode can be presented as a table such as Table 5.2.

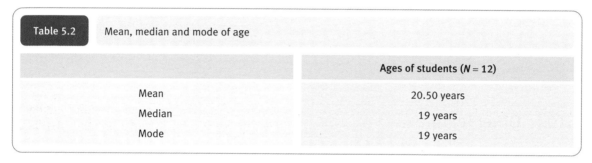

Table 5.2	Mean, median and mode of age	
		Ages of students (*N* = 12)
	Mean	20.50 years
	Median	19 years
	Mode	19 years

- Two decimal places are more than enough for most data. Most measurement is approximate, and the use of several decimal places tends to imply an unwarranted degree of precision.
- For the median, it is probably less confusing if you do not report values as 19.00 but as 19. However, if the decimal places were anything other than .00 then this should be reported since it indicates that the median is estimated and does not correspond to any actual scores for these particular data.

Summary of SPSS Statistics steps for descriptive statistics

Data

- In 'Variable View' of the 'Data Editor', 'name' the variables.
- In 'Data View' of the 'Data Editor' enter the data under the appropriate variable names.

Analysis

- Select 'Analyze', 'Descriptive Statistics' and 'Frequencies …'.
- Move the variables to be analysed to the right-hand box.
- Select 'Statistics …', 'Mean', 'Median', 'Mode', 'Variance' and 'Range'.

Output

- These statistics are presented in a single table.

For further resources including data sets and questions, please refer to the website accompanying this book.

CHAPTER 6

Shapes of distributions of scores

Overview

- It is important to study the shape of the distribution of scores for each variable. Ideally for most statistical techniques, a distribution should be symmetrical and normally distributed (bell-shaped).

- Some statistical techniques are at their most powerful when the distributions of the variables involved are normally distributed. Major deviations from normality should be avoided but, for relatively small sample sizes, visual inspection of frequency diagrams is the only practical way to assess this. The effects of failure to meet this criterion can be overstated. Sometimes it is possible to transform one's scores statistically to approximate a normal distribution but this is largely a matter of trial and error, using, for example, logarithmic scales.

- Nevertheless, researchers should be wary of very asymmetrical (skewed) distributions and distributions that contain a few unusually high or low scores (outliers). Histograms, for example, can be used to help detect asymmetry and outliers.

- Consider combining ranges of scores together (as opposed to tabulating each individual possible score) in order to clarify the distribution of the data. Small sample sizes, typical of much work in psychology and other social sciences, can lead to a sparse figure or diagram in which trends are not clear.

6.1 What are the different shapes of scores?

The initial stages of data analysis involve calculating and plotting statistics which describe the shape of the distribution of scores on each variable. Knowing how your data are distributed is obviously an intrinsic part of any study though it is frequently bypassed by those who see the primary aim of research as checking for statistical significance rather than understanding the characteristics of the things they are investigating. But there are other reasons for examining the distributions of scores. Statistical techniques are developed based on certain working assumptions. A common one is that scores on the variables follow the pattern of the bell-shaped curve

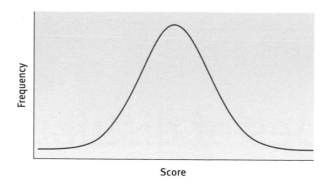

FIGURE 6.1   The normal distribution

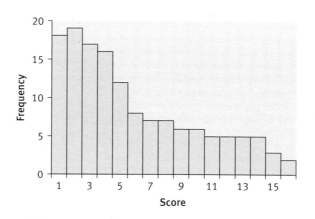

FIGURE 6.2   An example of a skewed distribution

or normal distribution (Figure 6.1). Since many statistical techniques are built on this assumption, if the data being analysed do not correspond to this ideal then one would expect, as is the case, that the statistical techniques would work less than perfectly. Some departures from the 'ideal' bell-shaped curve can affect how powerful (capable of giving statistically significant results) the test is. Distributions which are heavily skewed to the left or to the right may have this effect. Figure 6.2 is an example of a distribution skewed to the left.

Apart from skewness in a distribution, some distributions depart from the ideal normal distribution by being relatively steep or relatively flat (Figures 6.3 and 6.4). The degree of flatness and steepness is known as kurtosis. Kurtosis compares the shape of a distribution with that of the normal distribution using a special formula but SPSS will calculate it for you (see Section 5.10 in Chapter 5). If the curve is steep then the kurtosis has a positive value, if the curve is flat then the kurtosis has a negative value.

We discussed histograms in Chapter 4. There is more to be said about them – in particular the concept of the frequency curve. A frequency curve is basically a smoothed-out line which joins the peaks of the various bars of the histogram. This smoothed-out line fits very uncomfortably if there are just a few different values of the scores obtained. It is not easy to fit a smooth curve to the histogram that appears at the end of Chapter 4 (p. 54). The more data points the better from this point of view. Figure 6.5 is a histogram with many more data points – it is easier to fit a relatively smooth curve in this case as can be seen from the roughly drawn curve we have imposed on the figure.

As might be expected and assuming that all other things are equal, the more data points the better the fit the data can be to the normal curve. For this reason, often continuous variables

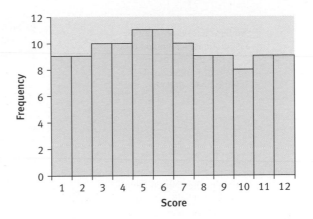

FIGURE 6.3 A flat frequency distribution

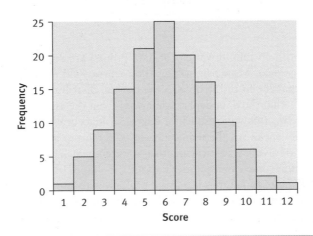

FIGURE 6.4 A steep frequency distribution

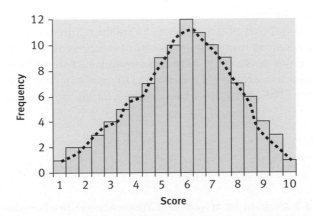

FIGURE 6.5 A histogram with more data points, making frequency curve fitting a little easier

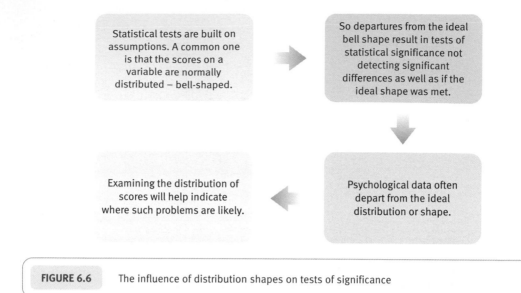

FIGURE 6.6	The influence of distribution shapes on tests of significance

(those with an infinite variety of data points) are regarded as ideal. Few psychological measures meet this ideal. Most psychological measures yield relatively bumpy frequency curves. Again this means that the fit of the data to the normal distribution is not as good as it could be. The consequence is that the data depart somewhat from the assumptions on which many statistical tests are built. Again this will result in a degree of inaccuracy though psychologists happily tolerate this. Generally speaking, these inaccuracies will tend to make the statistical test less powerful (i.e. less capable of giving statistical significant outcomes). This is illustrated in Figure 6.6. Figures 6.7 and 6.8 outline some essential ideas about the shapes of distributions.

Of course, sometimes it may be preferable to use frequency tables to look at the way in which the scores on a variable are distributed. Naturally, this makes it more difficult to identify a normal distribution but it does make it easier to spot errors in data entry (e.g. scores out of the range).

6.2 When to use histograms and frequency tables of scores

It is always highly desirable to obtain histograms and/or tables of frequencies of scores. This is an essential part of data analysis as conducted by a good researcher. Not to do so is basically incompetent as many important bits of information are contained within these simple statistics. Authorities on statistics have often complained that psychologists fail to use simple statistical techniques to explore their data prior to more complex analyses. Too often this failing is encouraged by SPSS instruction manuals which hardly mention descriptive statistics.

6.3 When not to use histograms and frequency tables of scores

Use a histogram in favour of a frequency table when the frequency table would contain so many categories of scores that it would be unwieldy. Frequency tables that are too big can be problematic.

Bell-shaped (normal) curves

- Many statistical techniques are based on the idea that the underlying frequency distribution of the scores is a symmetrical bell-shape.
- Few frequency distributions correspond to this bell-shape but often some deviation from this ideal makes little practical difference.
- The standard normal curve is one which has a mean of zero and a standard deviation of 1.

Skewed and flat/ steep distributions

- Some frequency distributions are markedly asymmetrical and have disproportionate numbers of scores towards either the left or the right. This is known as the degree of skew.
- Other distributions can be either very flat or very steep – this is known as the degree of kurtosis.
- Such distributions can introduce a degree of inaccuracy if the tendency is marked.

Non-normal distributions

- Where a frequency distribution departs markedly from the bell-shaped normal distribution, then there is a possibility that non-parametric statistical tests might be preferred. Some of these are discussed in Chapters 18 and 19.
- In practice, many psychologists pefer to use tests for normally distributed data in these circumstances since the inaccuracy introduced is usually small and there is a restricted range of non-parametric techniques available.

FIGURE 6.7 Essential ideas for shapes of distributions of scores

Frequency table

- This is a simple table which indicates how frequent each observation is.
- Where there are many different values then a frequency table can be unwieldy unless you use ranges of values rather than individual values.

Histogram

- This is a pictorial way of representing the frequencies of scores.
- Ideally, each bar of the histogram should be proportionate in width to the range of scores that the bar indicates. Unfortunately, this is not always the case.
- The heights of the bars indicate the relative frequencies of scores.

FIGURE 6.8 Ideas important in the shapes of distributions

6.4 Data requirements for histograms and frequency tables of scores

Histograms require score variables, but frequency tables can also be computed for nominal (category) data.

6.5   Problems in the use of histograms and frequency tables of scores

The major problem occurs when frequency tables are computed without consideration of the number of scores or categories that it will produce. Any variable can be used to generate a frequency table on SPSS but it needs to be remembered that for every different score then a new category will be created in the table, so if, for example, you measured people's heights in millimetres the resulting frequency table would be horrendous. This is a common problem with SPSS – it readily generates output whether or not the requested output is sensible. The adage 'junk in – junk out' is true but it is also true that thoughtless button pressing on SPSS means that a wheelbarrow may be needed to carry the printout home. Of course, it is possible to use SPSS procedures such as recode (see Chapter 42) to get output in a manageable and understandable form.

Be wary of the first charts and diagrams that SPSS produces. Often these need to be edited to produce something optimal.

Finally, it is common to find that student research is based on relatively few cases. This can result in somewhat difficult tables and diagrams unless some care is taken. It is very difficult to spot shapes of distributions if there are too many data points in the table or diagram. Consequently, sometimes it is better to use score ranges rather than the actual scores in order to generate good tables and diagrams. This will involve recoding your values (Chapter 42). Remember that in the worst cases SPSS will generate a different category for each different score to be found in your data. Recoding data would involve, for example, expressing age in ranges such as 15–19 years, 20–24 years, 25–39 years and so forth. Some experimentation may be necessary to achieve the best tables and diagrams.

You can find out more about distributions of scores in Chapter 4 of Howitt, D. and Cramer, D. (2011). *Introduction to Statistics in Psychology*, 5th edition. Harlow: Pearson.

6.6   The data to be analysed

We will compute a frequency table and histogram of the extraversion scores of the 50 airline pilots shown in Table 6.1 (*ISP*, Table 4.1).

Table 6.1	Extraversion scores of 50 airline pilots								
3	5	5	4	4	5	5	3	5	2
1	2	5	3	2	1	2	3	3	3
4	2	5	5	4	2	4	5	1	5
5	3	3	4	1	4	2	5	1	2
3	2	5	4	2	1	2	3	4	1

6.7   Entering the data

Step 1

In 'Variable View' of the 'Data Editor' name the first variable 'Extrav'.
Remove the two decimal places.
Save this data as a file to use for Chapter 47.

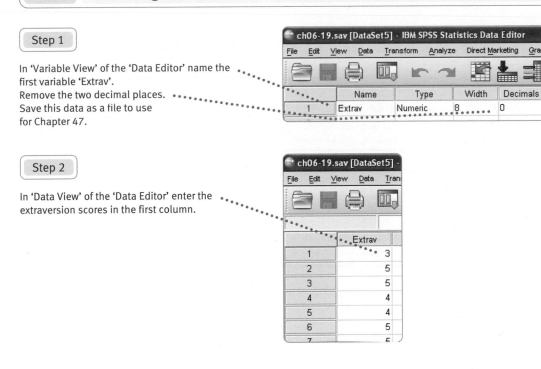

Step 2

In 'Data View' of the 'Data Editor' enter the extraversion scores in the first column.

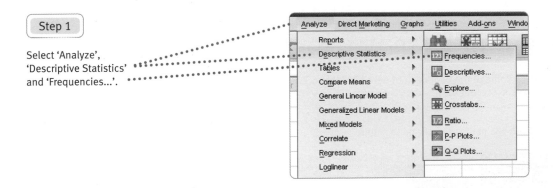

6.8   Frequency tables

Step 1

Select 'Analyze', 'Descriptive Statistics' and 'Frequencies...'.

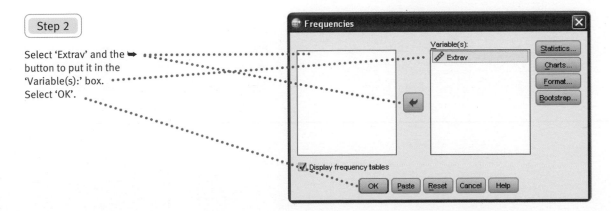

Step 2

Select 'Extrav' and the ➡ button to put it in the 'Variable(s):' box.
Select 'OK'.

6.9   Interpreting the output

The first column shows the five values of extraversion which are 1 to 5.

The second column shows the frequency of these values. There are seven cases with a value of 1.

Extrav

		Frequency	Percent	Valid Percent	Cumulative Percent
Valid	1	7	14.0	14.0	14.0
	2	11	22.0	22.0	36.0
	3	10	20.0	20.0	56.0
	4	9	18.0	18.0	74.0
	5	13	26.0	26.0	100.0
	Total	50	100.0	100.0	

The fifth column adds these percentages together cumulatively down the table. So 56% of the cases have values of 3 or less.

The third column expresses these frequencies as a percentage of the total number including missing data. Of all cases, 14% have a value of 1.

The fourth column expresses these frequencies as a percentage of the total number excluding missing data. As there are no missing cases the percentages are the same as in the third column.

REPORTING THE OUTPUT

Notice that we omitted some of the confusion of detail in Table 6.2. Tables and diagrams need to clarify the results.

Table 6.2   One style of reporting the table output

Extraversion Score	Frequency	Percentage frequency	Cumulative percentage frequency
1	7	14.0	14.0
2	11	22.0	36.0
3	10	20.0	56.0
4	9	18.0	74.0
5	13	26.0	100.0

6.10   Histograms

Select 'Graphs',
'Legacy Dialogs' and
'Histogram...'.
An alternative way is
to use 'Chart Builder...'.
This procedure is also
shown in Chapters 4 and 10.

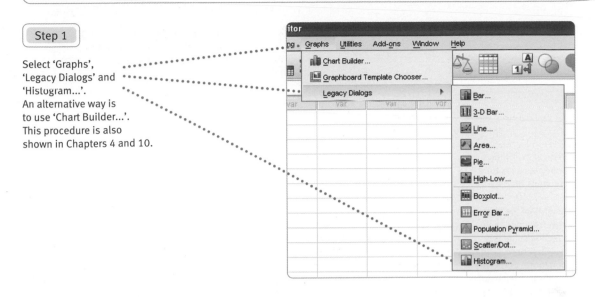

Select 'Extrav' and the ➡
button to put it in the
'Variable:' box.
Select 'OK'.

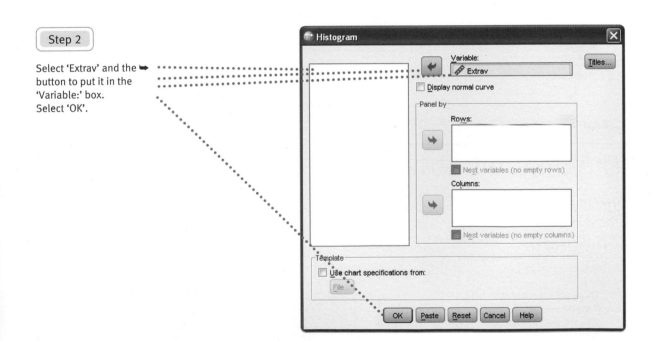

6.11   Interpreting the output

The mean extraversion score is 3.2.
The standard deviation is 1.414.
The number of cases is 50.

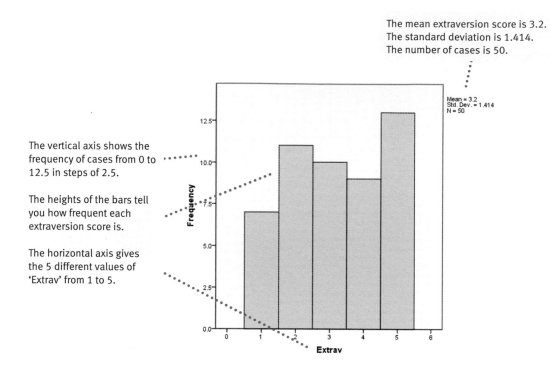

The vertical axis shows the frequency of cases from 0 to 12.5 in steps of 2.5.

The heights of the bars tell you how frequent each extraversion score is.

The horizontal axis gives the 5 different values of 'Extrav' from 1 to 5.

REPORTING THE OUTPUT

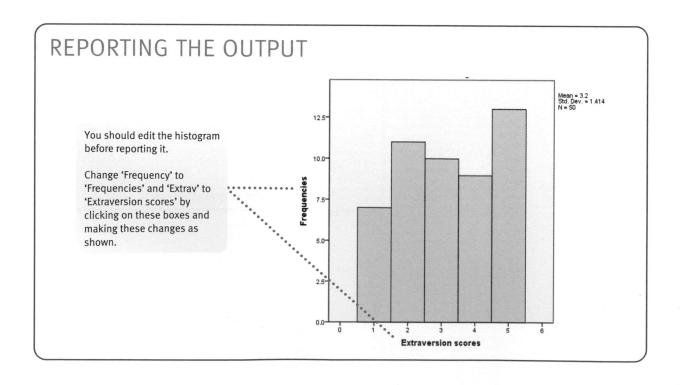

You should edit the histogram before reporting it.

Change 'Frequency' to 'Frequencies' and 'Extrav' to 'Extraversion scores' by clicking on these boxes and making these changes as shown.

Summary of SPSS Statistics steps for frequency distributions

Data

- In the 'Variable View' of the 'Data Editor', 'name' the variables.
- In the 'Data View' of the 'Data Editor' enter the data under the appropriate variable names.

Analysis

- For frequency tables, select 'Analyze', 'Descriptive Statistics' and 'Frequencies ...'.
- Move variables to be analysed to right hand side box.
- For histograms, select 'Graphs', and then either 'Chart Builder ...' (see Chapters 4 and 10) or 'Legacy Dialogs' (this chapter and Chapters 9 and 11)
- For 'Chart Builder ...', select 'OK' and 'Histogram'.
- Move the appropriate graph to the box above and the appropriate variable to the 'X-Axis?' box.
- For 'Legacy Dialogs', select 'Histogram' and then move the variable to the 'Variable:' box.

Output

- The frequency table shows the frequency as well as the percentage of the frequency for each value.
- If needed, use the 'Chart Editor' to edit the histogram.

For further resources including data sets and questions, please refer to the website accompanying this book.

CHAPTER 7

Standard deviation

The standard unit of measurement in statistics

7.1   What is standard deviation?

Whoever invented the standard deviation did not have psychology students in mind. Like a lot of statistical concepts, it was not designed to be easily comprehended by mere mortals but for its very special mathematical properties. It is actually fairly easy to say what the standard deviation is but much harder to understand what it is. Standard deviation is no more or less than the square root of the variance of a set of scores as depicted in Figure 7.1. If you understood the concept of variance (Chapter 5) then the standard deviation is simply the square root of the value of the variance. Unfortunately, the concept of variance is not that easy either since it is calculated by taking each score in a set of scores and subtracting the mean from that score, squaring each of these 'deviations' or differences from the mean, summing them up to give the total, and finally

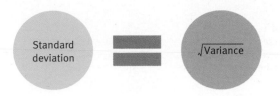

FIGURE 7.1   Calculating standard deviation from variance

dividing by the number of scores (or the number of scores – 1 to calculate the variance estimate). Easy enough computationally but it is a bit of a struggle to fix the concept in one's head.

Notice that the standard deviation is simply the square root of the variance, so it is the square root of something which involved squaring the deviation of each of the scores from the mean. Since it is the square root of squares the standard deviation gets us near to getting back to the original deviation anyway. And this is what conceptually the standard deviation is – a sort of average amount by which scores differ (deviate) from the mean. That's what it is but most of us would simply have taken the average deviation from the mean in the first place without all of this squaring and square rooting. The gods who invented statistics had worked out that the common-sense measure of average deviation from the mean did not have the mathematical advantages that standard deviation had.

Standard deviation comes into its own when our data have a normal distribution (the bell-shaped distribution we discussed in Chapter 6). In these circumstances, it so happens that if one counts the number of standard deviations a score is from the mean then one can say exactly what proportion of scores lie between the mean and that number of standard deviations from the mean. So if a score is one standard deviation from the mean then 34.13 per cent of scores lie between this score and the mean; if a score is two standard deviations from the mean then 47.72 per cent of scores lie between this score and the mean. How do we know this? Simply because these figures are a property of the normal curve. Most importantly, these figures apply to every normal curve. There are tables which give the percentages for every number of standard deviations from the mean.

These precise figures only apply to the normal distribution. If one does not have data which correspond to the normal curve then these figures are not accurate. The greater the deviation from the normal curve the more inaccurate become these figures. Perhaps now it is easier to see the importance of the normal distribution and the degree to which the normal distribution describes the frequency curve for your data.

One application of the standard deviation is something known as z-scores. This is simply a score re-expressed in terms of the number of standard deviations it is away from the mean score. So the z-score is simply the score minus the mean score of the set of data then divided by the standard deviation. In other words, the z-score is the number of standard deviations a score is away from the mean score. This is important since no matter the precise nature of a set of scores, they can always be converted to z-scores which then serve as a standard unit of measurement in statistics. It also means that one can quickly convert this number of z-scores to the proportion of scores which the score lies away from the mean of the set of scores.

You may not wish to do this for a particular set of data, but you need to know that many of the statistics described in this book can involve standardised measures which are very closely related to z-scores (e.g. multiple regression, log-linear analysis and factor analysis).

By the way, SPSS actually does not compute standard deviation but something called the estimated standard deviation. This is slightly larger than the standard deviation. Standard deviation applies when you simply wish to describe the characteristic of a set of scores whereas estimated standard deviation is used when one is using the characteristics of a sample to estimate the same characteristic in the population from which the sample came. It would be good to label estimated standard deviation as such despite what SPSS says in its output. Figure 7.2 outlines the main steps in calculating standard deviation.

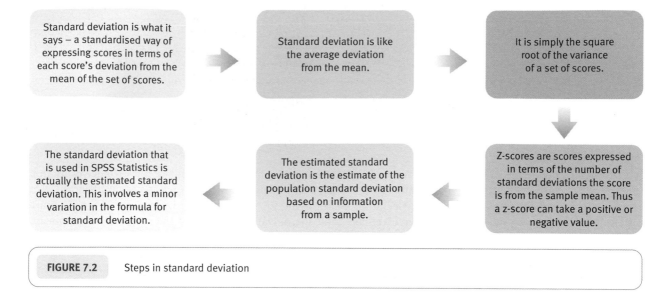

FIGURE 7.2   Steps in standard deviation

7.2   When to use standard deviation

Standard deviation is always valuable when considering any score variable. It is clearly much more informative when it is based on a normally distributed set of scores than where the distribution deviates substantially from that ideal. Variance, standard deviation and standard error (Chapter 5) are all used as measures of the variation in the scores on a variable. This is because there is such a close mathematical relation between each of them. The square root of variance is standard deviation and the variance divided by the square root of the sample size is the standard error. Of course, it is important to be consistent in terms of which one is used.

7.3   When not to use standard deviation

Do not use standard deviation when dealing with nominal (category variables). SPSS will do a calculation for you but the outcome is mathematically precise gibberish in this case.

7.4   Data requirements for standard deviation

Standard deviation can be calculated only on score data. Ideally, the variable in question should be normally distributed and, some would say, be measured on an equal-interval scale.

7.5   Problems in the use of standard deviation

Getting to understand standard deviation is a bit like a child trying to play a concerto on the piano before learning to play simple scales. That is, standard deviation, because of its abstract nature, is very difficult to grasp but invariably taught early in a statistics module.

Because of this, novices often stare blankly at SPSS output for standard deviation since it does not immediately tell a story to them. It's rather like, however, finding out that the average length of a plank of wood is 1.6 metres long – it is something to be taken for granted. Standard deviation is simply a sort of average amount by which scores deviate from the mean.

Remember that what SPSS calls standard deviation should really be labelled estimated standard deviation if one wishes to be precise.

You can find out more about standard deviation and *z*-scores in Chapter 5 of Howitt, D. and Cramer, D. (2011). *Introduction to Statistics in Psychology*, 5th edition. Harlow: Pearson.

7.6 The data to be analysed

The computation of the standard deviation and *z*-scores is illustrated with the nine age scores shown in Table 7.1 (based on *ISP*, Table 5.1).

Table 7.1	Data for the calculation of standard deviation							
Age	20	25	19	35	19	17	15	30	27

7.7 Entering the data

Step 1

In 'Variable View' of the 'Data Editor' name the first variable 'Age'. Remove the two decimal places.

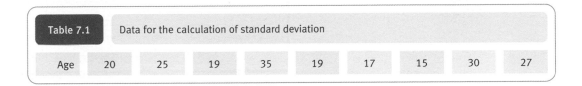

Step 2

In 'Data View' of the 'Data Editor' enter age in the first column.

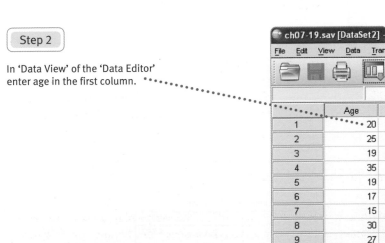

7.8   Standard deviation

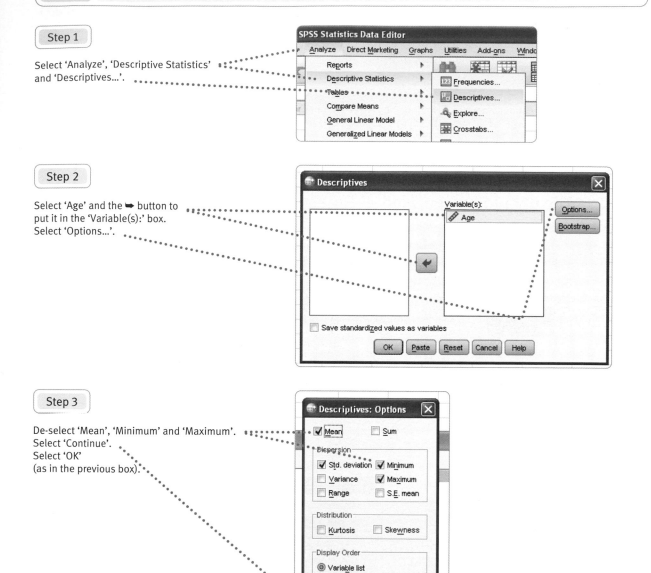

Step 1

Select 'Analyze', 'Descriptive Statistics' and 'Descriptives...'.

Step 2

Select 'Age' and the ➡ button to put it in the 'Variable(s):' box.
Select 'Options...'.

Step 3

De-select 'Mean', 'Minimum' and 'Maximum'.
Select 'Continue'.
Select 'OK'
(as in the previous box).

7.9   Interpreting the output

The number of cases is 9.
The standard deviation of 'Age' is 6.652.

Descriptive Statistics

	N	Std. Deviation
Age	9	6.652
Valid N (listwise)	9	

7.10   Z-scores

Step 1

In Step 2 of Section 7.8 select 'Save standardized values as variables'. Select 'OK'.

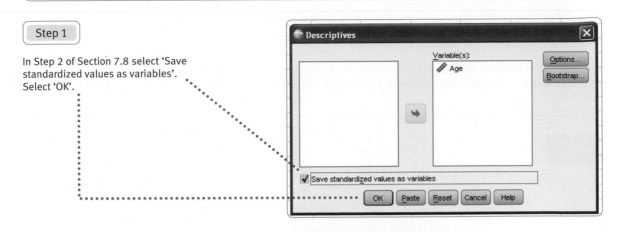

Step 2

The z or standardised scores are in the second column of 'Data View' in the 'Data Editor' and are called 'ZAge'.

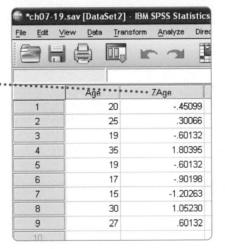

7.11   Other features

'Descriptives …' contains a number of statistical calculations which can easily be selected:

- Mean
- Sum
- Standard deviation (estimate)
- Range
- Minimum (score)
- Maximum (score)
- Standard error (S.E. mean)
- Kurtosis
- Skewness

These different statistical concepts are briefly explained in Chapter 5.

REPORTING THE OUTPUT

The standard deviation of just one variable can easily be mentioned in the text of your report:

The standard deviation of age was 6.65 years ($n = 9$).

However, it is more likely that you would wish to record the standard deviation alongside other statistics such as the mean and range, as illustrated in Table 7.2. You would probably wish to include these statistics for other numerical score variables that you have data on.

Table 7.2 The sample size, mean, range and standard deviations of age, IQ and verbal fluency

	n	Mean	Range	Standard deviation
Age	9	23.00	20.00	6.65
IQ	9	122.17	17.42	14.38
Verbal fluency	9	18.23	4.91	2.36

Summary of SPSS Statistics steps for standard deviation

Data

- In 'Variable View' of the 'Data Editor', 'name' the variables.
- In 'Data View' of the 'Data Editor' enter the data under the appropriate variable names.

Analysis

- Select 'Analyze', 'Descriptive Statistics' and 'Descriptives ...'.
- Move the variables to be analysed to the 'Variable(s):' box.
- Select 'Save standardised values as variables'.

Output

- The standard deviation is presented in a table with other default statistics unless these are de-selected.
- The standardised values are presented in the next free column of 'Data View' in the 'Data Editor' with the name of the original variable starting with a Z.

For further resources including data sets and questions, please refer to the website accompanying this book.

CHAPTER 8

Relationships between two or more variables

Tables

Overview

- A great deal of research explores the relationship between two or more variables. The univariate (single-variable) statistical procedures described so far have their place in the analysis of practically any data. Nevertheless, most research questions also require the interrelationships or correlations between different variables to be addressed.

- As with univariate statistics, a thorough bivariate statistical analysis of data requires an exploration of the basic trends in the data using cross-tabulation tables.

- Care has to be taken to make sure that the tables you obtain are useful and communicate well. In particular, ensure that your data for cross-tabulation tables contain only a small number of different data values. If they do not, SPSS Statistics will produce massive tables.

- Labelling tables in full is a basic essential, along with a clear title.

- The type of table which is most effective at communicating relationships in your data depends very much on the types of data involved. Two nominal variables are generally presented in terms of a cross-tabulation table.

- Remember that the effective presentation of basic descriptive statistics requires that researchers consider their tables carefully. Often researchers will have to 'tweak' things to make the finished tables effective forms of communication. That this can be done quickly is one of the advantages of using SPSS.

- It is recommended that the temptation to simply cut and paste tables from SPSS output into reports is avoided. There is virtually no SPSS output which cannot be improved upon and made clearer. The basic options of SPSS are not always ideal in this respect.

8.1   What tables are used to show relationships between variables?

Relationships between variables are the foundation stone of most psychological research. The idea of relationships imbues many common psychological research questions. Is intelligence related to income? Does depression cause suicide? Is there a correlation between age and memory deterioration? Is there an association between childhood attachment difficulties and adult attachment difficulties? These are all examples of questions which imply an association or, in other words, a relationship. There are other research questions which superficially may not imply a relationship but nevertheless do when examined more carefully. For example, 'Is there a difference between men and women in terms of their emotional intelligence?' This is a research question which can be put another way: Is there a relationship between a person's gender and the strength of their emotional intelligence? In other words, the conventional psychological notion of tests of difference and tests of association which is found in many current popular textbooks is fundamentally misleading.

There are many statistical techniques for showing that there are relationships between variables, as shown in Figure 8.1. But in this chapter we will concentrate on tables which do this.

Showing relationships between two variables needs an understanding of the difference between nominal and score data.

It also needs a decision about whether you want a figure (chart) or a table.

Consider your two variables. Are both nominal variables? Are both score variables? Or have you got one nominal variable and one score variable?

The number of values of the variable is a basic problem even for nominal data. You may decide to reduce the number of values by combining categories or by turning scores into ranges of scores.

If both of your variables are nominal category variables then a **compound bar chart** or a **stacked bar chart** is likely to be effective. As ever, too many categories lead to complex and uncommunicative charts. A **contingency table** may be a suitable alternative.

If you have two score variables, then the **scatterplot** may be the best way of showing this relationship. If you divide your variables into ranges of scores then you may be able to use a **cross-tabulation table** to show the relationship.

If one variable is a score variable and the other variable is a nominal variable then this could be simply presented as a **table of means** giving the mean score for each nominal category. Alternatively a bar chart which gives a pictorial representation of the mean score for the nominal categories would be appropriate.

It takes considerable effort to create a successful table or chart. Not only does it need to be titled and labelled properly and fully but it is important to be critical of your work and to try alternatives and modifications in order to produce something which communicates effectively.

FIGURE 8.1   Steps in showing relationships between two variables

The appropriate tabular methods used to investigate relationships depend on the sorts of data in question:

- If both variables are nominal (category) variables then a cross-tabulation table is suitable. The value of analysis becomes problematic if there are too many categories in one or both of the variables. One thing that might be done is to combine categories together into one if that is a reasonable thing to do given the detail of your data.

- If *both* variables are score variables then a table is usually not appropriate. The problem with tables is usually the large number of different values of scores on each variable. This can be dealt with by breaking each score variable into a small number of score ranges prior to producing the table using SPSS. How data can be recoded is described in Chapter 42. Apart from that, it is simply a matter of computing the cross-tabulation table of the two variables – or recoded variables. It sometimes is not quite so apparent that a crosstabulation table shows a relationship between two variables. Look at the table in Figure 8.2. There is a correlation indicated there since the frequencies near the diagonal from top left to bottom right tend to be larger than the frequencies found in other sections (cells) of the table.

		Knowledge					
		1.00	2.00	3.00	5.00	7.00	Total
Aptitude	1.00	6	0	0	0	0	6
	2.00	0	3	0	0	0	3
	3.00	0	0	3	0	0	3
	4.00	0	0	6	3	0	9
	5.00	0	3	0	0	0	3
	6.00	0	0	3	0	0	3
	8.00	0	0	0	0	3	3
Total		6	6	12	3	3	30

FIGURE 8.2 A cross-tabulation table

- If one variable is a score variable and the other is a nominal (category) variable then it might be appropriate to use a simple table which names the category and then gives the mean (and possibly other statistics such as the standard deviation, maximum and minimum scores and so forth). Tables such as Table 8.1 are quick and easy to interpret. Another advantage is that a lot of similar analyses can be presented in a single, compact table.

Table 8.1 Means presented as a table

	Mean aptitude score
Males	3.88
Females	3.69

You can find out more about methods of relating variables in Chapter 6 of Howitt, D. and Cramer, D. (2011). *Introduction to Statistics in Psychology*, 5th edition. Harlow: Pearson.

8.2   When to use tables to show relationships between variables

It is hard to conceive of circumstances in which a researcher would not wish to use a table to show relationships. Basic descriptive statistics are not kindergarten statistics there to ease students into their studies. They are vital to truly understanding not just the characteristics of one's variables but also clarifying exactly what the findings of one's research are. Too often novice researchers fail to understand their statistical analysis simply because they have not studied the essential descriptive statistics as part of their data analysis. So often do novice researchers leap to the stage of testing statistical significance which leaves them knowing that something is statistically significant but they do not quite know what that something is or what it means. SPSS speeds up the process of data analysis remarkably so there is little excuse for skipping vital stages such as computing means and standard deviations, for example. Put simply, descriptive statistics are the way in which you discover what your data have to say and tests of significance simply tell you the extent to which what the data say is statistically reliable enough to have faith in the trends you discover.

8.3   When not to use tables to show relationships between variables

The key thing is to select a method of graphical or tabular analysis which suits the data in question. We have given advice which will help you find the appropriate method for the majority of cases but, of course, there is always an element of judgement to apply in certain instances. Furthermore, one has to consider what to do when one has many similar analyses to report. In these circumstances a different approach may be preferable. Simply do not generate too many separate tables – try to keep them to an essential minimum, otherwise your report will be too cluttered. Also remember that the primary purpose of the descriptive statistics is to help you with your data analysis – it does not follow that you should stuff your reports with them. Again, it is a matter of judgement how many you use.

8.4   Data requirements for tables to show relationships between variables

There are different data requirements for the different methods. Two score or nominal variables are dealt with carefully planned cross-tabulation tables. Situations in which you have one score variable and one nominal variable can be most adequately dealt with by giving the means for each category of the nominal variable.

8.5   Problems in the use of tables to show relationships between variables

While the SPSS default versions of tables and charts are adequate for most aspects of data analysis, one should be very careful to consider what is needed for inclusion in research reports. Here the main criterion has to be how readily the table actually communicates to the reader. Consequently you may need to adjust things such as the labelling of the table to make it readily interpretable and meaningful to the reader.

8.6   The data to be analysed

We will illustrate the drawing up of a cross-tabulation table with the data shown in Table 8.2 (*ISP*, Table 6.4). This shows the number of men and women in a study who have or have not been previously hospitalised.

Table 8.2	Cross-tabulation table of gender against hospitalisation	
	Female	**Male**
Previously hospitalised	$F = 25$	$F = 20$
Not previously hospitalised	$F = 14$	$F = 30$

8.7   Entering the data

Step 1

The quickest way to enter the data in Table 8.2 is to create the four cells as three rows.
To do this we need three variables.
In 'Variable View' of the 'Data Editor' name the first variable 'Hospitalisation', the second variable 'Gender' and the third variable 'Freq'.
Remove the two decimal places.

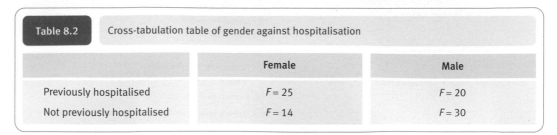

Step 2

Label the two values of 'Hospitalisation' (1 = 'Hospitalised'; 2 = 'Not hospitalised') and 'Gender' (1 = 'Females'; 2 = 'Males').
How to do this was discussed in Steps 5–10 in Section 2.8.

Step 3

Enter these numbers in 'Data View' of the 'Data Editor'.
The first row refers to 'Hospitalised' 'Females' of whom there are 25.
The second row to 'Hospitalised' 'Males' of whom there are 20.
The third row to 'Not hospitalised' 'Females' of whom there are 14.
The fourth row to 'Not hospitalised' 'Males' of whom there are 30.

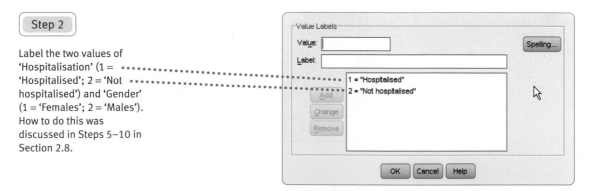

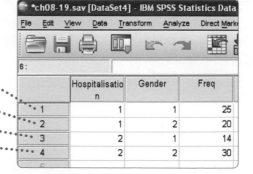

8.8 Weighting the data

Step 1

To weight the data so that the four cells have the appropriate number of cases in them, select 'Data' and 'Weight Cases...'.

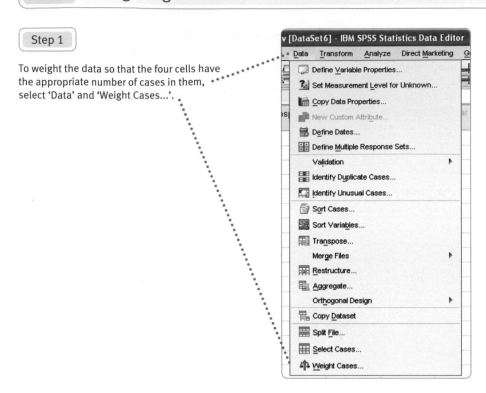

Step 2

Select 'Freq', 'Weight cases by' and the ➥ button to put it in the 'Frequency Variable:' box. Select 'OK'.

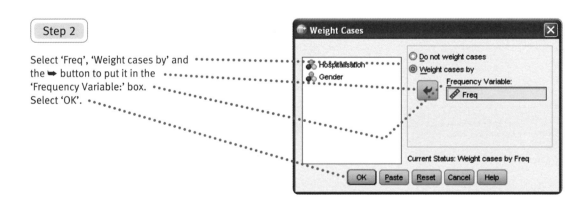

The cases are now weighted as shown by the 'Weight On' message in the lower right corner of the 'Data Editor'.

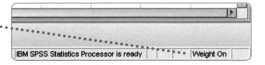

8.9   Cross-tabulation with frequencies

Step 1

Select 'Analyze',
'Tables' and 'Custom
Tables...'

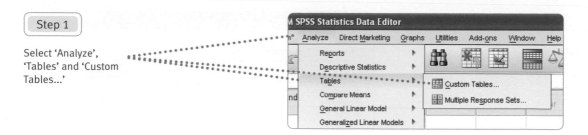

Step 2

If the variables have not been defined then do
so (see Step 3, Section 3.7). They should both
be nominal. Otherwise select 'OK'.

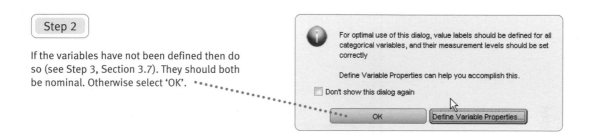

Step 3

Select 'Hospitalisation'
and drag it to 'Rows'.
Select 'Gender' and
drag it to 'Columns'.
Ensure 'nnnn' are in
the cells. If they are
not, go to Section 8.10
to see how to select
them.
Select 'OK'.

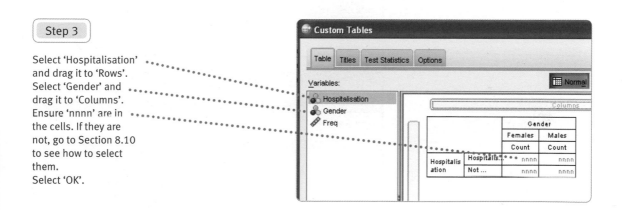

The table in the output is laid out
like Table 8.2. Unlike previous
versions of SPSS, presenting value
labels rather than numbers is the
default option.

		Gender	
		Females	Males
		Count	Count
Hospitalisation	Hospitalised	25	20
	Not hospitalised	14	30

8.10   Displaying frequencies as a percentage of the total number

Step 1

Select 'Hospitalisation' and
'Summary Statistics...'.

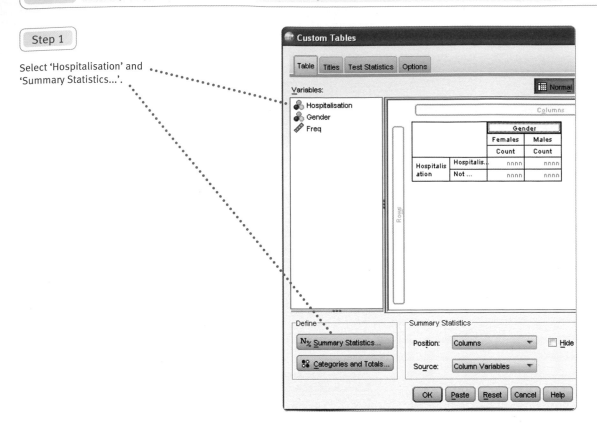

Step 2

Select 'Count' in
'Display:' and the ←
button to put it back
under 'Statistics:'
Select 'Table N%' and
the ➡ button to put it
under 'Display:'
Select 'Apply to
Selection' to return to
original dialog box.
Select 'OK'.

If you add the % in each
column they total 100.

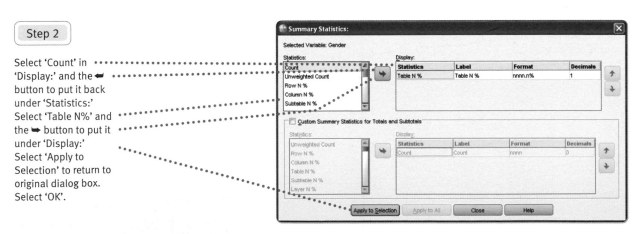

	Gender	
	1 Females	2 Males
	Table N %	Table N %
Hospitalisation 1 Hospitalised	28.1%	22.5%
2 Not hospitalised	15.7%	33.7%

8.11 Displaying frequencies as a percentage of the column total

Select 'Table N%' and the ← button to put it back under 'Statistics:' Select 'Col N%' and the → button to put it under 'Display:'
Select 'Apply to Selection' to return to original dialog box.
Select 'OK'.

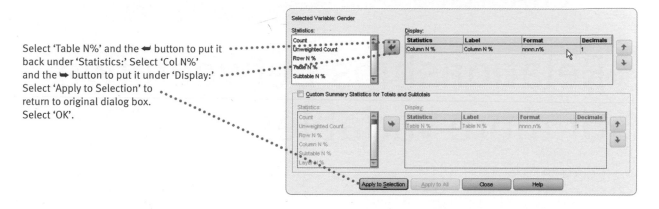

If you add the % in each column they total 100.

		Gender	
		Females	Males
		Column N %	Column N %
Hospitalisation	Hospitalised	64.1%	40.0%
	Not hospitalised	35.9%	60.0%

Summary of SPSS Statistics steps for contingency tables

Data

- In 'Variable View' of the 'Data Editor', 'name' the 'nominal' variables and select 'nominal' as the 'measure'.
- In 'Data View' of the 'Data Editor' enter the data under the appropriate variable names.

Analysis

- For a contingency table, select 'Analyze', 'Table' and 'Custom Tables ...'.
- Select 'OK', move the appropriate variables to the column and row boxes and select 'OK'.

Output

- Check tables are correct. They can be edited in the 'Chart Editor' if required.

For further resources including data sets and questions, please refer to the website accompanying this book.

CHAPTER 9

Relationships between two or more variables

Diagrams

- Another way of looking at and illustrating the relationship between two variables is through the use of diagrams and graphs such as compound (clustered) bar charts and scattergrams. These may be familiar to you already, but nevertheless can cause difficulties.

- Care has to be taken to make sure that the diagrams you obtain are useful and communicate well. In particular, ensure that your data for compound (clustered) bar charts only contain a small number of different data values. If they do not, SPSS Statistics will produce dense, unreadable graphs and diagrams.

- Labelling diagrams in full is essential, together with a clear title.

- Scattergrams work well when you have many different values for the scores on your variables.

- The type of diagram which is most effective at communicating relationships in your data depends very much on the types of data involved. Two score variables will generally be most effectively presented as a scattergram than a cross-tabulation table.

- Remember that the effective presentation of basic descriptive statistics requires that researchers consider their diagrams carefully. Often researchers will have to 'tweak' things to make the finished graphs effective forms of communication. This can be quickly done using SPSS.

- It is recommended that simply cutting and pasting diagrams from SPSS output into reports is avoided. There is virtually no SPSS output which cannot be improved upon and made clearer. The basic options of SPSS are not always ideal in this respect. Often the editing procedures available for SPSS charts can improve things enormously.

- Producing charts like those discussed in this chapter is one of the harder tasks using SPSS Statistics.

What diagrams are used to show relationships between variables?

As mentioned in the previous chapter, relationships between variables are the foundation stone of most psychological research. There are many statistical techniques for showing such relationships. In this chapter we will look at graphs which do this. The appropriate graphical methods used to investigate relationships depends on the sorts of data in question:

● If both variables are nominal (category) variables then the simplest graphical procedure is to produce a compound bar chart as in Figure 9.1 (or alternatively a stacked bar chart as in Figure 9.2). The value of analysis becomes problematic if there are too many categories in one or both of the variables. One thing that might be done is to combine categories into one if that is a reasonable thing to do given the detail of your data.

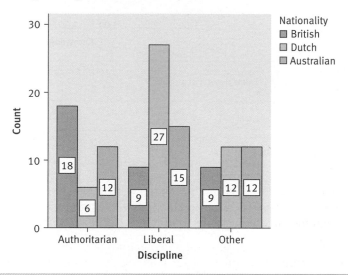

FIGURE 9.1   A compound bar chart

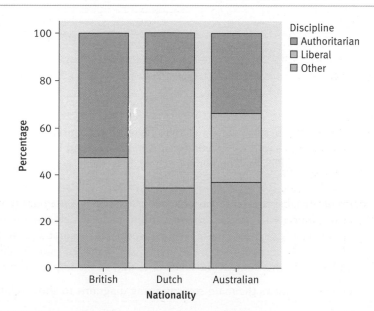

FIGURE 9.2   A stacked bar chart

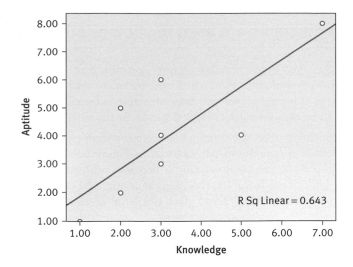

FIGURE 9.3   Example of a scatterplot

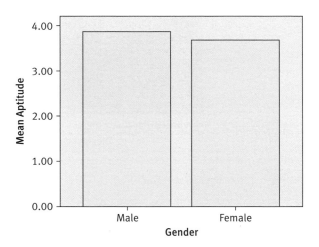

FIGURE 9.4   A bar chart representing means

- If *both* variables are score variables then the appropriate graphical method is the scattergram or scatterplot. An example of a scatterplot is given in Figure 9.3. We explain how to do a scatterplot on SPSS in Chapter 11 where it is dealt with in conjunction with the correlation coefficient.

- If one variable is a score variable and the other is a nominal (category) variable then it might be appropriate to produce the sort of bar chart which essentially displays the means for each nominal category (see Figure 9.4). However, there is not a great deal to be gained from the use of such a chart other than when giving, say, a PowerPoint presentation of some research to an audience. In this context the chart does add to the impact of the presentation.

Figure 9.5 shows the main steps in using diagrams to show relationships between variables.

You can find out more about methods of relating variables in Chapter 6 of Howitt, D. and Cramer, D. (2011). *Introduction to Statistics in Psychology*, 5th edition. Harlow: Pearson.

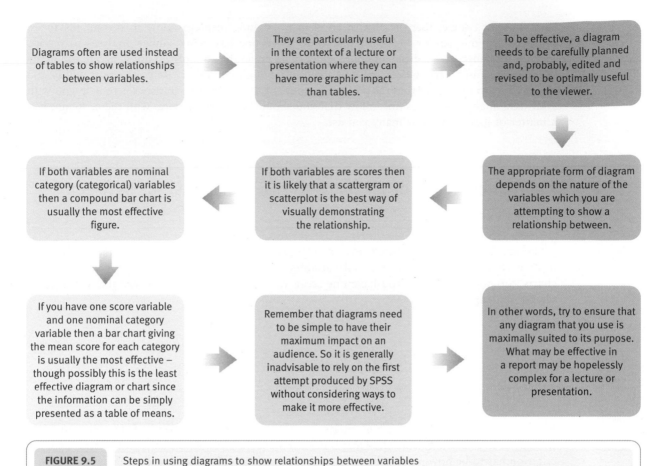

FIGURE 9.5   Steps in using diagrams to show relationships between variables

9.2   When to use diagrams to show relationships between variables

It is difficult to think of situations in which a researcher would not wish to use some of the graphical methods of showing relationships. Basic descriptive statistics are not only simple statistics there to help students into their studies. They are essential to truly understanding not just the characteristics of one's variables but also clarifying exactly what the findings of one's research are. Too often beginning researchers fail to understand their statistical analysis simply because they have not examined the essential descriptive statistics as part of their data analysis. Novice researchers often jump to the stage of testing statistical significance which leaves them knowing that something is statistically significant but they do not quite know what that something is or what it means. SPSS hastens the process of data analysis remarkably so there is little excuse for avoiding vital stages such as checking scattergrams. Put simply, descriptive statistics are the way in which you discover what your data have to say. Tests of significance simply tell you the extent to which what the data say is statistically reliable enough to have faith in the trends you discover.

9.3   When not to use diagrams to show relationships between variables

The essential thing is to select a method of graphical analysis which suits the data in question. We have given advice which will help you find the appropriate method for the majority of cases but, of course, there is always an element of judgement to apply in certain instances. Furthermore, one has to consider what to do when one has many similar analyses to report. In these circumstances a different approach may be preferable. Simply do not generate too many separate diagrams – try to keep them to an essential minimum otherwise your report will be too cluttered. Also remember that the primary purpose of the descriptive statistics is to help you with your data analysis – it does not follow that you should stuff your reports with them. Again, it is a matter of judgement how many you use.

9.4 Data requirements for diagrams to show relationships between variables

There are different data requirements for different types of diagrams. Two score variables are dealt with scattergrams; two nominal variables are dealt with using stacked or clustered bar charts. Situations in which you have one score variable and one nominal variable can be presented in a bar chart.

9.5 Problems in the use of diagrams to show relationships between variables

While the SPSS default versions of charts are adequate for most aspects of data analysis, one should be very careful to consider what is needed for inclusion in research reports. Here the main criterion has to be how readily the chart actually communicates to the reader. Consequently you may need to adjust things such as the labelling of the diagram to make it readily interpretable and meaningful to the reader.

Errors in the selection of appropriate charts can often be seen in the unusual look to them. In particular, scattergrams where a nominal variable has inadvertently been included as one of the variables tend to have the data points seemingly stacked vertically. Be on the look out for oddities like this. Figure 9.6 is a good example. Why does it look odd? Simply because gender has been included, which only has two categories. Hence the points on the plot pile up on top of each other.

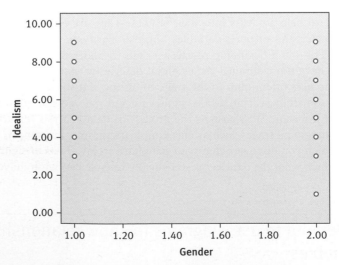

FIGURE 9.6   There are problems with this scatterplot, as explained in the text

9.6   The data to be analysed

We will illustrate the drawing up of a cross-tabulation table and compound bar chart with the data shown in Table 9.1 (*ISP*, Table 6.4). This shows the number of men and women in a study who have or have not been previously hospitalised. If your data are already entered into SPSS then Steps 1 to 5 may be ignored.

Table 9.1	Cross-tabulation table of gender against hospitalisation

	Female	Male
Previously hospitalised	$F = 25$	$F = 20$
Not previously hospitalised	$F = 14$	$F = 30$

9.7   Entering the data

Step 1

The quickest way to enter the data in Table 9.1 is to create the four cells as three rows. To do this we need three variables.
In 'Variable View' of the 'Data Editor' name the first variable 'Hospitalisation', the second variable 'Gender' and the third variable 'Freq'. Remove the two decimal places.

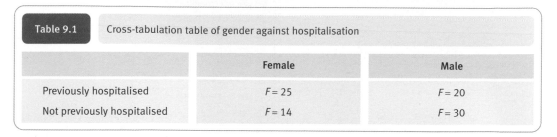

Step 2

Label the two values of 'Hospitalisation' (1 = 'Hospitalised'; 2 = 'Not hospitalised') and 'Gender' (1 = 'Females'; 2 = 'Males'). How to do this was discussed in Steps 5–10 in Section 2.8.

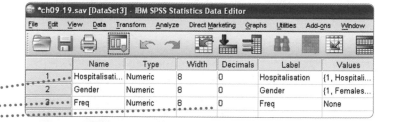

Step 3

Enter these numbers in 'Data View' of the 'Data Editor'. The first row refers to 'Hospitalised' 'Females' of whom there are 25. The second row to 'Hospitalised' 'Males' of whom there are 20. The third row to 'Not hospitalised' 'Females' of whom there are 14. The fourth row to 'Not hospitalised' 'Males' of whom there are 30.

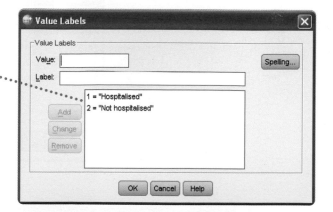
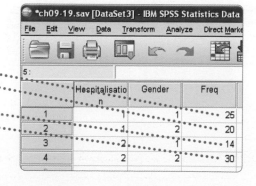

9.8   Weighting the data

Step 1

To weight the data so that the four cells have the appropriate number of cases in them, select 'Data' and 'Weight Cases...'.

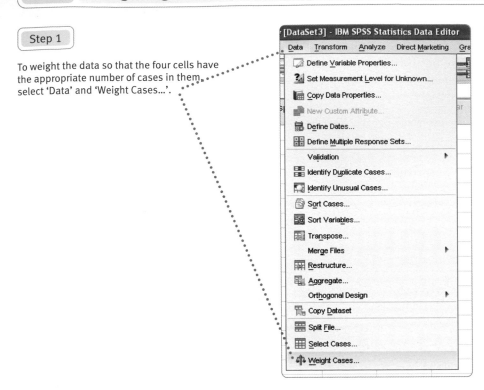

Step 2

Select 'Freq', 'Weight cases by' and the ➡ button to put it in the 'Frequency Variable:' box. Select 'OK'.

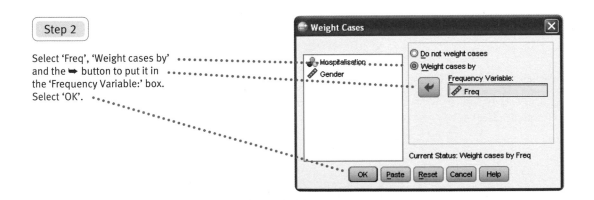

The cases are now weighted as shown by the 'Weight On' message in the lower right corner of the 'Data Editor'.

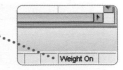

9.9   Compound (stacked) percentage bar chart

Step 1

To obtain a compound (stacked) percentage bar chart in which the bars represent 100 per cent you need to enter the percentage figures (called 'ColPerCent') for the two bars and weight them.

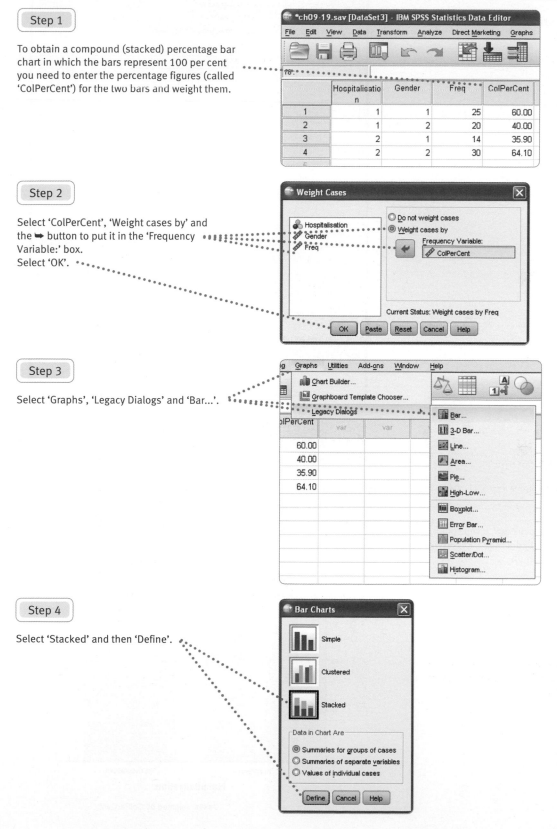

Step 2

Select 'ColPerCent', 'Weight cases by' and the ➡ button to put it in the 'Frequency Variable:' box.
Select 'OK'.

Step 3

Select 'Graphs', 'Legacy Dialogs' and 'Bar...'.

Step 4

Select 'Stacked' and then 'Define'.

Step 5

Select 'Hospitalisation' and the ➡ button
next to 'Category Axis:'.
Select 'Gender and the ➡ button next to
'Define Stacks by:'.
Select 'OK'.

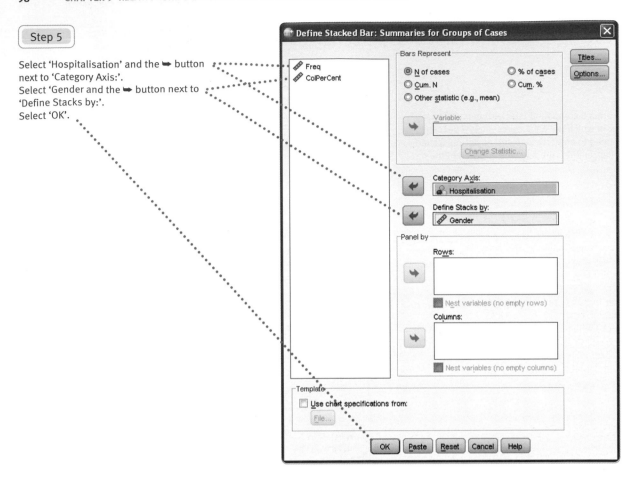

Step 6

'Count' refers to 'Per cent' which you could
change with the Chart Editor (See 'Reporting
histograms' in Chapter 6).

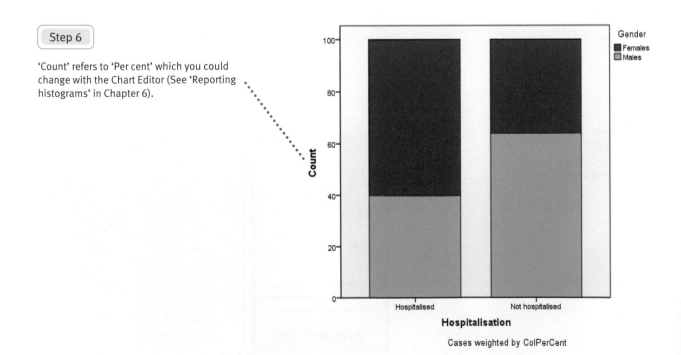

9.10   Compound histogram (clustered bar chart)

Step 1

Weight cases by 'Freq' instead of 'ColPerCent' by selecting 'Data' and 'Weight Cases...'.
You need to return 'ColPerCent' to the main box and then select 'Freq' to put it in the 'Frequency Variable:' box.
Select 'OK'.

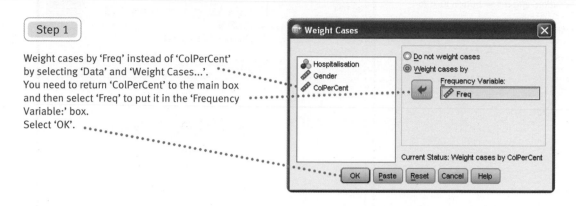

Step 2

Select 'Graphs', 'Legacy Dialogs' and 'Bar...'.

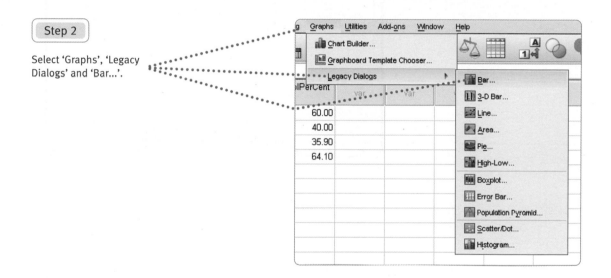

Step 3

Select 'Clustered' then 'Define'.

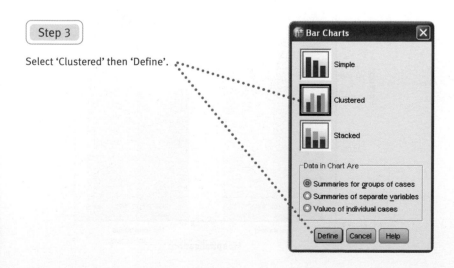

Step 4

Select 'Hospitalisation' and the ➡ button next to
'Category Axis:'.
Select 'Gender' and the ➡ button next to 'Define
Clusters by:'.
Select '% of cases'.
Select 'OK'.

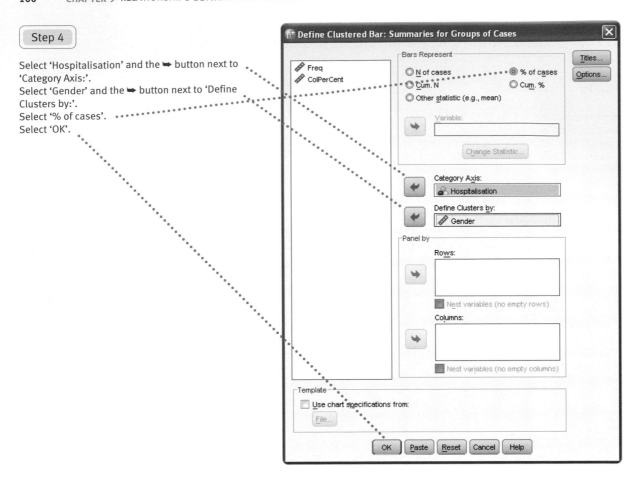

This is the bar chart produced.

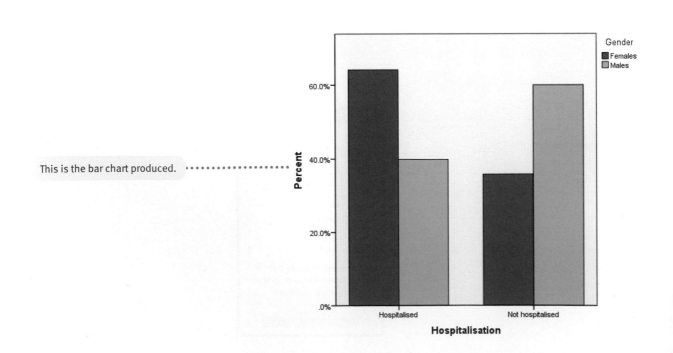

Summary of SPSS Statistics steps for bar charts

Data

- In 'Variable View' of the 'Data Editor', 'name' the variables.
- In 'Data View' of the 'Data Editor' enter the data under the appropriate variable names.

Analysis

- For bars, select 'Graphs', and then either 'Chart Builder ...' (see Chapters 4 and 10) or 'Legacy Dialogs' (this chapter and Chapters 6 and 11).
- For 'Chart Builder ...', select 'OK' and 'Bar'.
- Move the appropriate bar to the box above and the appropriate variable to the 'X-Axis?' box.
- For 'Legacy Dialogs', select 'Bar ...', the appropriate bar, and then 'Define' to move the appropriate variables to their respective boxes and select the appropriate statistic.

Output

- Check charts are correct. They can be edited in the 'Chart Editor' if required.

For further resources including data sets and questions, please refer to the website accompanying this book.

CHAPTER 10

Correlation coefficients
Pearson's correlation and Spearman's rho

Overview

- A correlation coefficient is a numerical index which indicates the strength and direction of a relationship between two variables.

- There are a number of different correlation coefficients. In general, the most common and most useful by far is the Pearson correlation coefficient. The phi, point biserial and Spearman's rho correlation coefficients are all merely variants of it.

- It is good practice to draw a scattergram as this represents the data included in a correlation coefficient. Not only will this give you a visual representation of the relationship but it also helps identify a number of problems such as a curved relationship or the presence of outliers.

- The Pearson correlation coefficient assumes a straight-line relationship between two variables. It is misleading if a curved relationship exists between the two variables. Outliers are extreme and unusual scores which distort the size of the correlation coefficient. Remedies include examining the relationship if the outliers are omitted. Alternatively, a Spearman correlation coefficient is less affected by outliers, and so one could compare the size of the Spearman correlation for the same data.

- A correlation coefficient is a numerical measure or index of the amount of association between two sets of scores. It ranges in size from a maximum of +1.00 through .00 to −1.00.

- The '+' sign indicates a positive correlation – that is, the scores on one variable increase as the scores on the other variable increase. A '−' sign indicates a negative correlation – that is, as the scores on one variable increase, the scores on the other variable decrease.

- A correlation of 1.00 indicates a perfect association between the two variables. In other words, a scattergram of the two variables will show that *all* of the points fit a straight line exactly. A value of .00 indicates that the points of the scattergram are essentially scattered randomly around any straight line drawn through the data or are arranged in a curvilinear manner. A correlation coefficient of −.5 would indicate a moderate negative relationship between the two variables.

- Spearman's rho is the Pearson correlation coefficient applied to the scores after they have been ranked from the smallest to the largest on the two variables separately. It is used when the basic assumptions of the Pearson correlation coefficient have not been met by the data – that is especially when the scores are markedly asymmetrical (skewed) on a variable.

- Since correlation coefficients are usually based on samples of data, it is usual to include a statement of the statistical significance of the correlation coefficient. Statistical significance is a statement of the likelihood of obtaining a particular correlation coefficient for a sample of data *if* there is no correlation (i.e. a correlation of .00) in the population from which the sample was drawn. SPSS Statistics can give statistical significance as an exact value or as one of the conventional critical significance levels (for example .05 and .01).

10.1   What is a correlation coefficient?

The simplest way to understand correlation coefficients is to conceive of them as a single numerical index which summarises some of the vital information in a scattergram or scatterplot. Figure 10.1 shows a scatterplot of the information to be found in Table 10.1 overleaf. That is to say, we have included a line of best fit through the various points on the scatterplot. This is a straight line. You will notice that the data points on the scatterplot are spread around the straight line. This means that the data do not fit the straight line perfectly. The correlation coefficient is really just an index of the spread of the data points around that best fitting straight line. Correlation coefficients range from absolute values of 1.00 at the most to .00 at the least. A correlation of 1.00 means that the data points *all* fit perfectly on the straight line whereas a correlation of .00 means that the data points fit the straight line very badly. Actually, a .00 correlation indicates that there is no straight line which fits the data points any better than on a chance basis. Values between .00 and 1.00 are indicative of increasingly strong relationships between the two variables – that is the fit of the data points to the straight line is getting closer and closer.

There is a little more to things than this. Some correlation coefficients have a negative sign in front of them such as –1.00 or –.50. The negative sign simply indicates that the slope of the scatterplot between two variables is negative – that is, the straight line points downwards from left to right rather than upwards. This negative slope can be seen in Figure 10.1. In other words, a negative sign in a correlation coefficient indicates that as scores get bigger on one of the two

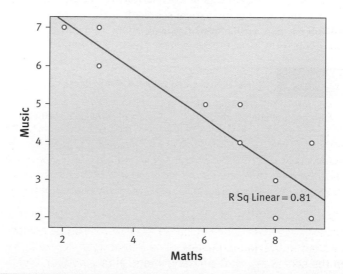

FIGURE 10.1   A scattergram showing the relationship between maths ability and musical ability

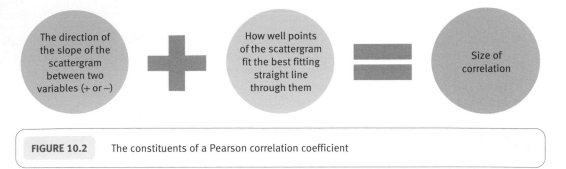

| **FIGURE 10.2** | The constituents of a Pearson correlation coefficient |

variables then they begin to get smaller on the other variable. Figure 10.2 illustrates the basic parts of a Pearson correlation coefficient.

A correlation coefficient can be spoken of as explaining or accounting for a certain amout of the variance between two variables. A correlation coefficient of 1.00 explains all of the variance between two variables – that is there is no variation at all of the data points from the best fitting straight line on the scatterplot so that all of the variation in data points is accounted for by the best fitting straight line. A correlation coefficient of .00 means that there is no variance accounted for by the best fitting straight line since the fit of the data points to the straight line is merely random. This idea of variance explained by a correlation coefficient is very important in psychology and needs to be understood in order to grasp the meaning of correlation coefficients adequately. The correlation coefficient is the proportion of variance shared between two variables. However, the variance explained in one variable by the other is calculated by squaring its value and expressing this as a proportion or percentage of the maximum value of a correlation coefficient possible, i.e. 1.00. Some examples of such values are shown in Table 10.1.

This table tells us that there is not a direct relationship between a correlation coefficient and the amount of variance that it accounts for. The relationship is exponential because it is based on squares. The percentage of variation explained is a better indicator of the strength of the relationship than the size of the correlation coefficient. So for example, although a correlation of .5 may seem to be half as good as a correlation of 1.00, this is not really the case. A correlation of .5 actually only explains .25 (25 per cent) of the variance. So a correlation of .5 is really a quarter as good as a correlation of 1.00.

The above comments apply directly to what is known as the Pearson product moment correlation coefficient – Pearson correlation for short. This is a correlation coefficient based on variables measured as numerical scores. It is the mother (father?) of a range of other correlation coefficients such as the Spearman rho.

Table 10.1	Correlations and proportion and percentage of variance explained in one of the variables	
Value of correlation coefficient *r*	**Square of correlation coefficient**	**% of variance in relationship between two variables explained (by best-fitting straight line)**
1.00	1.00	100.00
.70	.49	49.00
.50	.25	25.00
.25	.0625	6.25
.10	.01	1.00
.00	.00	0.00
−.50	.25	25.00
−1.00	1.00	100.00

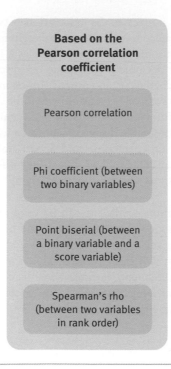
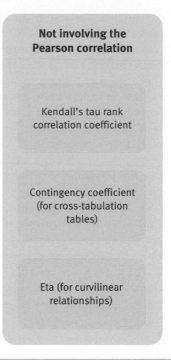

FIGURE 10.3   Some varieties of the correlation coefficient

The Spearman rho correlation coefficient is based on the Pearson correlation coefficient. It only differs in that the scores on the two variables are ranked separately from the smallest to the largest values. (Ranking simply puts scores in order from the smallest to the largest values. The smallest score would be given the rank 1, the next biggest score is given the rank 2, and so forth.) One has two sets of ranks to calculate the Spearman rho. Essentially if the Pearson correlation coefficient formula is applied to these ranks rather than the original scores, the resulting correlation coefficient is known as the Spearman correlation coefficient. It has traditionally been regarded as an alternative to the Pearson correlation coefficient when the variables involved are measured on a scale which does not have equal intervals. That is, the difference between 5 and 6 on the scale, say, is not the same quantity as the difference between 8 and 9 on the scale in terms of the 'psychological' units being measured. Modern psychologists tend to ignore this issue which once created fierce debates among researchers and statisticians. So the preference seems to be (although few would admit to it) to use the Pearson correlation coefficient. There are advantages to this since more powerful statistical techniques are available which are based on the Pearson correlation rather than on the Spearman rho.

Correlation coefficients are normally based on samples of data. As a consequence, it is necessary to test the statistical significance of correlation coefficients (see Box 1.1). The test of statistical significance assesses the possibility that the obtained correlation coefficient could come from a sample taken from a population in which the correlation is .00 (i.e. no correlation). SPSS displays significance levels routinely. Figure 10.3 shows some different kinds of correlation coefficient.

10.2   When to use Pearson's and Spearman's rho correlation coefficients

The Pearson correlation coefficient can be used in any circumstance in which it would be appropriate to draw a scattergram (scatterplot) between two score variables. In other words, the Pearson correlation coefficient can be used any time that the researcher wishes to know the extent to which one variable is related to another. It assesses the association between two variables – it is *not* a way of comparing means, for example. So many statistical techniques described later in this book are based on Pearson correlation in some form or another that it is impossible to avoid. Thus it is one of the key statistical ideas to understand before going on to study many advanced techniques.

The Spearman rho correlation coefficient, in comparison, is infrequently used. Its main function is to serve as an alternative to Pearson correlation when the assumptions of the Pearson correlation are not met by the data. Unfortunately, it is quite difficult to decide if these assumptions are violated so there would be considerable disagreement as to whether to use the Spearman rho correlation coefficient. If the Pearson correlation and Spearman's rho result in much the same conclusion as each other then there is no problem. But if they differ then one might wish to consider the possibility that the statistical assumptions underlying the Pearson correlation are not met by the data.

Versions of the Pearson correlation coefficient can be used when one is not considering the relationship between two sets of scores. For example, if one has a score variable and a binary variable (one with just two alternative categories such as yes and no or male and female) then the point biserial correlation coefficient may be calculated. This is the Pearson correlation coefficient applied to the score variable and the binary variable which is coded 1 and 2 for the two binary categories. If one has two binary variables then these can be correlated using the Pearson correlation coefficient though the outcome is conventionally known as the phi coefficient.

10.3 When not to use Pearson's and Spearman's rho correlation coefficients

Pearson correlation is based on linear (straight-line) relationships hence we draw a straight line through the points on a scattergram or scatterplot when illustrating Pearson correlation. But not all relationships are linear in psychological research. One should always examine the scattergram for the data to assess the extent to which the relationship is not linear – the best fitting line on a scatterplot might actually be a curve (i.e. a curvilinear relationship). If this is the case for your data then do not use Pearson correlation since it will be extremely misleading. In these circumstances, there is a correlation coefficient (eta) which may be calculated since it works with curved relationships.

> For a discussion of eta and its calculation see Chapter 34 in Howitt, D. and Cramer, D. (2011) *Introduction to Statistics in Psychology*, 5th edition. Harlow: Pearson Education.

The Spearman rho also assumes that the relationship is a more or less linear one and so this should not be used either where the relationship is a curve and not a straight line.

10.4 Data requirements for Pearson's and Spearman's rho correlation coefficients

The Pearson correlation coefficient requires two score variables as does Spearman's rho. However, the scores should ideally be normally distributed for the Pearson correlation and, some would argue, on an equal interval scale. One could consider using Spearman's rho where these requirements are not met, though the advantages of doing so may not balance out the costs for anything other than the most simple of research studies because it is a far less flexible statistic.

It is possible to use the Pearson correlation coefficient when one or both variables are in the form of binary variables – that is, there are only two possible response categories. Code these binary variables as 1 for one category and 2 for the other category in SPSS. If you then run the Pearson correlation procedure on these variables the resulting correlations are known as phi (if both variables are binary) and the point-biserial (if one variable is binary and the other a score variable).

10.5   Problems in the use of correlation coefficients

Most of the problems involved in using correlation can be identified by carefully looking at the scatterplots of the relationship. The problem of curvilinear relationships has already been discussed and the easiest way of identifying curvilinear relationships is simply by looking at the scatterplot. The alternative is to compare the size of the Pearson correlation on your data with the size of eta calculated on the same data.

Outliers are data points on the scatterplot which are at the extremes of the distribution. They are problematic because they can totally distort the correlation such is their impact. They can often be identified visually from the scatterplot. However, Spearman's rho is largely unaffected by outliers since they are changed into ranks along with the other scores. So one way of checking for outliers is to compare the Spearman's rho correlation with the Pearson correlation for the same data. If Spearman's rho is much smaller than the Pearson correlation, then suspect the influence of outliers which are making the Pearson correlation appear to be large.

You can find out more about the correlation coefficient in Chapter 7 of Howitt, D. and Cramer, D. (2011). *Introduction to Statistics in Psychology*, 5th edition. Harlow: Pearson.

10.6   The data to be analysed

We will illustrate the computation of Pearson's correlation, a scatter diagram and Spearman's rho for the data in Table 10.2 (*ISP*, Table 7.2), which gives scores for the musical ability and mathematical ability of 10 children.

Table 10.2	Scores on musical ability and mathematical ability for 10 children
Music score	**Mathematics score**
2	8
6	3
4	9
5	7
7	2
7	3
2	9
3	8
5	6
4	7

10.7   Entering the data

Step 1

In 'Variable View' of the 'Data Editor' name the first variable 'Music' and the second variable 'Maths'.
Remove the two decimal places by changing figure already here to zero.

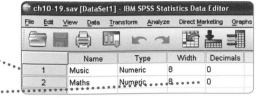

Step 2

In 'Data View' of the 'Data Editor' enter 'Music' scores in the first column and 'Maths' scores in the second column.
Save this data as a file to use for Chapter 11.

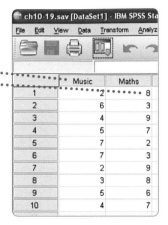

10.8   Pearson's correlation

Step 1

Select 'Analyze', 'Correlate' and 'Bivariate...'.

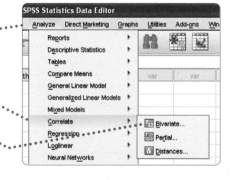

Step 2

Select 'Music' and 'Maths' either singly or together and the ➡ button to put them in the 'Variables:' box as they appear here.
Select 'OK'.

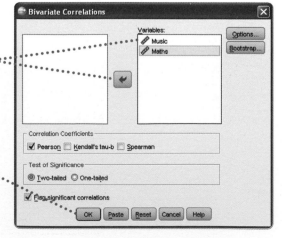

10.9   Interpreting the output

The correlation between 'Maths' and 'Music' is −.900. The two-tailed significance or probability level is .001 or less so the correlation is statistically significant.
The number of cases on which this correlation is based is 10. This information is also given in this cell.

Correlations

		Music	Maths
Music	Pearson Correlation	1	−.900**
	Sig. (2-tailed)		.000
	N	10	10
Maths	Pearson Correlation	−.900**	1
	Sig. (2-tailed)	.000	
	N	10	10

**. Correlation is significant at the 0.01 level (2-tailed).

REPORTING THE OUTPUT

- The correlation between music ability and mathematical ability is −.900. It is usual to round correlations to two decimal places, which would make it −.90. This is more than precise enough for most psychological measurements. Note that there is no need to put a 0 before the decimal point (e.g. −0.90) because a correlation cannot be bigger than ±1.00.
- The exact significance level to three decimal places is .000. This means that the significance level is less than 0.001. We would suggest that you do not use a string of zeros, as these confuse people. Always change the third zero to a 1. This means that the significance level can be reported as being $p < 0.001$.
- It is customary to present the degrees of freedom (df) rather than the number of cases when presenting correlations. The degrees of freedom are the number of cases minus 2, which makes them 8 for this correlation. There is nothing wrong with reporting the number of cases instead.
- In a report, we would write:

    There is a significant negative relationship between musical ability and mathematical ability, $r(8) = -.90$, $p < 0.001$. Children with more musical ability have lower mathematical ability.

The significance of the correlation coefficient is discussed in more detail in Chapter 10 of Howitt, D. and Cramer, D. (2011) *Introduction to Statistics in Psychology*, 5th edition. Harlow: Pearson.

10.10   Spearman's rho

Step 1

As for Pearson's correlation, select 'Analyze', 'Correlate', 'Bivariate' and the variables you want to correlate. Select 'Spearman'. If you don't want Pearson, deselect it.
Select 'OK'.

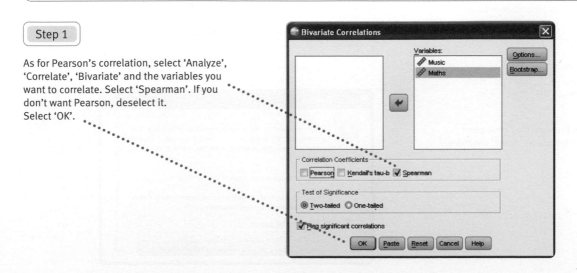

10.11 Interpreting the output

Correlations

Spearman's rho between 'Maths' and 'Music' is −.894. The two-tailed significance level of this correlation is .001 or less so the correlation is statistically significant. The number of cases is 10. This information is also given in this cell.

			Music	Maths
Spearman's rho	Music	Correlation Coefficient	1.000	−.894**
		Sig. (2-tailed)		.000
		N	10	10
	Maths	Correlation Coefficient	−.894**	1.000
		Sig. (2-tailed)	.000	.
		N	10	10

**. Correlation is significant at the 0.01 level (2-tailed).

REPORTING THE OUTPUT

- The correlation reported to two decimal places is −.89.
- The probability of achieving this correlation by chance is less than .001 (i.e. $p < .001$).
- We would report this in the following way:

  There is a statistically significant negative correlation between musical ability and mathematical ability, $\rho(8) = -.89$, $p < .001$. Those with the highest musical ability tend to be those with the lowest mathematical ability and vice versa.

  Rho should be represented by the Greek small letter ρ.

10.12 Scatter diagram

 Step 1

Select 'Graphs', and 'Chart Builder...'. The 'Legacy Dialog' procedure is used in Chapter 11.

Step 2

Select 'OK' as the variables have been appropriately defined as 'Scale' variables.

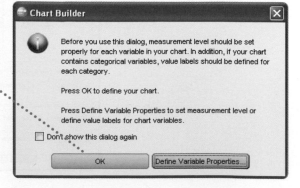

Step 3

Select 'Scatter/Dot' and move the 'Simple Scatter' image to the box above. Close 'Element Properties' box (not shown). Move 'Music' to the 'Y Axis?' and Maths' to the 'X Axis?'. Select 'OK'.

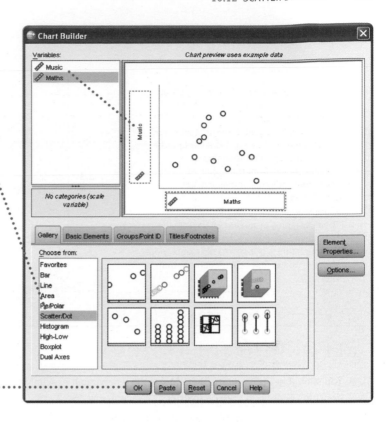

Step 4

To fit a correlation line to the scatterplot, double click anywhere in it which opens the 'Chart Editor'.
Select 'Elements' and 'Fit Line at Total'.
Select 'Close' in the 'Properties' box.
This is a regression line (Chapter 11).
To obtain a correlation line the two variables have to be standardised (Chapter 5).

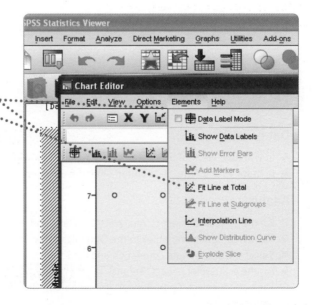

10.13   Interpreting the output

In this scattergram the scatter of points is relatively narrow, indicating that the correlation is high.

The slope of the scatter lies in a relatively straight line, indicating it is a linear rather than a curvilinear relationship.

The line moves from the upper left to the lower right, which signifies a negative correlation.

If the relationship is curvilinear, then Pearson's or Spearman's correlations may be misleading.

Although this is not so for these data, note that the points may represent more than one case.

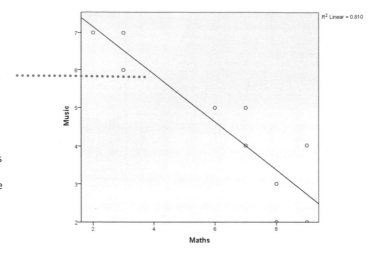

REPORTING THE OUTPUT

● You should never report a correlation coefficient without examining the scattergram for problems such as curved relationships or outliers (*ISP*, Chapter 7).

● In a student project it should always be possible to include graphs of this sort. Unfortunately, journal articles and books tend to be restricted in the figures they include because of economies of space and cost.

● We would write of the scattergram:

A scattergram of the relationship between mathematical ability and musical ability was examined. There was no evidence of a curvilinear relationship or the undue influence of outliers.

Summary of SPSS Statistics steps for correlation

Data

● In 'Variable View' of the 'Data Editor', 'name' the variables.
● In 'Data View' of the 'Data Editor' enter the data under the appropriate variable names.

Analysis

● For the correlation, select 'Analyze', 'Correlate' and 'Bivariate ...'.
● Move appropriate variables to the 'Variables:' box.
● Select the appropriate correlation and then 'OK'.
● For the scatterplot, select 'Graphs'.
● Then either select 'Chart Builder' (this chapter), 'OK', 'Scatter/Dot' and move the 'Simple Scatter' figure to the box above.
● Move appropriate variable names to the vertical and horizontal axes.
● Or select 'Legacy Dialogs' (Chapter 11), 'Scatter/Dot ...', 'Define', 'Y Axis' variable, 'X Axis' variable and 'OK'.

Output

● The correlation table shows the correlation, its significance level and the sample size.

For further resources including data sets and questions, please refer to the website accompanying this book.

Regression

Prediction with precision

Overview

- Where there is a relationship between two variables, it is possible to estimate or predict a person's score on one of the variables from their score on the other variable. The stronger the correlation, the better the prediction. This is known as *simple* regression because there are only two variables involved.

- Regression can be used on much the same data as the correlation coefficient. However, it is far less commonly used, partly because of the problem of comparability between values obtained from different sets of variables. (The beta weight can be used if such comparability is required.)

- The dependent variable in regression is the variable the value of which is to be predicted. It is also known as the criterion variable, the predicted variable or the Y-variable.

- The independent variable is the variable being used to make the prediction. It is also known as the predictor variable or the X-variable.

- Great care is needed not to get the independent variable and the dependent variable confused. This can easily happen with simple regression. The best way of avoiding problems is to examine the scatterplot or scattergram of the relationship between the two variables. Make sure that the horizontal x-axis is the independent variable and that the vertical y-axis is the dependent variable. One can then check what the cut point is approximately from the scattergram as well to get an idea of what the slope should be. The cut point is where the slope meets the vertical axis. These estimates may be compared with their calculated values to ensure that an error has not been made. If problems are found, the most likely reason is that the independent and dependent variables have been confused.

- The simple regression technique described in this chapter expresses relationships in terms of the original units of measurement of the variables involved. Thus, if two different studies use slightly different variables it is difficult to compare the outcomes of the studies using this form of regression.

- In regression, the relationship between two variables is described mathematically by the slope of the best fitting line through the points of the scattergram together with the point at which this regression line cuts the (vertical) axis of the scattergram. Therefore, the relationship between two variables requires the value of the slope (usually given the symbol B or b) and the intercept or cut point in the vertical axis (usually given the symbol a or described as the constant).

- Regression becomes a much more important technique when one is using several variables to predict values on another variable. These techniques are known as multiple regression (see Chapters 32–35). When the dependent variable is a nominal category variable, then the appropriate statistical analysis will be a form of logistic regression (see Chapters 37 and 38).

11.1   What is simple regression?

One of the most difficult statistical techniques for novices to understand is that of simple regression. It is one of the earliest statistical methods, predating the correlation coefficient. The first thing to understand is that regression is applied to exactly the same data as the Pearson correlation coefficient. But it does something different. Look at the scatterplot in Section 11.11. When we discussed Pearson correlation in Chapter 10, we explained how the correlation coefficient is a measure of how closely the data points fit the straight line through the data points – the bigger the correlation coefficient, the closer the data points tend to be to the straight line. Regression is different in that it describes the characteristics of the straight line itself. In order to describe the best fitting straight line we need to know two things: (1) where the line cuts the vertical axis of the scatterplot and (2) what the slope of the line is. The point at which the vertical axis is cut by the straight line is known as the cut point, intercept or constant whereas the slope is the regression weight. The slope is merely the amount that the line goes up (or down) for every unit that one goes along the horizontal axis. The slope can be negative, which indicates that the line goes downwards towards the right of the scatterplot.

The good news is that all of the hard work is done for you on SPSS Statistics. The best fitting straight line is calculated by SPSS so it is not a matter of trial and error.

However, it needs to be understood that regression is affected by which variable you put on the horizontal (or X axis) and which variable you put on the vertical (or Y) axis. Quite different figures emerge in regression depending on your choice. Why does this matter? One of the functions of regression is to allow the user to make predictions from the value of one variable to the other variable. This is possible if it is established from a sample of data that there is a good correlation between the two variables. For example, if you measured the heights and weights of a sample of participants you would find a correlation between the two. On the basis of this information, you would assume that someone who is tall is likely to be heavier than someone who is small. The only problem is that this is not very precise.

Figure 11.1 gives a scattergram for the relationship between musical and mathematical ability which is the data in Table 11.1 on page 117, identical to that in Table 10.2 in the previous chapter. The best fitting straight line (also known as the regression line) has been drawn in. We have added in some extra things: (1) the cut point of the regression line on the vertical axis (which, by the way, can be negative, i.e. below the horizontal axis, and (2) the words independent and dependent variable which refer to the horizontal and vertical axes respectively. You could predict

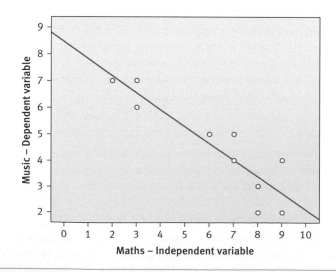

FIGURE 11.1   Scattergram of the relationship between music and maths ability

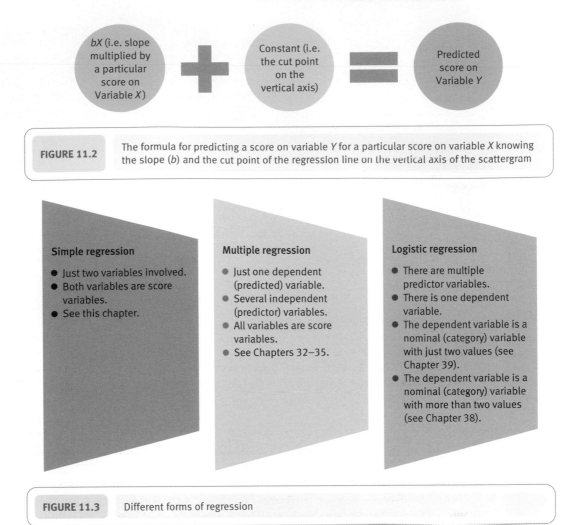

FIGURE 11.2   The formula for predicting a score on variable *Y* for a particular score on variable *X* knowing the slope (*b*) and the cut point of the regression line on the vertical axis of the scattergram

Simple regression

● Just two variables involved.
● Both variables are score variables.
● See this chapter.

Multiple regression

● Just one dependent (predicted) variable.
● Several independent (predictor) variables.
● All variables are score variables.
● See Chapters 32–35.

Logistic regression

● There are multiple predictor variables.
● There is one dependent variable.
● The dependent variable is a nominal (category) variable with just two values (see Chapter 39).
● The dependent variable is a nominal (category) variable with more than two values (see Chapter 38).

FIGURE 11.3   Different forms of regression

music ability scores using this scatterplot if you wished simply by drawing a vertical line from the relevant point on the horizontal axis (i.e. a particular individual's maths ability score) to the regression line then horizontally to the vertical axis, The point that the vertical axis is cut is the predicted score on musical ability. It would be more accurate, though, to use the formula in Figure 11.2, where *b* is the slope of the regression line and *X* is a particular score on the *x*-axis from which the score on the *y*-axis is to be predicted. You would obtain the figures for the slope of the regression line and the constant from the SPSS output for regression. The value of *X* you choose would be the score on the *X*-variable of a particular individual that you are interested in. Such predictions are rarely made in research settings. Figure 11.3 shows some different forms of regression.

11.2   When to use simple regression

Generally, in psychology, simple regression of this sort would be rarely used. Usually psychologists would use the correlation coefficient in preference to regression when doing simple regression. Indeed, the regression weight expressed in standardised form is the same as the Pearson correlation coefficient. In terms of learning statistics, it is vital because it introduces the basic ideas of regression. Regression comes into its own when extended into what is termed multiple regression (see Chapters 32–35) in which there are several predictor variables involved. It is also closely related conceptually to logistic regression.

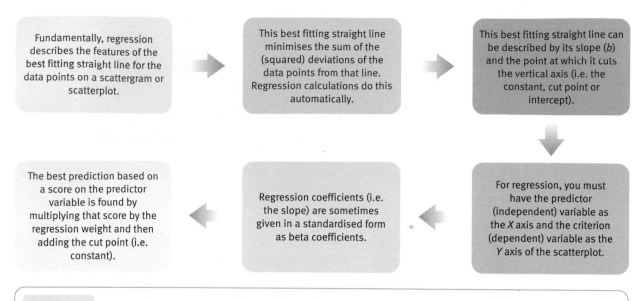

Fundamentally, regression describes the features of the best fitting straight line for the data points on a scattergram or scatterplot.

This best fitting straight line minimises the sum of the (squared) deviations of the data points from that line. Regression calculations do this automatically.

This best fitting straight line can be described by its slope (*b*) and the point at which it cuts the vertical axis (i.e. the constant, cut point or intercept).

The best prediction based on a score on the predictor variable is found by multiplying that score by the regression weight and then adding the cut point (i.e. constant).

Regression coefficients (i.e. the slope) are sometimes given in a standardised form as beta coefficients.

For regression, you must have the predictor (independent) variable as the *X* axis and the criterion (dependent) variable as the *Y* axis of the scatterplot.

FIGURE 11.4   Conceptual stages in understanding regression

However, it can be used on any pair of score variables especially when the data approximate equal-interval measurement, which is the assumption for most score variables in psychology.

11.3   When not to use simple regression

Few psychologists use simple regression in their analyses. For most analyses, it generally has no advantages over the Pearson correlation coefficient with which it has a lot in common.

In addition, do not use simple regression where one has one or more binary variables (two values or two category variables) since the output may be fairly meaningless – it certainly will not look like a typical scatterplot and so may cause confusion.

11.4   Data requirements for simple regression

Two score variables are needed. It is best if these are normally distributed but some deviation from this ideal will not make much difference.

11.5   Problems in the use of simple regression

Simple regression is beset with the same problems as the Pearson correlation coefficient (see Chapter 10). So non-linear relationships are a problem as are outliers. These can be checked for using a scatterplot of the data.

From a learner's point of view, there is one extremely common mistake – putting the predictor variable in the analysis wrongly so that it is treated as the dependent variable by SPSS. Making a scatterplot should help with this since it is possible to estimate the likely values from this – especially the constant or cut point on the vertical axis. This can be checked against the computed values since this may reveal such a confusion of the independent and dependent variables.

Sometimes the way in which SPSS gives the constant can cause confusion. The constant appears under the column for *B* weights or coefficients which is not what it is. You should have no trouble if you follow the instructions in Section 11.9.

You can find out more about simple regression in Chapter 8 of Howitt, D. and Cramer, D. (2011). *Introduction to Statistics in Psychology*, 5th edition. Harlow: Pearson.

11.6   The data to be analysed

We will illustrate the computation of simple regression and a regression plot with the data in Table 11.1 (*ISP*, Table 8.2), which gives a score for the musical ability and mathematical ability of 10 children. These data are identical to those used in the previous chapter on correlation. In this way, you may find it easier to appreciate the differences between regression and correlation.

The music scores are the criterion or the dependent variable, while the mathematics scores are the predictor or independent variable. With regression, it is essential to make the criterion or dependent variable the vertical axis (*y*-axis) of a scatterplot and the predictor or independent variable the horizontal axis (*x*-axis).

Table 11.1	Scores on musical ability and mathematical ability for 10 children	
Music score		**Mathematics score**
2		8
6		3
4		9
5		7
7		2
7		3
2		9
3		8
5		6
4		7

11.7   Entering the data

If you have saved the data select the file. Otherwise enter the data again.

Step 1

In 'Variable View' of the 'Data Editor' name the first variable 'Music' and the second variable 'Maths'.
Remove the two decimal places.

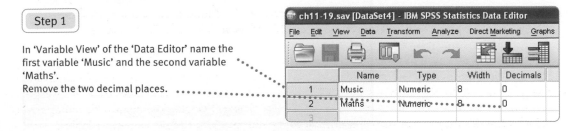

Step 2

In 'Data View' of the 'Data Editor' enter 'Music' scores in the first column and 'Maths' scores in the second column.

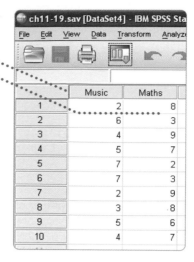

11.8   Simple regression

Step 1

Select 'Analyze', 'Regression' and 'Linear...'.

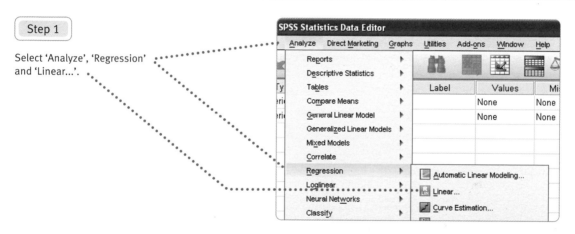

Step 2

Select 'Music' and the ➡ button beside 'Dependent:' to put 'Music' in that box.
Select 'Maths' and the ➡ button beside 'Independent(s):' to put 'Maths' in that box.
Select 'Statistics...'.

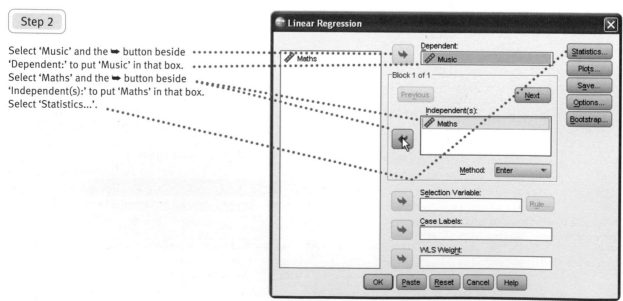

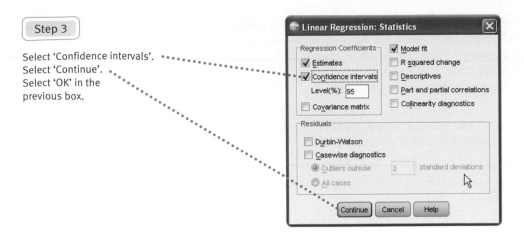

Step 3

Select 'Confidence intervals'.
Select 'Continue'.
Select 'OK' in the
previous box.

11.9 Interpreting the output

The table below is the last table of the output which has the essential details of the regression analysis. It is very easy to reverse the independent variable and dependent variable accidentally. Check the table titled Coefficients[a]. Under the table the name of the dependent variable is given. In this case it is Music, which is our dependent variable. If it read Maths then we would have made a mistake and the analysis would need to be redone as the regression values would be incorrect.

The intercept or constant is 8.425. This is the point at which the regression line cuts the vertical axis.

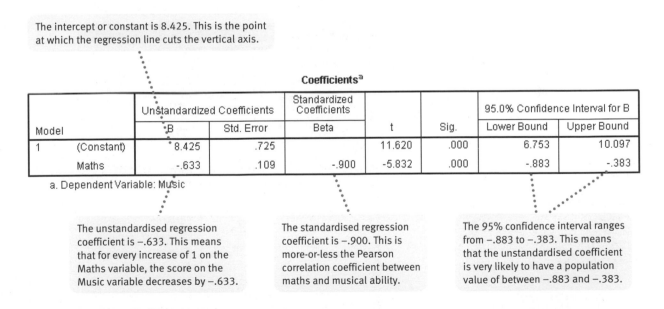

Coefficients[a]

Model		Unstandardized Coefficients B	Std. Error	Standardized Coefficients Beta	t	Sig.	95.0% Confidence Interval for B Lower Bound	Upper Bound
1	(Constant)	8.425	.725		11.620	.000	6.753	10.097
	Maths	-.633	.109	-.900	-5.832	.000	-.883	-.383

a. Dependent Variable: Music

The unstandardised regression coefficient is –.633. This means that for every increase of 1 on the Maths variable, the score on the Music variable decreases by –.633.

The standardised regression coefficient is –.900. This is more-or-less the Pearson correlation coefficient between maths and musical ability.

The 95% confidence interval ranges from –.883 to –.383. This means that the unstandardised coefficient is very likely to have a population value of between –.883 and –.383.

In simple regression involving two variables, it is conventional to report the regression equation as a slope (*b*) and an intercept (*a*) as explained in *ISP* (Chapter 8). SPSS does not quite follow this terminology. Unfortunately, at this stage the SPSS output is far more complex and detailed than the statistical sophistication of most students:

● B is the slope. The slope of the regression line is called the unstandardised regression coefficient in SPSS. The unstandardised regression coefficient between 'Music' and 'Maths' is displayed under B and is –.633, which rounded to two decimal places is –0.63. What this means is that for every increase of 1.00 on the horizontal axis, the score on the vertical axis changes by –.633.

- The 95 per cent confidence interval for this coefficient ranges from −.88 (−.883) to −.38 (−.383). Since the regression is based on a sample and not the population, there is always a risk that the sample regression coefficient is not the same as that in the population. The 95 per cent confidence interval gives the range of regression slopes within which you can be 95 per cent sure that the population slope will lie.

- The intercept (*a*) is referred to as the constant in SPSS. The intercept is presented as the (Constant) and is 8.425, which rounded to two decimal places is 8.43. It is the point at which the regression linc cuts the vertical (*y*) axis.

- The 95 per cent confidence interval for the intercept is 6.753 to 10.097. This means that, based on your sample, the intercept of the population is 95 per cent likely to lie in the range of 6.75 to 10.10.

- The column headed 'Beta' gives a value of −.900. This is actually the Pearson correlation between the two variables. In other words, if you turn your scores into standard scores (*z*-scores) the slope of the regression and the correlation coefficient are the same thing.

11.10   Regression scatterplot

It is generally advisable to inspect a scattergram of your two variables when doing regression. This involves the steps involved in plotting a scattergram as described in Chapter 10.

Step 1

Select 'Graphs', 'Legacy Dialogs' and 'Scatter/Dot...'.

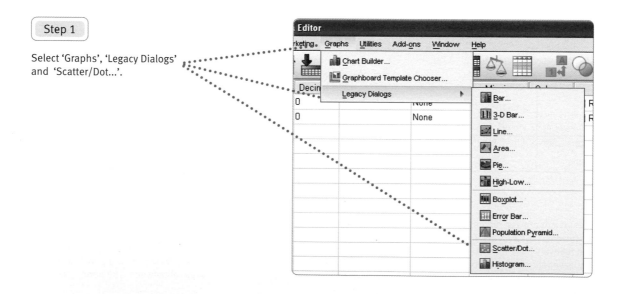

Step 2

Select 'Define' as 'Simple Scatter' is the preselected default.

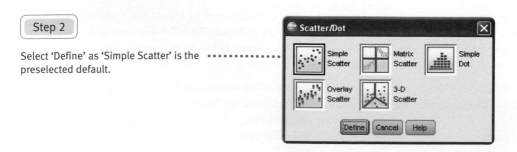

Step 3

Select 'Music' and the ➡ button
beside 'Y Axis:' to put it in this
box as it is the criterion.
Select 'Maths' and the ➡ button
beside 'X Axis:' to put it in this
box as it is the predictor.
Select 'OK'.
The output is the same as that for
the correlation in Section 10.9.

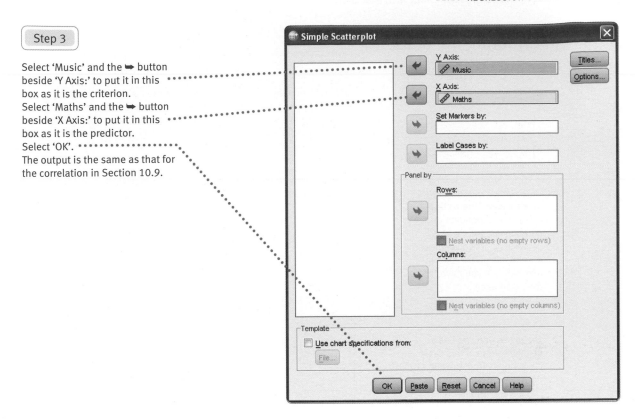

Step 4

To fit a regression line to the scatterplot,
double click anywhere in it which opens
the 'Chart Editor'. Select 'Elements'
and 'Fit Line at Total'.
Select 'Close' in 'Properties' box.

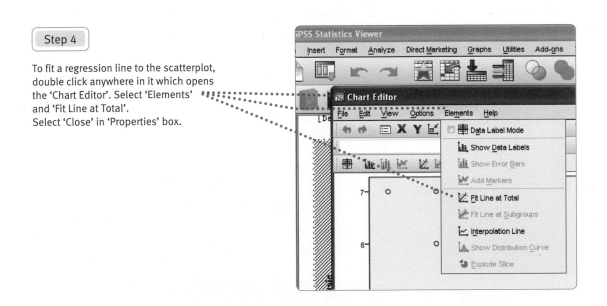

11.11   Interpreting the output

The points on the scatterplot are close to the regression line (i.e. the line is a good fit to the data). Furthermore, the points seem to form a straight line (i.e. the relationship is not curvilinear) and there is no sign of outliers (i.e. especially high or low points).

In **regression**, the vertical axis is the dependent or criterion variable; in this case Music. It is a common error to mix up the axes in regression. If you have, then start again.

The regression line has a negative slope in this case, i.e. it slopes from top left down to bottom right. The B weight therefore has a − value.

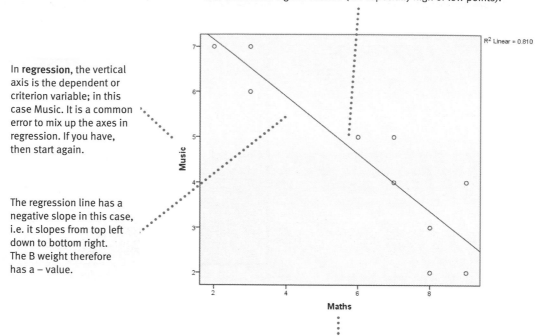

In **regression**, the horizontal axis is the independent or predictor variable; in this case Maths. It is a common error to mix up the axes in regression. If you have then re-do the analysis reversing the axes.

- The regression line sloping from the top left down to bottom right indicates a negative relationship between the two variables.

- The points seem relatively close to this line, which suggests that Beta (correlation) should be a large (negative) numerical value and that the confidence interval for the slope should be relatively small.

REPORTING THE OUTPUT

Although all of the output from SPSS is pertinent to a sophisticated user, many users might prefer to have the bare bones at this stage.

● With this in mind, we would write about the analysis in this chapter:

The scatterplot of the relationship between mathematical and musical ability suggested a linear negative relationship between the two variables. It is possible to predict accurately a person's musical ability from their mathematical ability. The equation is $Y' = 8.43 + (-0.63X)$ where X is an individual's mathematics score and Y' is the best prediction of their musical ability score.

● An alternative is to give the scatterplot and to write underneath $a = 8.43$ and $B = -0.63$.
● One could add the confidence intervals such as:

The 95 per cent confidence interval for the slope of the regression line is $-.88$ to $-.38$. Since this confidence interval does not include 0.00 the slope differs significantly from a horizontal straight line.

However, this would be a relatively sophisticated interpretation for novices in statistics.

Summary of SPSS Statistics steps for simple regression

Data

● In 'Variable View' of the 'Data Editor', 'name' the variables.
● In 'Data View' of the 'Data Editor' enter the data under the appropriate variable names.

Analysis

● For the correlation, select 'Analyze', 'Regression' and 'Linear …'.
● Move the dependent variable to the 'Dependent:' box and the independent variable to the 'Independent(s):' box and then select 'OK'.
● For the scatterplot, select 'Graphs',
● Then either select 'Chart Builder' (Chapter 10), 'OK', 'Scatter/Dot' and move the 'Simple Scatter' figure to the box above.
● Move appropriate variable names to the vertical and horizontal axes.
● Or select 'Legacy Dialogs' (this chapter), 'Scatter/Dot …', 'Define', 'Y Axis' variable, 'X Axis' variable and 'OK'.

Output

● The 'Coefficients' table shows the unstandardised and standardised regression coefficient and its significance level.
● For the scattergram, a regression line can be fitted by double clicking anywhere on the scattergram to bring up the Chart Editor.
● Select 'Elements' and 'Fit Line at Total'.

For further resources including data sets and questions, please refer to the website accompanying this book.

PART 3

Significance testing and basic inferential tests

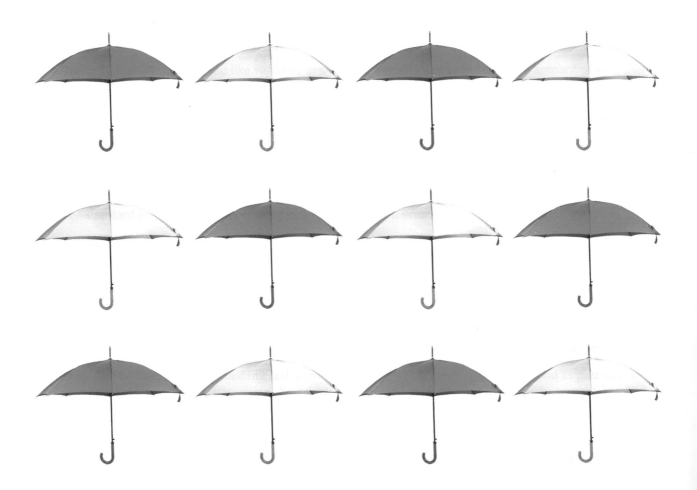

13.6 The data to be analysed

The computation of a related *t*-test is illustrated with the data in Table 13.1, which shows the number of eye-contacts made by the same babies with their mothers at six and nine months (*ISP*, Table 12.6). The purpose of the analysis is to see whether the amount of eye-contact changes between these ages.

Table 13.1	Number of one-minute segments with eye-contact at different ages	
Baby	**Six months**	**Nine months**
Clara	3	7
Martin	5	6
Sally	5	3
Angie	4	8
Trevor	3	5
Sam	7	9
Bobby	8	7
Sid	7	9

13.7 Entering the data

Step 1

In 'Variable View' of the 'Data Editor' label the first variable 'Six_mths' and the second variable 'Nine_mths'.
Remove the two decimal places by changing the figure for the decimals to 0.

Step 2

In 'Data View' of the 'Data Editor' enter the data in the first two columns.
(Save this data file to use in Chapter 18.)

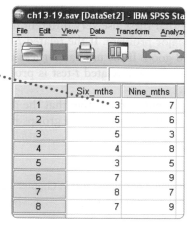

13.8   Related *t*-test

Step 1

Select 'Analyze', 'Compare Means' and 'Paired-Samples T Test...'.

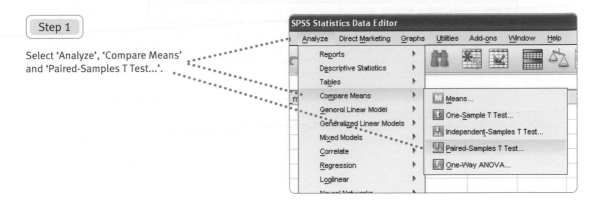

Step 2

Select 'Six_mths' and the ➡ button to put it in the 'Paired Variables:' box. Do the same for 'Nine_mth'. Select 'OK'.

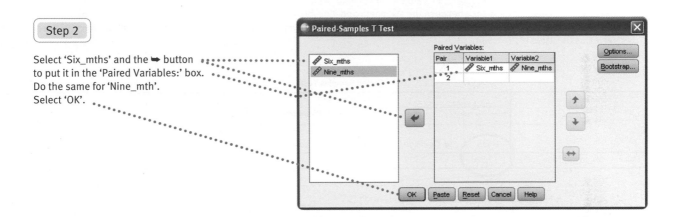

13.9   Interpreting the output

Paired Samples Statistics

		Mean	N	Std. Deviation	Std. Error Mean
Pair 1	Six_mths	5.25	8	1.909	.675
	Nine_mths	6.75	8	2.053	.726

The first table shows the mean, the number of cases and the standard deviation of the two groups. The mean for 'Six_mths' is 5.25 and its standard deviation is 1.909. The two standard deviations are very similar which is an advantage.

Paired Samples Correlations

		N	Correlation	Sig.
Pair 1	Six_mths & Nine_mths	8	.419	.301

The second table shows the degree to which the two sets of scores are correlated. The correlation between them is .419. This is a moderate correlation although it is not significant as the significance is greater than .05. Correlated tests such as the related *t*-test should have a substantial correlation between the two sets of scores. Be careful: this table is *not* the test of significance. It is a common mistake, among novices, to confuse this correlation between two variables with the significance of the difference between the two variables.

The first three columns containing figures are the basic components of the calculation of the related *t*-test. The 'Mean' of −1.500 is actually the difference between the six month and nine month means, so it is really the mean difference. The value of *t* is based on this mean difference (−1.500) divided by the standard error of the mean (.756). This calculation gives us the value of *t* (−1.984).

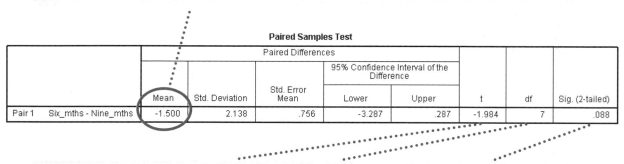

Paired Samples Test

		Paired Differences					t	df	Sig. (2-tailed)
					95% Confidence Interval of the Difference				
		Mean	Std. Deviation	Std. Error Mean	Lower	Upper			
Pair 1	Six_mths - Nine_mths	-1.500	2.138	.756	-3.287	.287	-1.984	7	.088

The third and last table shows the *t* value (−1.984), the degrees of freedom (7) and the two-tailed significance level (.088). As the significance level is greater than .05 this difference is not significant. The one-tailed level is obtained by dividing it by 2 which is .044 and significant. (However, unless the difference has been predicted in advance of data collection on the basis of strong theoretical and/or empirical reasons, only the two-tailed test is appropriate.)

- In the first table of the output, the mean number of eye-contacts at six months ('Six_mths') and at nine months ('Nine_mths') is displayed under Mean. Thus the mean amount of eye contact is 5.25 at 6 months and 6.75 at nine months.

- In the second table of the output is the (Pearson) correlation coefficient between the two variables (eye-contact at six months and eye-contact at nine months). *Ideally*, the value of this should be sizeable (in fact it is .419) and statistically significant (which it is not with a two-tailed significance level of .301). The related *t*-test assumes that the two variables are correlated, and you might consider an unrelated *t*-test (Chapter 14) to be more suitable in this case.

- In the third table of the output the difference between these two mean scores is presented under the 'Mean' of 'Paired Differences' and the standard error of this mean under 'Std. Error Mean'. The difference between the two means is −1.50 and the estimated standard error of means for this sample size is .76.

- The *t*-value of the difference between the sample means, its degrees of freedom and its two-tailed significance level are also shown in this third table. The *t*-value is −1.984, which has an exact two-tailed significance level of .088 with seven degrees of freedom.

REPORTING THE OUTPUT

- We could report these results as follows:

    The mean number of eye-contacts at six months ($M = 5.25$, SD = 1.91) and at nine months ($M = 6.75$, SD = 2.05) did not differ significantly, $t(7) = -1.98$, two-tailed $p = .088$.

- In this book, to be consistent, we will report the exact probability level for non-significant results as above. However, it is equally acceptable to report them as '$p > .05$' or '*ns*' (which is short for non-significant).
- Notice that the findings would have been statistically significant with a one-tailed test. However, this would have to have been predicted with sound reasons prior to being aware of the data. In this case one would have written to the effect 'The two means differed significantly in the predicted direction, $t(7) = -1.98$, one-tailed $p = .044$.' A one-tailed test should only be used if, prior to data collection, the direction of the difference between means has been predicted on the basis of strong theoretical reasons or a strong consistent trend in the previous research. These requirements are rarely met in student research so, generally, two-tailed significance testing should be the norm.
- Once again, to be consistent throughout this book, we will report the exact probability level for significant findings where possible. Note that when SPSS displays the significance level as '.000', we need to present this as '$p < .001$' since the exact level is not given. It is equally acceptable to report significant probabilities as '$p < .05$', '$p < .01$' and '$p < .001$' as appropriate.
- If you prefer to use confidence intervals (see Chapter 15), you could report your findings as:

    The mean number of eye-contacts at six months was 5.25 (SD = 1.91) and at nine months was 6.75 (SD = 2.05). The difference was 1.50. The 95 per cent confidence interval for this difference is -3.29 to .29. Since the confidence interval passes through 0.00, the difference is not statistically significant at the two-tailed 5 per cent level.

- Some statisticians advocate the reporting of confidence intervals rather than significance levels. However, it remains relatively uncommon to give confidence intervals.

Summary of SPSS Statistics steps for related *t*-test

Data

- In 'Variable View' of the 'Data Editor', 'name' the variables.
- In 'Data View' of the 'Data Editor' enter the data under the appropriate variable names.

Analysis

- Select 'Analyze', 'Compare Means' and 'Paired-samples T test …'.
- Move the variables to be compared to the 'Paired variable(s):' box and then select 'OK'.

Output

- The *t* value, degrees of freedom (*df*) and exact two-tailed significance level are reported in the last three columns of the 'Paired Samples' table.

For further resources including data sets and questions, please refer to the website accompanying this book.

CHAPTER 14

The *t*-test

Comparing two groups of unrelated/ uncorrelated scores

Overview

- The uncorrelated or unrelated *t*-test is used to calculate whether the means of two sets of scores are significantly different from each other. It is the most commonly used version of the *t*-test. Statistical signficance indicates that the two samples differ to an extent which is unlikely to be due to chance factors as a consequence of sampling. The variability inherent in the available data is used to estimate how likely it is that the difference between the two means would be if, in reality, there is no difference between the two samples.

- The unrelated *t*-test is used when the two sets of scores come from two different samples of people. (Refer to the previous chapter on the related *t*-test if your scores come from just one set of people or if you have employed a matching procedure.)

- Data entry for related and unrelated variables are very different in SPSS Statistics. So take care to plan your analysis before entering your data in order to avoid problems and unnecessary work. SPSS, however, is very flexible and errors of this sort are usually straightforward to correct using copy-and-paste and similar features.

- SPSS procedures for the unrelated *t*-test are very useful and go beyond usual textbook treatments of the topic. That is, they include an option for calculating the *t*-test when the variances of the two samples of scores are significantly different from each other. Most textbooks erroneously suggest that the *t*-test is too inaccurate to use when the variances of the two groups are unequal. This additional version of the unrelated *t*-test is rarely mentioned in statistics textbooks but is extremely valuable.

- If you have more than two sets of scores to compare, then refer to Chapter 21 on the unrelated analysis of variance.

14.1   What is the unrelated *t*-test?

The unrelated *t*-test simply tests the statistical significance of the difference between two different group (i.e. sample) means (see Box 1.1 for a discussion of statistical significance). That is, it addresses the question of whether the average score for one group (e.g. males) is significantly different from the average score for the other group (females). It is much more commonly used than the related *t*-test (discussed in Chapter 13). It is called the *t*-test because the computation is based on a mathematical distribution known as the *t*-distribution.

The unrelated *t*-test basically involves the difference between the two sample means. Of course, scores vary and this variation may be responsible for the apparent difference between the sample means. So a formula based on combining the variances of the two separate samples is calculated. This is known as the standard error of the differences between two sample means. The smaller this is compared to the size of the difference between the two sample means then the more likely it is for the two means to be statistically significantly different from each other. These two versions of the *t*-test are illustrated in Figure 14.1.

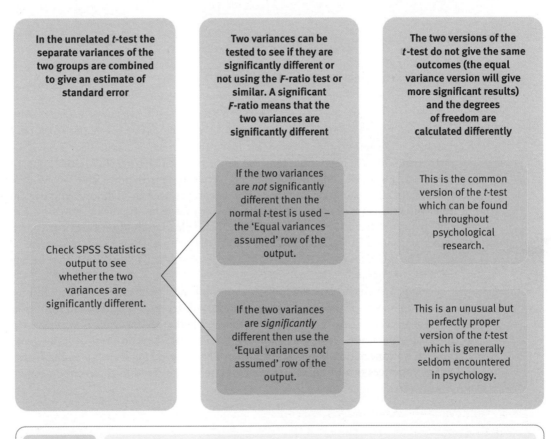

FIGURE 14.1   The two types of unrelated *t*-test

Table 14.1	Emotionality scores in two-parent and lone-parent families
Two-parent family X_1	**Lone-parent family** X_2
12	6
18	9
14	4
10	13
19	14
8	9
15	8
11	12
10	11
13	9
15	
16	
Mean = 13.42	Mean = 9.50

Table 14.1 gives data from a study for which the unrelated *t*-test is an appropriate test of significance. The data consist of scores on a single variable coming from two groups of participants – the important criterion for using the unrelated *t*-test. If you have just one group of participants then it is possible that the related *t*-test (Chapter 13) is the appropriate technique. Group 1 consists of children brought up in two-parent families and Group 2 consists of those brought up in lone-parent families. The researcher wishes to know whether these two groups differ on a measure of emotionality. It is easy to see from the mean scores for each group shown in Table 14.1 that the two groups differ. Quite clearly, children from two-parent families have higher emotionality scores (mean = 13.42) than those from the lone-parent families (mean = 9.50). The question is whether this difference is statistically significant, that is, hard to explain on the basis of chance.

There is a slight problem in the calculation of the standard error of the difference between the two group means. If the two variances are very different, then combining them is problematic in terms of statistical theory. It is possible to test whether two variances are statistically significantly different using the *F*-ratio (see Chapter 20). SPSS uses Levene's test of homogeneity to do this. If the two variances differ significantly then SPSS also provides a version of the unrelated *t*-test 'Equal variances not assumed'. This should be used instead in these circumstances.

14.2   When to use the unrelated *t*-test

The unrelated *t*-test can be used where there are two separate groups of participants for which you have scores on any score variable. It is best if the scores are normally distributed and correspond to an equal interval level of measurement, though any score data can be used with a loss of accuracy. SPSS provides two versions of the unrelated *t*-test: one where the variances of the two sets of scores are equal and the other where the two variances are significantly different.

14.3   When not to use the unrelated *t*-test

Do not use the unrelated *t*-test where you have three or more groups of participants – the mean of each group you wish to compare with the mean of each other group. ANOVA is a more suitable statistic in these circumstances since multiple comparisons using an unrelated *t*-test increases the likelihood of having one or more comparisons significant by chance.

Some researchers would prefer to use a non-parametric or distribution-free test of significance (Chapter 18) if the scores are not measured on an equal interval scale of measurement. Mostly, however, considerable violations of the assumptions of the unrelated *t*-test are tolerated.

If your two sets of scores come from one group of participants, then you might need the related *t*-test in preference (see Chapter 13).

14.4   Data requirements for the unrelated *t*-test

One single score variable and two distinct groups of participants are needed.

14.5   Problems in the use of the unrelated *t*-test

SPSS causes users some problems. The common problem is confusing the significance of the *F*-ratio comparing the variances of the two groups of scores with the significance of the *t*-test. They are quite distinct but very close together in SPSS output.

There are two versions of the unrelated *t*-test. One is used if the two sets of scores have similar variances. This is described as 'Equal variances assumed' in SPSS. Where the *F*-ratio shows a statistically significant difference in the variances then the output for 'Equal variances not assumed' should be used.

As is often the case in statistical analysis, it is useful to be clear what the difference in the two group means indicates prior to the test of significance. The test of significance then merely confirms whether this is a big enough difference to be regarded as statistically reliable. It is very easy to get confused by SPSS output giving mean differences otherwise.

You can find out more about unrelated *t*-test in Chapter 13 of Howitt, D. and Cramer, D. (2011). *Introduction to Statistics in Psychology*, 5th edition. Harlow: Pearson.

14.6   The data to be analysed

The computation of an unrelated *t*-test is illustrated with the data in Table 14.1, which shows the emotionality scores of 12 children from two-parent families and 10 children from single-parent families (*ISP*, Table 13.8). In SPSS this sort of *t*-test is called an independent samples *t*-test. The purpose of the analysis is to assess whether emotionality scores are different in two-parent and lone-parent families.

14.7 Entering the data

Step 1

In 'Variable View' of the 'Data Editor' label the first variable 'Family'. This defines the two types of family.
Label the second variable 'Emotion'. These are the emotionality scores.
Remove the two decimal places by changing the value here to 0 if necessary.

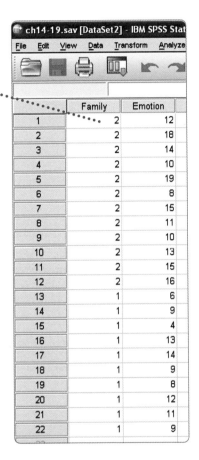

Step 2

In 'Data View' of the 'Data Editor' enter the values of the two variables in the first two columns.
Save the data as a file to use in Chapter 18.

	Family	Emotion
1	2	12
2	2	18
3	2	14
4	2	10
5	2	19
6	2	8
7	2	15
8	2	11
9	2	10
10	2	13
11	2	15
12	2	16
13	1	6
14	1	9
15	1	4
16	1	13
17	1	14
18	1	9
19	1	8
20	1	12
21	1	11
22	1	9

Take a good look at Step 2. Notice that there are two columns of data. The second column ('emotion') consists of the 22 emotionality scores from *both* groups of children. The data are not kept separate for the two groups. In order to identify to which group the child belongs, the first column ('family') contains lots of 1s and 2s. These indicate, in our example, children from a lone-parent family (they are the rows with 1s in 'family') and children from two-parent families (they are the rows with 2s in 'family'). Thus a single column is used for the dependent variable (in this case, emotionality, 'emotion') and another column for the independent variable (in this case, type of family, 'family'). So each row is a particular child, and their independent variable and dependent variable scores are entered in two separate columns in the 'Data Editor'.

14.8   Unrelated *t*-test

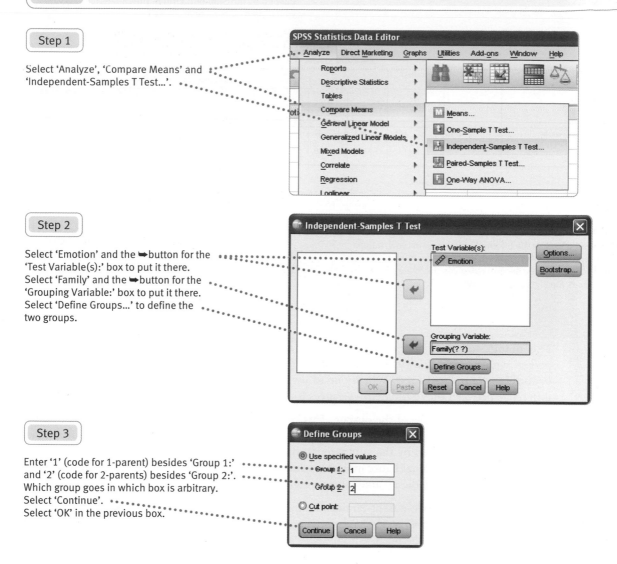

Step 1

Select 'Analyze', 'Compare Means' and 'Independent-Samples T Test...'.

Step 2

Select 'Emotion' and the ➡button for the 'Test Variable(s):' box to put it there.
Select 'Family' and the ➡button for the 'Grouping Variable:' box to put it there.
Select 'Define Groups...' to define the two groups.

Step 3

Enter '1' (code for 1-parent) besides 'Group 1:' and '2' (code for 2-parents) besides 'Group 2:'.
Which group goes in which box is arbitrary.
Select 'Continue'.
Select 'OK' in the previous box.

14.9   Interpreting the output

Group Statistics

	Family	N	Mean	Std. Deviation	Std. Error Mean
Emotion	1	10	9.50	3.100	.980
	2	12	13.42	3.370	.973

This first table shows for each group the number of cases, the mean and the standard deviation. The mean for the lone parents is 9.50. There is obviously a difference, therefore, between the two types of family. The question next is whether the two means differ significantly.

The value of *t* is simply the mean difference (−3.917) divided by the Standard Error of the Difference (1.392) which yields the value of −2.813.

Independent Samples Test

		Levene's Test for Equality of Variances		t-test for Equality of Means					95% Confidence Interval of the Difference	
		F	Sig.	t	df	Sig. (2-tailed)	Mean Difference	Std. Error Difference	Lower	Upper
Emotion	Equal variances assumed	.212	.650	-2.813	20	.011	-3.917	1.392	-6.821	-1.013
	Equal variances not assumed			-2.836	19.768	.010	-3.917	1.381	-6.800	-1.034

If the significance of Levene's test is greater than .05, which it is here at .650, use the information on this first row. If the significance of Levene's test is .05 or less, use the information on the second row. The second row gives the figures for when the variances are significantly different.

For equal variance, *t* is −2.813 which with 20 degrees of freedom is significant at .011 for the two-tailed level. To obtain the one-tailed level, divide this level by 2 which gives .006 rounded to three decimal places.

The output for the uncorrelated/unrelated *t*-test on SPSS is particularly confusing even to people with a good knowledge of statistics. The reason is that there are two versions of the uncorrelated/unrelated *t*-test. Which one to use depends on whether or not there is a significant difference between the (estimated) variances for the two groups of scores.

- Examine the first table of the output. This contains the means and standard deviations of the scores on the dependent variable (emotionality) of the two groups. Notice that an additional figure has been added by the computer to the name of the column containing the dependent variable. This additional figure indicates which of the two groups the row refers to. If you had labelled your values, these value labels would be given in the table.

- For children from two-parent families ('family 2') the mean emotionality score is 13.42 and the standard deviation of the emotionality scores is 3.37. For the children of lone-parent families ('family 1') the mean emotionality score is 9.50 and the standard deviation of emotionality is 3.10.

- In the second table, read the line 'Levene's Test for Equality of Variances'. If the probability value is statistically significant then your variances are *unequal*. Otherwise they are regarded as equal.

- Levene's test for equality of variances in this case tells us that the variances are equal because the *p* value of .650 is not statistically significant.

- Consequently, you need the row for 'Equal variances assumed'. The *t*-value, its degrees of freedom and its probability are displayed. The *t*-value for equal variances is −2.813, which with 20 degrees of freedom has an exact two-tailed significance level of .011.

- Had Levene's test for equality of variances been statistically significant (i.e. .05 or less), then you should have used the second row of the output which gives the *t*-test values for unequal variances.

REPORTING THE OUTPUT

- We could report the results of this analysis as follows:

  The mean emotionality scores of children from two-parent families (M = 13.42, SD = 3.37) is significantly higher, $t(20) = -2.81$, two-tailed $p = .011$, than that of children in lone-parent families (M = 9.50, SD = 3.10).

- It is unusual to see the t-test for unequal variances in psychological reports. Many psychologists are unaware of its existence. So what happens if you have to use one? In order to clarify things, we would write:

  Because the variances for the two groups were significantly unequal, $F = 8.43$, $p < .05$, a t-test for unequal variances was used.

  Knowing that variances are unequal may be of interest in itself. As the degrees of freedom for an unequal variance t-test involves decimal places, this information tells you which version of the t-test has been used.

- If you prefer to use the confidence intervals (see Chapter 15), you might write:

  The difference between the emotionality scores for the children from two-parent families (M = 13.42, SD = 3.37) and lone-parent families (M = 9.50, SD = 3.10) is −3.92. The 95 per cent confidence interval for this difference is −6.82 to −1.01. Since this interval does not include 0.00, the difference is statistically significant at the two-tailed .05 level.

Summary of SPSS Statistics steps for unrelated t-test

Data

- In 'Variable View' of the 'Data Editor', 'name' the variables.
- In 'Data View' of the 'Data Editor' enter the data under the appropriate variable names.

Analysis

- Select 'Analyze', 'Compare Means' and 'Independent Samples T test …'.
- Move the dependent variable to the 'Test variable(s):' box and the independent variable to the 'Grouping Variable;' box.
- Select 'Define Groups …', enter the values for the two groups, select 'Continue' and then 'OK'.

Output

- In the 'Independent Samples Test' table, first check whether Levene's test is significant. If not significant, use the t value, degrees of freedom (df) and significance level for the first row (Equal variances assumed). If Levene's test is significant, use the values in the second row (Equal variances not assumed).

For further resources including data sets and questions, please refer to the website accompanying this book.

CHAPTER 15

Confidence intervals

Overview

- Confidence intervals are advocated as an alternative and, perhaps, improved way of conceptualising inferential statistics since they better reflect the fact that statistical analyses are subject to a certain amount of uncertainty.

- A confidence interval is basically a range of a statistic within which the population value is likely to lie. Conventionally single point statistics such as the mean are given, which fail to reflect the likely variability of the estimate. The confidence interval would suggest that the population mean is likely to lie between two points or limits (e.g. 2.3 and 6.8).

- The most common confidence interval is the 95 per cent value. This gives the interval between the largest and smallest values which cut off the extreme 2.5 per cent of values in either direction. That is, the 95 per cent confidence interval includes the 95 per cent of possible population values in the middle of the distribution.

15.1 What are confidence intervals?

The idea of confidence intervals is an important one but too little used in psychological reports. Professional statistical practice demands that they are used though, as yet, they are far from universally employed. They can be regarded as an alternative to the point estimate approach which is very familiar since it is the conventional approach. Point estimates can be made of any statistic including the mean, the slope of a regression line, and so forth. This book is full of examples of such point statistics. But point estimates fail to emphasise appropriately the extent to which statistical estimates are imprecise. To estimate the mean of a population from a sample as 8.5 suggests a degree of precision which is a little misleading. So the confidence interval approach does something different – that is, it gives the most likely range of the population values of the statistics.

Although expert statisticians recommend the greater use of confidence intervals in psychology they still tend to be supplementary rather than central to most analyses if, indeed, they appear at all. Nevertheless, the underlying concept is relatively easy to understand and, thanks to SPSS, involve little or no labour. In inferential statistics, we use all sorts of statistics from a sample or samples to estimate the characteristics of the population from which the sample was drawn. Few of us will have any difficulty understanding that if the mean age of a sample of participants is 23.5 years then our best estimate of the mean age of the population is exactly the same at 23.5 years. This estimate is known as a point estimate simply because it consists of a single figure (or point). Most of us will also understand that this single figure is simply the best estimate and that it is likely that the population mean will be, to a degree, different. We know this because we know that when we randomly sample from a population that the sample mean is likely to be similar to the population mean but generally will be a little different. So it follows that if we use a sample to estimate a population characteristic then our estimate will be subject to the variation due to sampling.

Confidence intervals simply are the most likely spread of values of a population characteristic (parameter) when it is estimated from a sample characteristic (statistic). So instead of an estimate of the mean population age of 23.5 one would get a confidence interval of, say, 20.0 years to 27.0 years. Actually the confidence interval gives the spread of the middle 95 per cent of means in the case of the 95 per cent confidence interval or the middle 99 per cent of means in the case of the 99 per cent confidence interval. Figure 15.1 outlines some key facts about confidence intervals.

The value of the confidence interval is essentially dependent on the spread of scores in the original sample of data: the bigger the spread of scores then the wider will be the confidence interval. So the variability in the scores is used to estimate the standard deviation of the population. This standard deviation can be used to estimate the standard error of the population. Then it is simply a matter of working out the number of standard errors that cover the middle 95 per cent of the distribution of sample means which could be drawn from that population. The number of standard errors can be obtained from something known as the *t*-distribution for any sample size but, as we are using SPSS Statistics, then we can leave the computer to do the calculation.

SPSS calculates confidence intervals automatically as part of certain statistical procedures. Although we have dwelt on the confidence interval for the mean in this chapter, there are other statistics for which confidence intervals are provided as listed at the end of this chapter.

Sometimes the concept of *confidence limits* is used. Confidence limits are merely the extreme values of the *confidence interval*. In the above example, the 95 per cent confidence limits are 4.64 and 15.36. Figure 15.2 shows the key steps in understanding confidence intervals.

Confidence intervals

- It is increasingly recommended that researchers report confidence intervals and give less emphasis to point estimates.
- They are preferred because they acknowledge the basic variability of research data.

99% and 95% confidence intervals

- Confidence intervals may be more or less cautious as is the case with other statistics.
- The 95% confidence interval is less cautious than the 99% confidence interval.

Confidence limits

- Confidence intervals can be expressed in terms of the upper and lower bounds (values) of the confidence interval.
- Otherwise the confidence interval can be expressed as a value above or below the mean.

FIGURE 15.1   Some key facts about confidence intervals

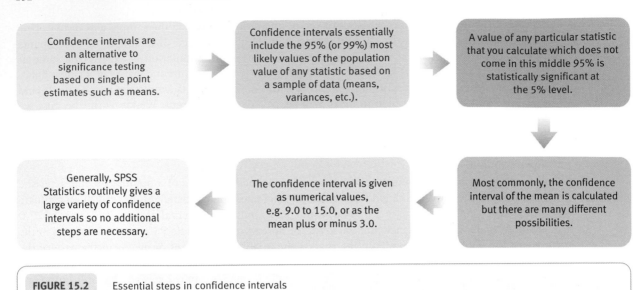

FIGURE 15.2	Essential steps in confidence intervals

15.2 The relationship between significance and confidence intervals

At first sight, statistical significance and confidence intervals appear dissimilar concepts. This is incorrect since they are both based on much the same inferential process. Remember that in significance testing we usually test the null hypothesis of no relationship between two variables. This usually boils down to a zero (or near-zero) correlation or to a difference of zero (or near-zero) between sample means. If the confidence interval does not contain this zero value then the obtained sample mean is statistically significant at 100 per cent minus the confidence level. So if the 95 per cent confidence interval is 2.30 to 8.16 but the null hypothesis would predict the population value of the statistic to be 0.00, then the null hypothesis is rejected at the 5 per cent level of significance. In other words, confidence intervals contain enough information to judge statistical significance. However, statistical significance alone does not contain enough information to calculate confidence intervals.

You can find out more about confidence intervals in Chapter 37 of Howitt, D. and Cramer, D. (2011). *Introduction to Statistics in Psychology*, 5th edition. Harlow: Pearson.

15.3 Confidence intervals and limits in SPSS Statistics

Confidence intervals have been presented in the output for the following tests described in this book:

- Regression *B*: see the table on page 119.
- Related *t*-test: see the table on page 140.
- Unrelated *t*-test: see the table on page 148.

- One-way unrelated ANOVA: see the table on page 208.

- Multiple-comparison tests: see the table on page 236.

- One-way ANCOVA: see the table on page 257.

  Confidence intervals can also be readily obtained for the following statistics when using SPSS:

- One-sample *t*-test.

- One-way related ANOVA.

- Two-way unrelated ANOVA.

- Two-way mixed ANOVA.

- Regression – predicted score.

For further resources including data sets and questions, please refer to the website accompanying this book.

CHAPTER 16

Chi-square

Differences between unrelated samples of frequency data

Overview

- Chi-square is generally used to assess whether two or more samples each consisting of frequency data (nominal data) differ significantly from each other. In other words, it is the usual statistical test to analyse cross-tabulation or contingency tables based on two nominal category variables.

- It can also be used to test whether a single sample differs significantly from a known population. The latter application is the least common because population characteristics are rarely known in research.

- It is essential to remember that chi-square analyses frequencies. These should *never* be converted to *percentages* for entry into SPSS Statistics as they will give misleading outcomes when calculating the value and significance of chi-square. This should be clearly distinguished from the use of percentages when one is trying to interpret what is happening in a contingency table.

- Also remember that a chi-square analysis needs to include the data from every individual only once. That is, the total frequencies should be the same as the number of people used in the analysis.

- The analysis and interpretation of 2×2 contingency tables are straightforward. However, interpretation of larger contingency tables is not quite so easy and may require the table to be broken down into a number of smaller tables. Partitioning chi-square, as this is known, usually requires adjustment to the significance levels to take into account the number of sub-analyses carried out.

- This chapter also includes the Fisher exact test, which can be useful in some circumstances when the assumptions of the chi-square are not met by your data (especially when the expected frequencies are too low).

- Versions of chi-square are used as measures of goodness-of-fit in some of the more advanced statistical techniques discussed later in this book such as logistic regression. A test of goodness-of-fit simply assesses the relationship between the available data and the predicted data based on a set of predictor variables. Consequently, it is essential to understand chi-square adequately, not simply because of its simple application but because of its role in more advanced statistical techniques.

16.1 | What is chi-square?

Chi-square is a test of statistical significance which is used when the data consist of only nominal (category) variables (see Figure 16.1). It can be used with a single nominal variable (one-sample chi-square) but it is much more usual to use it with two nominal variables. If you have three or more nominal variables then it is likely that log-linear analysis (Chapter 37) is the solution, though it is a much more complex statistic. It has to be said that many psychologists never collect data in the form of nominal category data rather than score data. As a consequence, despite its familiarity among psychologists, chi-square is not so commonly used as this might imply. Figure 16.2 outlines different types of chi-square.

The basic principle underlying chi-square is the comparison of the frequencies of cases of members of samples with what might be expected to be the frequencies in the population from which the samples are drawn:

- In the one-sample chi-square, the population distribution is usually based on some known population (e.g. the national statistics on the number of males and females) or some theoretical distribution (e.g. the population distribution of heads and tails for tosses of a coin can be said to be 50 per cent heads and 50 per cent tails). (If you only have one sample then your best estimate of the population characteristics is therefore exactly the same as that sample which is not very helpful.) There are very few circumstances in which a researcher would need to carry out a one-sample chi-square.

- In the two-sample chi-square, the population distribution is estimated from the available data since you will have two or more separate samples. This is simply done by combining the characteristics of two or more samples of data.

This is illustrated by the data in Table 16.1 which is sometimes known as a cross-tabulation table or contingency table. This table shows the data from a study in which a sample of males and a sample of females (i.e. the nominal variable gender) have been asked about their favourite television programme (i.e. the nominal variable favourite type of programme). Some mention soap operas, some crime dramas, but others mention neither of these. The researcher wants to know if males and females differ in terms of their favourite types of programme – which is exactly the

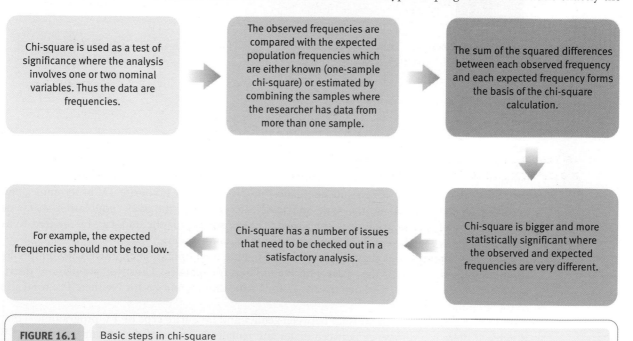

FIGURE 16.1   Basic steps in chi-square

One-way chi-square	• There is just one independent variable with two or more categories.
	• Usually the expected frequencies are based on known population distribution or theoretical expectations.

Two-way chi-square	• There are two independent variables each with two or more categories – try to avoid too many as interpretation becomes difficult.
	• Expected frequencies are based on the marginal totals of the row and column corresponding to each cell of the cross-tabulation table.

Fisher's exact probability tests	• This is a test which can be used for a chi-square with two independent variables each with two different categories or where there are two categories for one independent variable and three categories for the other.
	• It is not based on calculating expected frequencies.

McNemar's test of significance of changes	• This is a special version of chi-square which examines changes over time in terms of membership of two categories of the independent variable (see Chapter 17).
	• The basic assumption is that if there is no change then the same proportions will change to one condition as changes to the other condition.

Log-linear analysis	• This is a complex statistic which is used when there are three or more independent variables with two or more categories in each (see Chapter 37 for details).
	• The calculation of expected frequencies is very complex/protracted and is not feasible other than using a computer program such as SPSS Statistics.

FIGURE 16.2   Different types of chi-square

Table 16.1   Relationship between favourite TV programme and gender

Respondents	Soap opera	Crime drama	Neither
Males	27	14	19
Females	17	33	9

same thing as asking whether there is a relationship between a respondent's gender and their favourite type of TV programme. In chi-square, the assumption is that there is no relationship between the two nominal variables. In terms of our example this means that there is no difference between one group of respondents (males) in terms of their favourite TV programmes and the other (females). That is, the male and females come from the same population so far as favourite type of TV programme is concerned. The chi-square for the data can be referred to as a 2×3 chi-square as there are two rows and three columns of data.

What this amounts to is that it is possible to estimate the population characteristics simply by combining the findings for the males with the findings for females. Thus the 'population' distribution is 44 (i.e. 27 + 17) for Soap opera, 47 for Crime drama, and 28 for Neither. The total size of the 'population' is 119.

But what if there is a relationship between gender and favourite type of TV programme – that is, what if there is a difference between males and females in terms of their favourite type of TV programme? If this were the case then the 'population' distribution will be very different from the distribution of favourite types of TV programme for males and females separately. If males and females are no different in terms of their favourite type of TV programme then they should show the same distribution as the estimated 'population' distribution (which, after all, was simply the separate samples combined). In the terminology of chi-square, the cell frequencies are called the observed frequencies and the population frequencies are called the expected frequencies (since they are what we would expect on the basis that there is no difference between the samples).

The only complication is that the population frequencies have to be scaled down to the sample sizes in question which do not have to be equal. So the expected frequency of males in the soap opera condition in Table 16.1 is calculated on the basis that in the population there are 44/119 who prefer Soap operas and that there is a total of 60 (i.e. 27 + 14 + 19) males. So the expected frequency of males who like Soap operas is 44/119 of 60 which equals 22.185. Notice that this expected value is different from the obtained value of 27 which can be seen in Table 16.1. The more that the obtained and expected values differ for all of the entries in the table, then the more that the male participants and the female participants differ. Chi-square is based on the sum of the squared differences between the obtained frequencies and the expected frequencies.

Chi-square is subject to a range of restrictions which mean that the data may need to be modified to make it suitable for analysis or a different analysis performed. The Fisher exact probability can be used, for example, instead of a 2 × 2 chi-square in some circumstances – that is where there are two rows and two columns of data. It is also available for the 2 × 3 case.

16.2 When to use chi-square

There are few alternatives to the use of chi-square when dealing solely with nominal (category) variables. Chi-square can only be used with nominal (category) data. It can be used where there is just one nominal variable but also where there are two different nominal variables. There may be any *practical* number of categories (values) of each of the nominal variables though you will probably wish to restrict yourself to just a few categories as otherwise the analysis may be very difficult to interpret. Chi-square is not used for three or more nominal variables where log-linear analysis may be the appropriate statistic.

16.3 When not to use chi-square

There are a number of limitations on the use of chi-square. The main problem is when there are too many expected frequencies which are less than five. Any more than 20–25 per cent of the expected frequencies below five indicates that chi-square would be problematic. Sometimes a Fisher exact test can be used instead because low expected frequencies do not invalidate this test in the same way. An alternative is to combine categories of data together if this is appropriate, thus increasing the expected frequencies to five or above (hopefully). However, it is not always meaningful to combine categories with each other. Sometimes the alternative is simply to delete the small categories. This is all based on judgement and it is difficult to give general advice other than to be as thoughtful as possible (see Section 16.5).

16.4 Data requirements for chi-square

Each participant in the research contributes just one to the frequencies of each value of the nominal variable. It is difficult to violate this requirement using SPSS but even then it is wise to check that the number of participants you have in your analysis is the same as the total number of frequencies in the chi-square if you are imputting your own contingency or cross-tabulation table into SPSS.

16.5 Problems in the use of chi-square

Chi-square is not an easy statistical test to use well. Fortunately, psychologists tend to prefer score data so the problems with chi-square are not routinely experienced by researchers. The main difficulties with chi-square are as follows:

- Usually it is necessary to have a fairly substantial numbers of participants to use chi-square appropriately. This is because having small expected frequencies invalidates the test. Usually if you have more than about 20 or 25 per cent expected frequencies of less than five then you should not use chi-square. You may be able to use the Fisher exact test instead in the 2×2 case. Otherwise it is a case of trying to re-jig the contingency table in some way. But the main implication is to avoid categories of the nominal variable which attract few cases. These could be deleted or put with other categories into a combined 'other' category.

- A major problem in using chi-square is what happens when the analysis is of something larger than a 2×2 contingency table. The problem is that an overall significant chi-square tells you that the samples differ, it does not entirely tell you in what way the samples differ. So, for example, do women tend to prefer Soap operas and men tend to prefer Crime drama? It is possible to analyse the data as a 2×2 chi-square by dropping the category 'Neither', for example. Indeed it is possible to form as many 2×2 analyses out of a bigger cross-tabulation table as you wish though it would be usual to adjust the significance level for the number of separate chi-squares carried out (see *ISP*).

- One thing that gets students confused is it is possible to enter a cross-tabulation table directly in SPSS using a weighting procedure. This would be useful in circumstances where you have carried out a study and already have your data in the form of a simple table. However, if your data are part of a bigger SPSS spreadsheet then this weighting procedure would not be used. It all depends on what form your data are in.

You can find out more about chi-square in Chapter 14 of Howitt, D. and Cramer, D. (2011). *Introduction to Statistics in Psychology*, 5th edition. Harlow: Pearson.

16.6 The data to be analysed

The computation of chi-square with two or more samples is illustrated with the data in Table 16.1 (*ISP*, Table 14.8). This table shows which one of three types of television programme is favoured by a sample of 119 teenage boys and girls. To analyse a table of data like this one with SPSS, first we have to input the data into the 'Data Editor' and weight the cells by the frequencies of cases in them.

- As we are working with a ready-made table, it is necessary to go through the 'Weighting Cases' procedure first (see Section 16.7). Otherwise, you would enter Table 16.1 case by case, indicating which category of the row and which category of the column each case belongs to (see Section 16.8). We need to identify each of the six cells in Table 16.1. The rows of the table represent the gender of the participants, while the columns represent the three types of television programme. We will then weight each of the six cells of the table by the number of cases in them.

- The first column, called 'gender' in Step 1 of Section 16.7, contains the code for males (1) and females (2). (These values have also been labelled.)

- The second column, called 'program', holds the code for the three types of television programme: soap opera (1), crime drama (2) and neither (3). (These values have also been labelled.)

16.7 Entering the data using the 'Weighting Cases' procedure

Step 1

In 'Variable view' of the 'Data Editor' label the first three variables 'gender', 'program' and 'freq' respectively.
Remove the two decimal places. Label the values of 'gender' and 'program'.

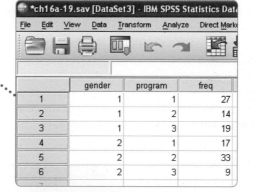

Step 2

In 'Data view' of the 'Data Editor' enter the appropriate values. Each row represents one of the six cells in Table 16.1.

	gender	program	freq
1	1	1	27
2	1	2	14
3	1	3	19
4	2	1	17
5	2	2	33
6	2	3	9

Step 3

To weight these cells, select 'Data' and 'Weight Cases...'.

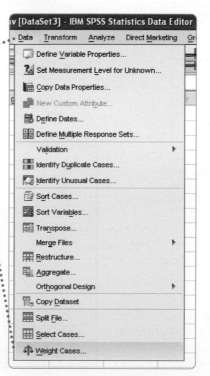

Step 4

Select 'freq', 'Weight cases by' and the ➡ button to put 'freq' into the 'Frequency Variable:' box.
Select 'OK'.

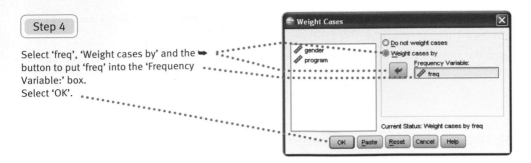

16.8   Entering the data case by case

Enter the values for the two variables for each of the 119 cases.

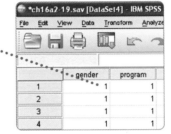

16.9   Chi-square

Step 1

Select 'Analyze', 'Descriptive Statistics' and 'Crosstabs...'.

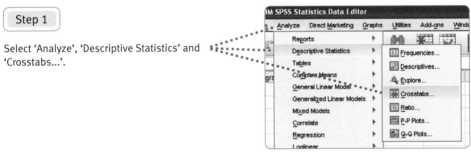

Step 2

Select 'gender' and the ➡ button for 'Row(s):' to put it there.
Select 'program' and the ➡ button for 'Column(s):' to put it there.
Select 'Statistics...'.

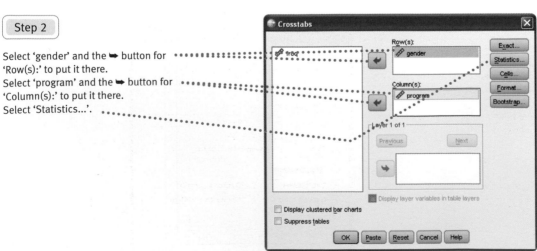

Step 3

Select 'Chi-square' and 'Continue'. ••••••••••

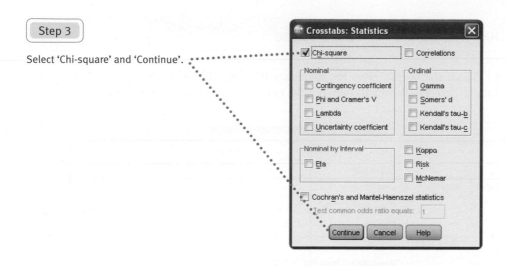

Step 4

Select 'Cells'. ••••••••••

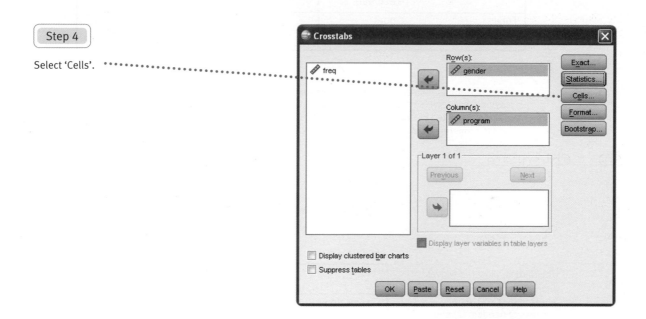

Step 5

Select 'Expected' under 'Counts'. •••••••
Select 'Unstandardized' under •••••
'Residuals'. Residuals mean
differences.
Select 'Continue' and then
'OK' in the previous box.

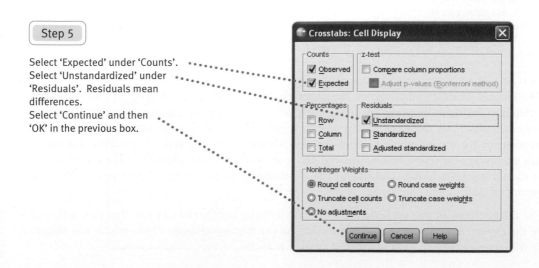

16.10 Interpreting the output for chi-square

The second table of the output gives the frequency (Count), the expected frequency (Expected Count) and the difference (Residual) between the two for each of the six cells of the table.

gender * program Crosstabulation

		Statistics	program Soap	program Crime	program Neither	Total
gender	Males	Count	27	14	19	60
		Expected Count	22.2	23.7	14.1	60.0
		Residual	4.8	-9.7	4.9	
	Females	Count	17	33	9	59
		Expected Count	21.8	23.3	13.9	59.0
		Residual	-4.8	9.7	-4.9	
Total		Count	44	47	28	119
		Expected Count	44.0	47.0	28.0	119.0

E.g. the count or number of females saying they prefer 'soaps' is 17, the number expected by chance is 21.8 and the difference between these is −4.8.

You must understand what the figures for count for the gender by program cross-tabulation mean. This is equivalent to the original data table (Table 16.1). You could turn the figures into percentages. (See the screenshot in Step 5 which gives three different options – Row, Column and Total percentages.) Choosing column percentages would probably help you see that males tend to prefer soap operas and neither more than females. Females tend to prefer crime programmes more than do males.

Chi-Square Tests

	Value	df	Asymp. Sig. (2-sided)
Pearson Chi-Square	13.518[a]	2	.001
Likelihood Ratio	13.841	2	.001
Linear-by-Linear Association	.000	1	.987
N of Valid Cases	119		

The third and last table gives the value of (Pearson's) chi-square (13.518), the degrees of freedom (2) and the two-tailed significance (.001). As this value is less than .05, this chi-square is significant.

a. 0 cells (.0%) have expected count less than 5. The minimum expected count is 13.88.

- The second (middle) table shows the observed and expected frequencies of cases and the difference (residual) between them for each cell. The observed frequency (called Count) is presented first and the expected frequency (called Expected Count) second. The observed frequencies are always whole numbers, so they should be easy to spot. The expected frequencies are always expressed to one decimal place, so they are easily identified. Thus the first cell of the table (males liking soap opera) has an observed frequency of 27 and an expected frequency of 22.2.

- The final column in this table (labelled 'Total') lists the number of cases in that row followed by the expected number of cases in the table. So the first row has 60 cases, which will always be the same as the expected number of cases (i.e. 60.0).

- Similarly, the final row in this table (labelled 'Total') first presents the number of cases in that column followed by the expected number of cases in the table for that column. Thus the first column has 44 cases, which will always be equal to the expected number of cases (i.e. 44.0).

- The chi-square value, its degrees of freedom and its significance level are displayed in the third table on the line starting with the word Pearson, the person who developed this test. The chi-square value is 13.518 which, rounded to two decimal places, is 13.52. Its degrees of freedom are two and its exact two-tailed probability is .001.

- Also shown underneath this table is the 'minimum expected count' of any cell in the table, which is 13.88 for the last cell (females liking neither). If the minimum expected frequency is less than 5.0 then we should be wary of using chi-square. If you have a 2 × 2 chi-square and small expected frequencies occur, it would be better to use the Fisher exact test which SPSS displays in the output in these circumstances.

REPORTING THE OUTPUT

There are two alternative ways of describing these results. To the inexperienced eye they may seem very different but they amount to the same thing:

- We could describe the results in the following way:

  There was a significant difference between the observed and expected frequency of teenage boys and girls in their preference for the three types of television programme, $\chi^2(2) = 13.51$, $p = .001$.

- Alternatively, and just as accurate:

  There was a significant association between gender and preference for different types of television programme, $\chi^2(2) = 13.51$, $p = .001$.

- In addition, we need to report the direction of the results. One way of doing this is to state that:

  Girls were more likely than boys to prefer crime programmes and less likely to prefer soap operas or both programmes.

16.11 Fisher's exact test

The chi-square procedure computes Fisher's exact test for 2 × 2 tables when one or more of the four cells has an expected frequency of less than 5. Fisher's exact test would be computed for the data in Table 16.2 (*ISP*, Table 14.14).

Table 16.2	Photographic memory and gender	
	Photographic memory	**No photographic memory**
Males	2	7
Females	4	1

16.12 Interpreting the output for Fisher's exact test

gender * memory Crosstabulation

			memory		
		Statistics	Photographic	Non-photographic	Total
gender	Males	Count	2	7	9
		Expected Count	3.9	5.1	9.0
		Residual	-1.9	1.9	
	Females	Count	4	1	5
		Expected Count	2.1	2.9	5.0
		Residual	1.9	-1.9	
Total		Count	6	8	14
		Expected Count	6.0	8.0	14.0

This is the second of three tables in the output showing the observed (Count) and expected (Expected Count) frequencies of the four cells.

Chi-Square Tests

	Value	df	Asymp. Sig. (2-sided)	Exact Sig. (2-sided)	Exact Sig. (1-sided)
Pearson Chi-Square	4.381[a]	1	.036		
Continuity Correction[b]	2.340	1	.126		
Likelihood Ratio	4.563	1	.032		
Fisher's Exact Test				.091	.063
Linear-by-Linear Association	4.068	1	.044		
N of Valid Cases	14				

a. 3 cells (75.0%) have expected count less than 5. The minimum expected count is 2.14.
b. Computed only for a 2x2 table

This is the third and last table in the SPSS output. It shows the values of the statistics, their degrees of freedom and their significance levels. The significance of Fisher's exact test for this table is .091 at the two-tailed level and .063 at the one-tailed level.

REPORTING THE OUTPUT

- We would write:

  There was no significant relationship between gender and the possession of a photographic memory, two-tailed Fisher exact $p = .091$.

  or

  Males and females do not differ in the frequency of possession of a photographic memory, two-tailed Fisher exact $p = .091$.

- However, with such a small sample size, the finding might best be regarded as marginally significant and a strong recommendation made that further studies should be carried out in order to establish with more certainty whether girls actually do possess photographic memories more frequently.

16.13 One-sample chi-square

We will illustrate the computation of a one-sample chi-square with the data in Table 16.3 (*ISP*, Table 14.16), which shows the observed and expected frequency of smiling in 80 babies. The expected frequencies were obtained from an earlier large-scale study.

Table 16.3	Data for a one-sample chi-square		
	Clear smilers	**Clear non-smilers**	**Impossible to classify**
Observed frequency	35	40	5
Expected frequency	40	32	8

Step 1

Enter the data in 'Data View' of the 'Data Editor' having named the variables and removed the two decimal places. Label the three categories. Weight the cells or cases with 'freq'.

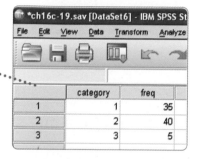

Step 2

Select 'Analyze', 'Nonparametric Tests', 'Legacy Dialogs' and 'Chi-square...'.

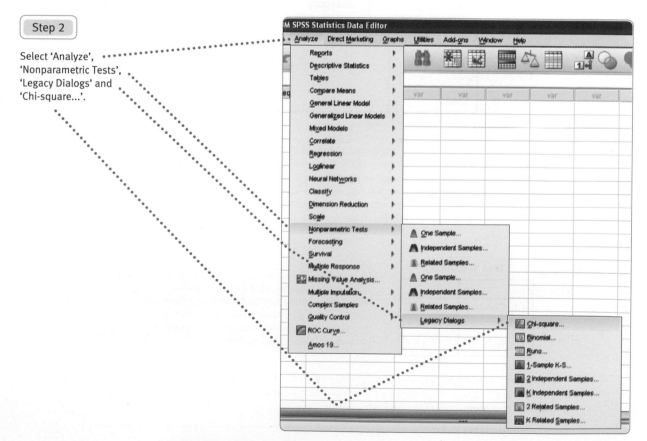

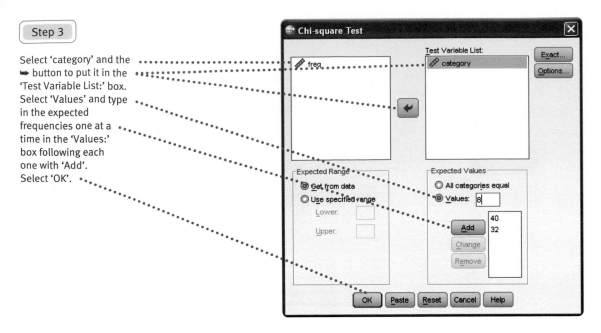

Step 3

Select 'category' and the
➡ button to put it in the
'Test Variable List:' box.
Select 'Values' and type
in the expected
frequencies one at a
time in the 'Values:'
box following each
one with 'Add'.
Select 'OK'.

16.14 Interpreting the output for a one-sample chi-square

category

	Observed N	Expected N	Residual
Smilers	35	40.0	-5.0
Non-smilers	40	32.0	8.0
Unclassifiable	5	8.0	-3.0
Total	80		

The first of the two tables in the output shows the observed and the expected frequencies
for the three categories together with the difference or residual between them. The first
column shows the three categories, the second column the observed N or frequencies,
the third column the expected N or frequencies and the fourth column the residual or
difference between the observed and the expected frequencies. The observed frequency
of smilers is 35 and the expected frequency 40.0.

Test Statistics

	category
Chi-Square	3.750[a]
df	2
Asymp. Sig.	.153

a. 0 cells (.0%)
have expected
frequencies less
than 5. The
minimum
expected cell
frequency is 8.0.

The second table shows the value of chi-square (3.750),
the degrees of freedom (2) and the significance level
(.153). As the significance level is greater than .05,
the observed frequencies do not differ significantly
from those expected by chance.

REPORTING THE OUTPUT

We could describe the results of this analysis as follows:

There was no statistical difference between the observed and expected frequency for the three categories of smiling in infants, $\chi^2(2) = 3.75$, $p = .153$.

16.15   Chi-square without ready-made tables

In this chapter we have concentrated on how one can analyse data from pre-existing contingency tables. This is why we need the weighting procedure. However, you will not always be using ready-made tables. Any variables which consist of just a small number of nominal categories can be used for chi-square. For example, if one wished to examine the relationship between gender (coded 1 for male, 2 for female) and age (coded 1 for under 20 years, 2 for 20 to 39 years, and 3 for 40 years and over), the procedure is as follows. (a) Enter the age codes for your, say, 60 cases in the first column of the Data Editor. (b) Enter the age categories for each of these cases in the equivalent row of the next column. You can then carry out your chi-square as follows. You do not go through the weighting procedure first. The frequencies in the cells are calculated for you by SPSS.

Summary of SPSS Statistics steps for chi-square

Data

- Name the variables in 'Variable View' of the 'Data Editor'.
- Enter the data under the appropriate variable names in 'Data View' of the 'Data Editor'.

Analysis

- For a one-sample chi-square, select 'Analyze', 'Nonparametric Tests', 'Legacy Dialogs' and 'Chi-Square Test …'.
- Move category variable to 'Test Variable List:' box and enter 'Expected Values' of frequencies for the categories of that variable.
- For a more than one-sample chi-square, select 'Analyze', 'Descriptive Statistics' and 'Crosstabs …'.
- Move appropriate variables to Row and Column box.
- Select 'Statistics …' and 'Chi-square'.
- Select 'Cells …' and 'Expected' and 'Unstandardized'.

Output

- Check that the minimum expected frequencies have been met.
- Check that the test is statistically significant with $p = .05$ or less.
- Determine direction of significant difference.

For further resources including data sets and questions, please refer to the website accompanying this book.

CHAPTER 17

McNemar's test
Differences between related samples of frequency data

Overview

● McNemar's test is generally used to assess whether two related samples each consisting of frequency data (nominal data) differ significantly from each other. In other words, it is the usual statistical test to analyse cross-tabulation or contingency tables based on two nominal category variables which are related.

● It is sometimes known as a test of the significance of change since it involves comparing two related samples which might be the same group of participants measured at two different points in time.

17.1 What is McNemar's test?

McNemar's test of significance of changes is a test of statistical significance which is used when the data consist of only nominal (category) variables. This is a test of the change over time of the categories of a nominal variable in which a sample is placed. The nominal variable can only have two different categories so the data may look superficially like a 2×2 chi-square or contingency table but the calculation is actually based on the difference numbers between the two categories for the two samples. The expectation is, if the null hypothesis of no difference (or no change) is true, that the differences will be the same for the two categories. The steps in understanding McNemar's test are illustrated in Figure 17.1. Notice that only two out of the four cells refer to change. McNemar's test can be thought of as the chi-square equivalent of a related test.

17.2 When to use McNemar's test

There are few alternatives to the use of McNemar's test when dealing solely with two related nominal (category) variables. McNemar's test can only be used with nominal (category) data.

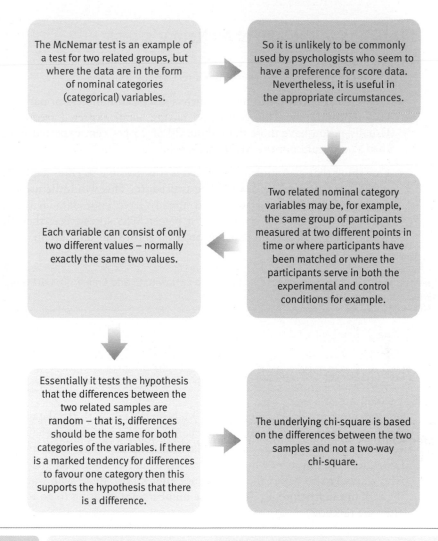

The McNemar test is an example of a test for two related groups, but where the data are in the form of nominal categories (categorical) variables.

So it is unlikely to be commonly used by psychologists who seem to have a preference for score data. Nevertheless, it is useful in the appropriate circumstances.

Two related nominal category variables may be, for example, the same group of participants measured at two different points in time or where participants have been matched or where the participants serve in both the experimental and control conditions for example.

Each variable can consist of only two different values – normally exactly the same two values.

Essentially it tests the hypothesis that the differences between the two related samples are random – that is, differences should be the same for both categories of the variables. If there is a marked tendency for differences to favour one category then this supports the hypothesis that there is a difference.

The underlying chi-square is based on the differences between the two samples and not a two-way chi-square.

FIGURE 17.1   Steps in understanding the McNemar test

17.3 When not to use McNemar's test

The main limitation on the use of McNemar's test is when there are too many expected frequencies which are less than five. Any more than 20–25 per cent of the expected frequencies below five indicates that McNemar's test would be problematic.

17.4 Data requirements for McNemar's test

Each participant in the research contributes to two of the frequencies of each value of the nominal variable.

17.5   Problems in the use of McNemar's test

The main difficulties with McNemar's test are as follows:

- Usually it is necessary to have a fairly substantial numbers of participants to use McNemar's test appropriately. This is because having small expected frequencies invalidates the test. Usually if you have more than about 20 or 25 per cent expected frequencies of less than five then you should not use McNemar's test.

- One thing that gets students confused is it is possible to enter a cross-tabulation table directly in SPSS Statistics using a weighting procedure. This would be useful in circumstances where you have carried out a study and already have your data in the form of a simple table. However, if your data are part of a bigger SPSS spreadsheet then this weighting procedure would not be used. It all depends on what form your data are in.

You can find out more about McNemar's test in Chapter 14 of Howitt, D. and Cramer, D. (2011). *Introduction to Statistics in Psychology*, 5th edition. Harlow: Pearson.

17.6   The data to be analysed

We will illustrate the computation of McNemar's test with the data in Table 17.1, which shows the number of teenage children who changed or did not change their minds about going to university after listening to a careers talk favouring university education (*ISP*, Table 14.17). The table gives the numbers who wanted to go to university before the talk and after it (30), those who wanted to go before the talk but not after it (10), those who wanted to go to university after the talk but not before it (50), and the numbers not wanting to go to university both before and after the talk (32).

Table 17.1	Students wanting to go to university before and after a careers talk	
	1 Before talk 'yes'	**2 Before talk 'no'**
1 After talk 'yes'	30	50
2 After talk 'no'	10	32

17.7   Entering the data using the 'Weighting Cases' procedure

Step 1

In 'Variable view' of the 'Data Editor' label the first three variables 'Before', 'After' and 'Freq' respectively.
Remove the two decimal places.
Label the values of 'Before' and 'After' as 'Yes' for 1 and 'No' for 2.

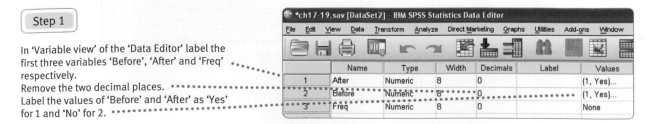

Step 2

Enter the data in 'Data View' of
the 'Data Editor'.

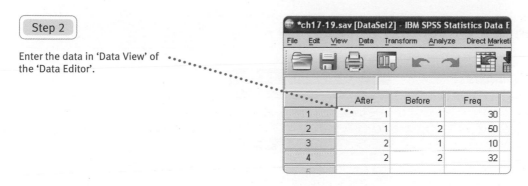

Step 3

To weight these cells,
select 'Data' and 'Weight Cases...'.

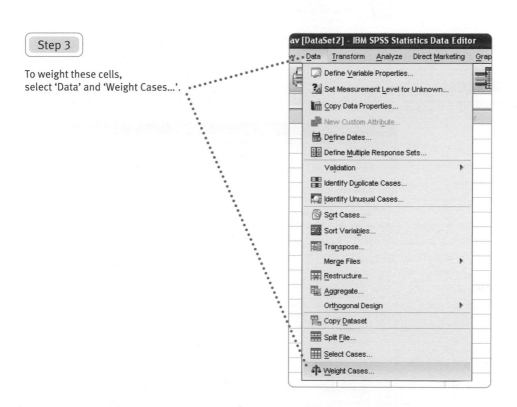

Step 4

Select 'Freq', 'Weight cases by'
and the ➡ button to put 'Freq'
in the 'Frequency Variable:' box.
Select 'OK'.

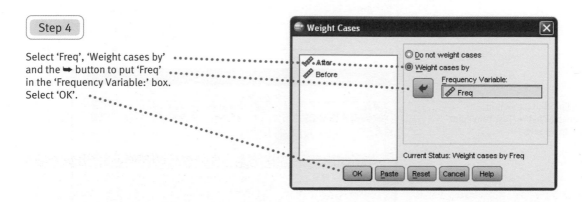

17.8   Entering the data case by case

Enter the values for the two variables for
each of the 119 cases.

17.9   McNemar's test

Step 1

Select 'Analyze',
'Nonparametric Tests',
'Legacy Dialogs' and '2
Related Samples...'.

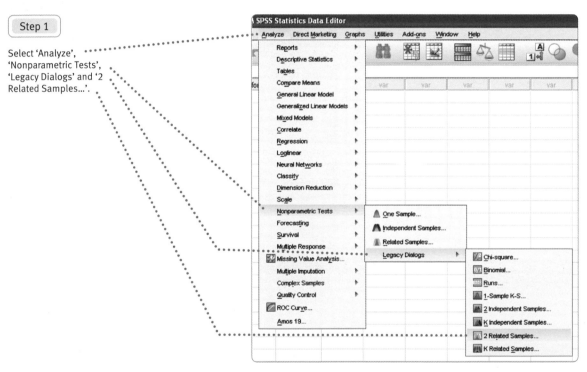

Step 2

Select 'After', 'Before' and the ➡ button to
put these two variables in the 'Test Pairs:' box.
Select 'Wilcoxon' to de-select it.
Select 'McNemar'.
Select 'OK'.

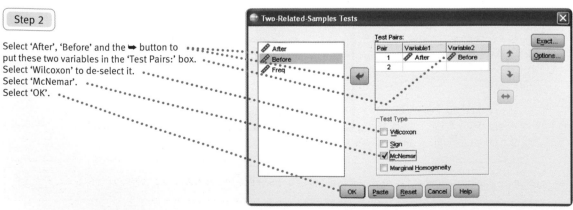

17.10 Interpreting the output for McNemar's test

The first of the two tables shows the frequencies of cases in the four cells as in Table 17.1. The two values, 1 and 2, have not been labelled.

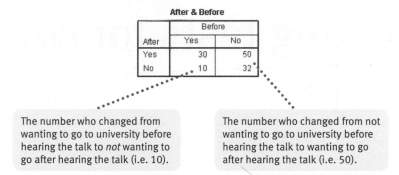

After & Before

After	Before	
	Yes	No
Yes	30	50
No	10	32

The number who changed from wanting to go to university before hearing the talk to *not* wanting to go after hearing the talk (i.e. 10).

The number who changed from not wanting to go to university before hearing the talk to wanting to go after hearing the talk (i.e. 50).

The second table in the output shows the total number (N) of cases (122), the value of chi-square (25.350) and the significance level (.000). Technically the significance level can never be 0. It is less than .001. As it is less than .05 it is significant. This means that there has been a significant change in the number of teenagers changing their minds about going to university after listening to a careers talk favouring university education.

Test Statistics[b]

	After & Before
N	122
Chi-Square[a]	25.350
Asymp. Sig.	.000

a. Continuity Corrected
b. McNemar Test

REPORTING THE OUTPUT

We can report the results of this analysis as follows:

There was a significant increase in the number of teenagers who wanted to go to university after hearing the talk, $\chi^2(1) = 25.35$, $p < .001$.

Summary of SPSS Statistics steps for McNemar's test

Data

- Name the variables in 'Variable View' of the 'Data Editor'.
- Enter the data under the appropriate variable names in 'Data View' of the 'Data Editor'.

Analysis

- Select 'Analyze', 'Nonparametric Tests', 'Legacy Dialogs' and '2-Related Samples ...'.
- Move appropriate pair of variables to the 'Test Pairs:' box and select 'McNemar'.

Output

- Check that the minimum expected frequencies have been met.
- Check that the test is statistically significant with p .05 or less.
- Determine the direction of significant difference.

For further resources including data sets and questions, please refer to the website accompanying this book.

CHAPTER 18

Ranking tests for two groups

Non-parametric statistics

Overview

- Sometimes you may wish to know whether the means of two different sets of scores are significantly different from each other but feel that the requirement that the scores on each variable are roughly normally distributed (bell-shaped) is not fulfilled. Non-parametric tests can be used in these circumstances.

- Non-parametric tests are ones which make fewer assumptions about the characteristics of the population from which the data came. This is unlike parametric tests (such as the *t*-test) which makes more assumptions about the nature of the population from which the data came. The assumption of normality (bell-shaped frequency curves) is an example of the sort of assumptions incorporated into parametric statistics.

- Strictly speaking, non-parametric statistics do not test for differences in means. They cannot, since, for example, they use scores turned into ranks. Usually they test whether the ranks in one group are typically larger or smaller than the ranks in the other groups.

- We have included the sign test and Wilcoxon's test for related data. In other words, they are the non-parametric equivalents to the related *t*-test. Wilcoxon's test should be used in preference to the sign test when comparing score data.

- The Mann–Whitney *U*-test is used for unrelated data. That is, it is the non-parametric equivalent to the unrelated *t*-test.

18.1 What are non-parametric tests?

There is a whole variety of statistical techniques which were developed for use in circumstances in which data do not meet the standards required of other more commonly used statistical techniques. These are known as non-parametric or distribution-free methods. By this is meant that they do not depend mathematically on standard distributions such as the normal distribution, the *t*-distribution and so forth unlike the *t*-test and similar techniques. Many statistical techniques were developed on the assumption that the data are normally distributed. To the extent that this is not the case, the results of the analysis using these tests may be to a degree misleading. Non-parametric tests are often referred to as ranking tests since many of them are based on turning the scores into ranks (where 1, 2, 3, etc. refer to the smallest score, second smallest score and third smallest score, etc.). However, not all non-parametric tests use ranking.

The other issue which results in the use of non-parametric tests is the scale of measurement involved. The interval scale of measurement assumes that the distances between each point on the scale are equal. If you think of a ruler (or rule which carpenters use) then the marks on the scale (e.g. millimetres) are of equal size no matter where you look on the ruler. That is basically what an interval (or equal interval) scale is. It is possible to do things like work out the average length of things using such a scale. Imagine that the ruler is made of plastic and it is heated such that it can be pulled and stretched. Now following this, the millimetres will not all be the same size anymore. Strictly speaking, the ruler is now inaccurate and some would say useless. It is no longer possible to work out the average length of things because the ruler has stretched more in some places than others. If the stretching had been constant then the ruler would still be able to measure lengths though no longer in millimetres but some other unit. Since the stretching is not even, can the ruler serve any purpose at all? Well, it could still be used to say which things are longer than which other things, for example.

What has this to do with psychological measurements? Psychologists have in the past argued that it has a lot to do with such measures as intelligence, extraversion and other measures. So unless they felt that their measures were on an interval scale then they would use non-parametric tests which do not assume equal intervals but only that a higher score on the scale would indicate greater intelligence, for example. The trouble with this is how do we know that psychological measures have equal intervals? The answer is that we do not.

So does that mean that we should never use statistics which assume equal intervals? Well some would argue that. On the other hand, psychologists regularly seem to ignore the equal interval requirement. One reason is that they might feel that psychological measures reflect more or less roughly equal intervals so it does not matter that much. This is not much more than an article of faith and it has no factual basis. Others argue that the underlying psychological scale of measurement does not matter since mathematically it is the numbers (scores) which we do the statistical analysis on. So we do not need to bother about equal intervals and psychological measurement. This is the view that we take.

The parametric tests which are built on the equal interval assumption are among the most powerful statistical techniques available for psychologists to use. It would be extremely restricting not to use them.

If you wish to use a non-parametric test then there is no particular problem with this in simple analyses. Unfortunately for complex data it may result in a weaker analysis being employed and the statistical question may not be so well addressed. Some may regard this as statistical heresy but overwhelmingly it is clear that psychologists largely proceed on the assumption that their measures are on an interval scale – more or less.

SPSS Statistics has a range of non-parametric statistical tests for one sample cases, two sample cases and multiple sample cases. These are generally analogous to a parametric equivalent such as the *t*-tests or ANOVA.

● The Mann–Whitney *U*-test is equivalent to the unrelated *t*-test and can be applied to exactly the same data.

● The Wilcoxon matched-pairs test is equivalent to the related *t*-test and can be applied to exactly the same data.

● The sign test is a weaker test than the Wilcoxon matched-pairs test and cannot generally be recommended as a substitute for the related *t*-test.

SPSS has these under 'Analyze', 'Nonparametric Tests' and 'Legacy Dialogs'. That is also where the one-sample chi-square is to be found, though this is not strictly a non-parametric test. Figure 18.1 shows the key steps in using non-parametric statistics.

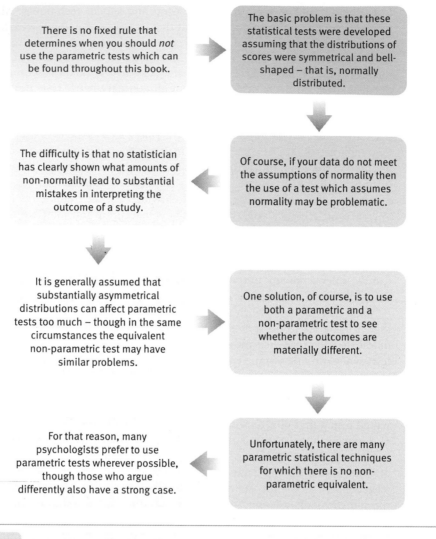

FIGURE 18.1   Key steps in non-parametric statistics

18.2   When to use non-parametric tests

Generally speaking, use non-parametric tests only when you have a very clear and strong reason for doing so. Quite what would be sufficient reason is not obvious. Before computers non-parametric tests would often be recommended as they are easier to compute by hand than parametric tests in general, but these circumstances are unlikely to occur nowadays. You might find it a useful exercise to compare the results of equivalent parametric and non-parametric analyses on the same data. It is unlikely that your interpretation will be very different and, where it is, these will be marginal circumstances close to the dividing line between statistical significance and non-significance.

However, you might find non-parametric tests useful in circumstances where you suspect that outliers might be affecting your results. These are uncommonly high or low scores which may disproportionately affect your findings. Because of the way they work, non-parametric tests are less prone to the influence of outliers. So, if you suspect the influence of outliers, then it would be useful to compare the results of a parametric and the equivalent non-parametric test on the same data. The non-parametric test will be likely to be much less statistically significant if outliers are confusing the parametric test.

18.3   When not to use non-parametric tests

Generally speaking, non-parametric tests are less powerful than their parametric equivalents. Basically this means that when applied to the same data, the non-parametric test is somewhat less likely to give statistically significant results. Unfortunately, it is not possible to give a rule of thumb which indicates the extent of this. If significance is being strictly interpreted as the 5 per cent or .05 level then this may make a difference. However, it is our observation that psychologists are increasingly prepared to consider marginally significant findings (e.g. the 6 per cent and .06 levels). This implies that the loss of power from using non-parametric tests may not be quite so important as it once was.

18.4   Data requirements for non-parametric tests

The data requirements for non-parametric tests are similar to their parametric equivalents. However, assumptions that the scores are normally distributed or that the measurement is equal interval need not be met.

18.5   Problems in the use of non-parametric tests

There are few problems in using non-parametric tests other than their reduced power compared with their parametric equivalents. In the past, before SPSS, they were considered student-friendly techniques because they were so easy to calculate. SPSS has put an end to that advantage and replaced them in the affections of teachers of statistics.

You can find out more about non-parametric tests for two groups in Chapter 18 of Howitt, D. and Cramer, D. (2011). *Introduction to Statistics in Psychology*, 5th edition. Harlow: Pearson.

18.6   The data to be analysed

The computation of two non-parametric tests for related scores is illustrated with the data in Table 18.1, which was also used in Chapter 13 and which shows the number of eye-contacts made by the same babies with their mothers at six and nine months. Notice that the sign test (Section 18.8) and the Wilcoxon matched-pairs test (Section 18.10) produce different significance levels. The sign test seems rather less powerful at detecting differences than the Wilcoxon matched-pairs test.

Table 18.1	Number of one-minute segments with eye-contact at different ages	
Baby	**Six months**	**Nine months**
Clara	3	7
Martin	5	6
Sally	5	3
Angie	4	8
Trevor	3	5
Sam	7	9
Bobby	8	7
Sid	7	9

18.7 Entering the data

Select the file of the data if you saved it. Otherwise enter the data.

Step 1

In 'Variable View' of the 'Data Editor' label the first variable 'Six_mths' and the second variable 'Nine_mths'.
Remove the two decimal places by changing the figure for the decimals to 0.

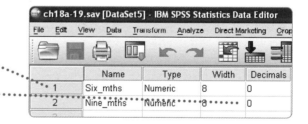

Step 2

In 'Data View' of the 'Data Editor' enter the data in the first two columns.

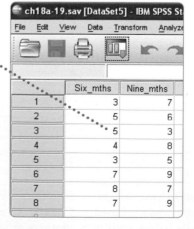

18.8 Related scores: sign test

Step 1

Select 'Analyze', 'Nonparametric Tests', 'Legacy Dialogs' and '2 Related Samples...'.

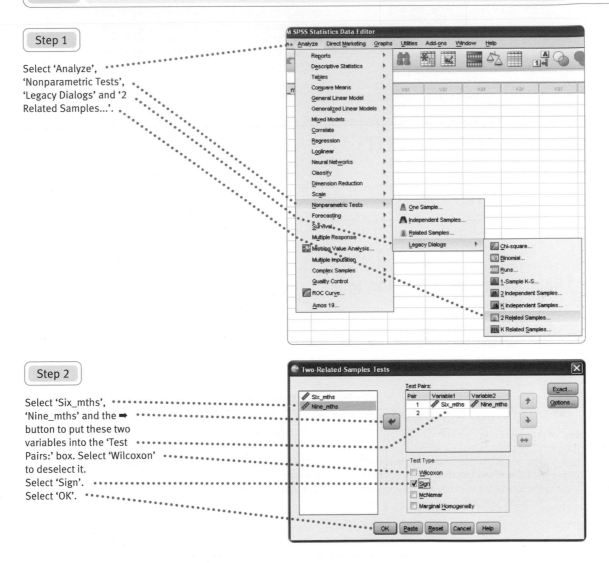

Step 2

Select 'Six_mths', 'Nine_mths' and the ➡ button to put these two variables into the 'Test Pairs:' box. Select 'Wilcoxon' to deselect it.
Select 'Sign'.
Select 'OK'.

18.9 Interpreting the output for the sign test

The first of the two tables of output can be ignored. It shows the number of negative (2), positive (8) and no (0) differences in smiling between the two ages.

Frequencies

		N
Nine_mths - Six_mths	Negative Differences[a]	2
	Positive Differences[b]	6
	Ties[c]	0
	Total	8

a. Nine_mths < Six_mths
b. Nine_mths > Six_mths
c. Nine_mths = Six_mths

Test Statistics[b]

	Nine_mths - Six_mths
Exact Sig. (2-tailed)	.289[a]

a. Binomial distribution used.
b. Sign Test

The second table shows the significance level of this test. The two-tailed probability is .289 or 29%, which is clearly not significant at the 5% level. The binomial distribution refers to the statistical technique by which probabilities can be found for samples consisting of just two different possible values, as is the case with the sign test (given that we ignore ties).

REPORTING THE OUTPUT

We could report these results as follows:

There was no significant change in the amount of eye-contact between six and nine months, Sign test ($n = 8$) $p = .289$.

18.10 Related scores: Wilcoxon test

The Wilcoxon test is the default option on the '2 Related Samples' tests dialog box. If you have previously deselected it, reselect it. Then 'OK' the analysis to obtain the Wilcoxon test output.

18.11 Interpreting the output for the Wilcoxon test

The first of the two tables of output can be ignored. It shows the number of negative (2), positive (6) and no (0) differences in the ranked data for the two ages as well as the mean and the sum of the negative and positive ranked data. The values of 'Nine_mths' are bigger than those for 'Six_mths'.

Ranks

		N	Mean Rank	Sum of Ranks
Nine_mths - Six_mths	Negative Ranks	2[a]	3.00	6.00
	Positive Ranks	6[b]	5.00	30.00
	Ties	0[c]		
	Total	8		

a. Nine_mths < Six_mths
b. Nine_mths > Six_mths
c. Nine_mths = Six_mths

Test Statistics[b]

	Nine_mths - Six_mths
Z	-1.706[a]
Asymp. Sig. (2-tailed)	.088

a. Based on negative ranks.
b. Wilcoxon Signed Ranks Test

The second table shows the significance level of this test. Instead of using tables of critical values, the computer uses a formula which relates to the z distribution. The z value is −1.706, which has a two-tailed probability of .088. This means that the difference between the two variables is not statistically significant at the 5% level.

REPORTING THE OUTPUT

We could report these results as follows:

There was no significant difference in the amount of eye-contact by babies between six and nine months, Wilcoxon, $z(n = 8) = -1.71$, two-tailed $p = .088$.

18.12   Unrelated scores: Mann–Whitney U-test

We will illustrate the computation of a non-parametric test for unrelated scores with the data in Table 18.2, which shows the emotionality scores of 12 children from two-parent families and 10 children from single-parent families.

Table 18.2	Emotionality scores in two-parent and lone-parent families	

Two-parent family X_1	Lone-parent family X_2
12	6
18	9
14	4
10	13
19	14
8	9
15	8
11	12
10	11
13	9
15	
16	

18.13   Entering the data

Select the file of the data if you saved it. Otherwise enter the data.

Step 1

In 'Variable View' of the 'Data Editor' label the first variable 'Family'. This defines the two types of family.
Label the second variable 'Emotion'.
These are the emotionality scores.
Remove the two decimal places by changing the value here to 0 if necessary.

ch18b-19.sav [DataSet6] - IBM SPSS Statistics Data Editor

File   Edit   View   Data   Transform   Analyze   Direct Marketing   Gra

	Name	Type	Width	Decimals
1	Family	Numeric	8	0
2	Emotion	Numeric	8	0

Step 2

In 'Data View' of the 'Data Editor' enter the values of the two variables in the first two columns.

18.14 Mann–Whitney *U*-test

Step 1

Select 'Analyze', 'Nonparametric Tests', 'Legacy Dialogs' and '2 Independent Samples...'.

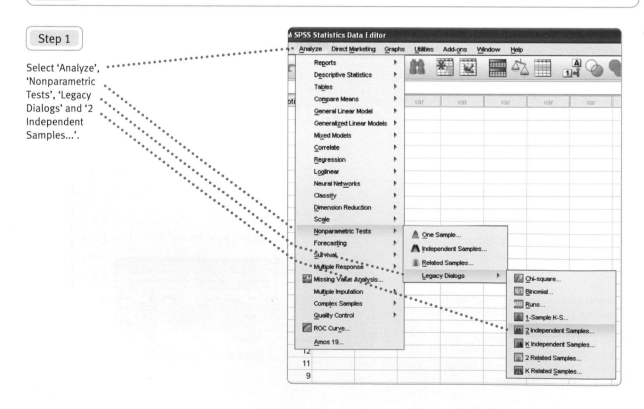

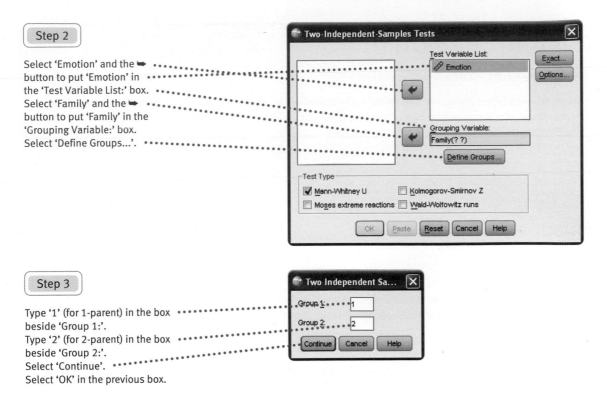

Step 2

Select 'Emotion' and the ➡ button to put 'Emotion' in the 'Test Variable List:' box.
Select 'Family' and the ➡ button to put 'Family' in the 'Grouping Variable:' box.
Select 'Define Groups...'.

Step 3

Type '1' (for 1-parent) in the box beside 'Group 1:'.
Type '2' (for 2-parent) in the box beside 'Group 2:'.
Select 'Continue'.
Select 'OK' in the previous box.

18.15 Interpreting the output for the Mann–Whitney *U* test

The first of the two tables of the output can be ignored. It shows that the average rank given to 'Emotion' for the first group (i.e. value = 1) is 7.85 and the average rank given to the second group (i.e. value = 2) is 14.54. This means that the scores for Group 2 tend to be larger than those for Group 1.

Ranks

	Family	N	Mean Rank	Sum of Ranks
Emotion	1	10	7.85	78.50
	2	12	14.54	174.50
	Total	22		

Test Statistics[b]

	Emotion
Mann-Whitney U	23.500
Wilcoxon W	78.500
Z	-2.414
Asymp. Sig. (2-tailed)	.016
Exact Sig. [2*(1-tailed Sig.)]	.014[a]

a. Not corrected for ties.
b. Grouping Variable: Family

The second table shows the basic Mann–Whitney statistic, the *U*-value, is 23.500, which is statistically significant at the .014 level.

In addition, the computer has printed out a *z*-value of −2.414, which is significant at the .016 level. This is the value of the Mann–Whitney test when a correction for tied ranks has been applied. As can be seen, this has only altered the significance level marginally to .016 from .014.

REPORTING THE OUTPUT

We could report the results of this analysis as follows:

The Mann–Whitney U-test found that the emotionality scores of children from two-parent families were significantly higher than those of children in lone-parent families, $U(n_1 = 10, n_2 = 12) = 23.5$, two-tailed $p = 0.016$.

Summary of SPSS Statistics steps for non-parametric tests for two groups

Data

● Name the variables in 'Variable View' of the 'Data Editor'.
● Enter the data under the appropriate variable names in 'Data View' of the 'Data Editor'.

Analysis

● For the two related data tests, select 'Analyze', 'Nonparametric Tests', 'Legacy Dialogs' and '2-Related Samples …'.
● Move the appropriate pair of variables to the 'Test Pairs:' box and select 'Sign'. Wilcoxon is already selected.
● For the Mann–Whitney test, select 'Analyze', 'Nonparametric Tests' and '2 Independent Samples …'.
● Move the dependent variable to the 'Test Variables List:' box and the grouping (independent) variable to the 'Grouping Variable:' box.
● Select 'Define Groups …', define the two groups, select 'Continue' and then 'OK'.

Output

● Check to see if the p-value is significant at .05 or less.
● If it is determine the direction of the difference.

For further resources including data sets and questions, please refer to the website accompanying this book.

CHAPTER 19

Ranking tests for three or more groups

Non-parametric statistics

Overview

- Sometimes you may wish to know whether the means of three or more different sets of scores are significantly different from each other but feel that the requirement that the scores on each variable are roughly normally distributed (bell-shaped) is not fulfilled. Non-parametric tests can be used in these circumstances.

- Non-parametric tests are ones which make fewer assumptions about the characteristics of the population from which the data came. This is unlike parametric tests (such as the t-test) which make fairly stringent assumptions about the distribution of the population from which the data came. The assumption of normality (bell-shaped frequency curves) is an example of the sort of assumptions incorporated into parametric statistics.

- Very few sets of data would meet the requirements of parametric testing exactly. However, some variation makes little difference. The problem is knowing just how much the assumptions of parametric tests can be violated without it having a major impact on the outcome of the study.

- Strictly speaking, non-parametric statistics do not test for differences in means. They cannot, since, for example, they use scores turned into ranks. Usually they test whether the ranks in one group are typically larger or smaller than the ranks in the other groups.

- We have included the Friedman test for related data. In other words, it is the non-parametric equivalent to the related ANOVA.

- The Kruskal–Wallis test is used for unrelated data. That is, it is the non-parametric equivalent to the unrelated ANOVA.

19.1 What are ranking tests?

Several non-parametric tests were described in Chapter 18. However, these dealt with circumstances in which only two sets of scores were compared. If you have three or more sets of scores there are other tests of significance which can be used. These are nowhere near so flexible and powerful as the analyses of variance described in Chapters 21–27 although they could be applied to the same data as some of the ANOVAs discussed in subsequent chapters. Nevertheless, in circumstances in which you have three or more sets of data to compare, then these ranking tests might be appropriate. There are unrelated and related versions of these ranking methods so they can be applied to the data in Chapter 21 on unrelated ANOVA and Chapter 22 on related ANOVA.

The decision to use non-parametric tests, then, does have an impact in that the analysis is less likely to be statistically significant than where the parametric equivalent is used – all other things being equal. Generally speaking, in simple research designs the non-parametric test will be adequate. The problem is that there is no non-parametric equivalent to many of the variants of ANOVA discussed later in this book – so SPSS Statistics does not have non-parametric equivalents of MANOVA, ANCOVA, mixed-design ANOVA and so forth. The researcher may therefore have no choice but to use the parametric test. If it were possible to clearly specify the circumstances in which parametric tests are too risky to apply then it would be less of a problem but, unfortunately, such information is generally not available. The best advice is to try to ensure that the data collected are of as good a quality as possible. Problems such as highly skewed data distributions may reflect that the researcher's measures have been poorly designed, for example.

It needs to be understood that the non-parametric techniques generally use ranks rather than the original scores. This results in a situation in which it is not strictly speaking true to discuss the differences in group means though, of course, the ranks do reflect differences in the means.

The statistical tests discussed in this chapter are equivalent to the parametric ANOVAs described later in this book. So the Kruskal–Wallis test is used as an alternative to the unrelated one-way ANOVA (Chapter 21) and the Friedman test is the alternative to the related one-way ANOVA (Chapter 22). Figure 19.1 outlines parametric and their non-parametric equivalents in this book.

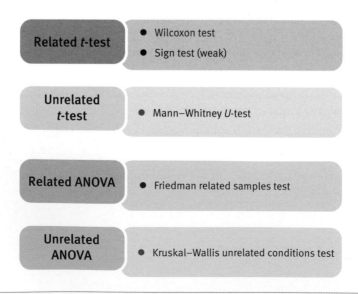

FIGURE 19.1   Parametric and non-parametric equivalents in this book

19.2   When to use ranking tests

The tests described in this chapter can be used in simple one-way unrelated and related ANOVA designs. They can be employed when the data do not meet the requirements of the parametric equivalents. However, this is a very subjective matter to assess and no hard-and-fast rules can be given. There is a loss of power when choosing the non-parametric version and often a loss of the flexibility offered by the parametric versions. For example, you will have to carry out multiple comparisons tests using non-parametric versions of the *t*-test (see Chapter 18) with adjustments for the number of comparisons made. It is also important to remember that there are no non-parametric versions of some variants of ANOVA.

19.3   When not to use ranking tests

We would avoid using the ranking (non-parametric) version of the test wherever possible. As a way of learning about the practical limitations of these, we would suggest that you compare the outcomes of the parametric and non-parametric analyses on your data. You may be surprised how little difference it makes.

19.4   Data requirements for ranking tests

Given the loss of power when ranking tests are used, then the researcher should satisfy themselves that the data do not correspond to a normal distribution and that the data cannot be regarded as on an equal interval scale. Other than that, the data requirements are the same as for the parametric equivalent.

19.5   Problems in the use of ranking tests

One major problem is that student research is often carried out on relatively small samples. This means that there is insufficient data to decide whether the data are normally distributed, skewed and so forth.

The limited variety of ranking tests on SPSS is a further limitation to their use.

You can find out more about non-parametric tests for three or more groups in Appendix B2 of Howitt, D. and Cramer, D. (2011). *Introduction to Statistics in Psychology*, 5th edition. Harlow: Pearson.

19.6   The data to be analysed

The computation of the non-parametric test for related scores is illustrated with the data in Table 19.1, which shows the recall of the number of pairs of nonsense syllables for the same participants under the three conditions of high, medium and low distraction. The computation of the non-parametric test for related scores is illustrated with the data in Table 19.2 (see p. 191), which shows the reading ability of three separate groups of children under the three conditions of high, medium and low motivation.

19.7   Friedman three or more related samples test

This test is used in circumstances in which you have three or more *related* samples of scores. The scores for each participant in the research are ranked from smallest to largest separately. In other words the scores for Joe Bloggs are ranked from 1 to 3 (or however many conditions there are), the scores for Jenny Bloggs are also ranged from 1 to 3, and so forth for the rest. The test essentially examines whether the average ranks in the several conditions of the experiment are more or less equal, as they should be if the null hypothesis is true.

Table 19.1 gives the scores in an experiment to test the recall of pairs of nonsense syllables under three conditions – high, medium and low distraction. The same participants were used in all conditions of the experiment.

Table 19.1	Scores on memory ability under three different levels of distraction		
	Low distraction	**Medium distraction**	**High distraction**
John	9	6	7
Mary	15	7	2
Shaun	12	9	5
Edmund	16	8	2
Sanjit	22	15	6
Ann	8	3	4

19.8   Entering the data for the Friedman test

Step 1

In 'Variable View' of the 'Data Editor' label the first variable 'Low', the second variable 'Medium' and the third variable 'High'.
Remove the two decimal places by changing the figure for the decimals to 0.

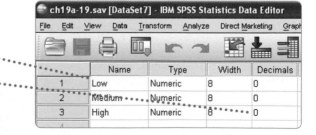

Step 2

In 'Data View' of the 'Data Editor' enter the data in the first three columns.

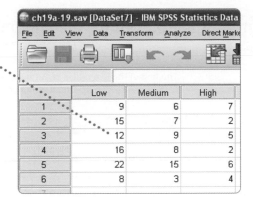

19.9 Friedman test

Step 1

Select 'Analyze', 'Nonparametric Tests', 'Legacy Dialogs' and 'K Related Samples...'.

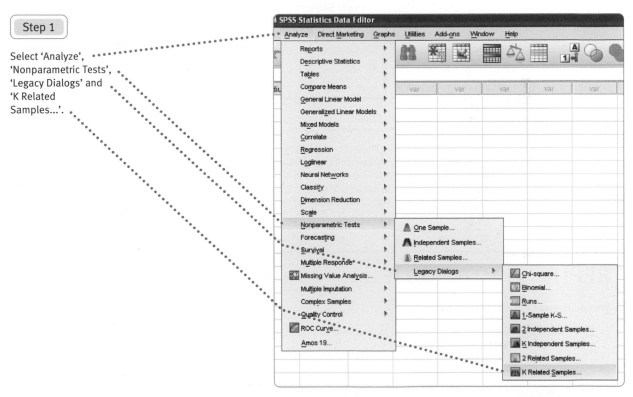

Step 2

Select 'Low', 'Medium', 'High' and the ➡ button to put these three variables into the 'Test Variables:' box. Make sure that 'Friedman' is selected. Select 'OK'.

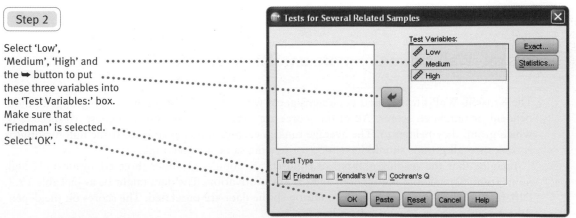

19.10 Interpreting the output for the Friedman test

The first of the two tables of output can be ignored.
It shows the mean rank for the three conditions.

Ranks

	Mean Rank
Low	3.00
Medium	1.67
High	1.33

Test Statistics[a]

N	6
Chi-Square	9.333
df	2
Asymp. Sig.	.009

a. Friedman Test

The second table shows the significance level of this test. The two-tailed probability is .009 or 0.9%, which is significant at the 5% level. We would then need to determine which of these three groups differed significantly. We could use a Wilcoxon matched-pairs signs-test (Chapter 18) to do this.

REPORTING THE OUTPUT

We could report these results as follows:

There was a significant difference in recall in the three conditions, Friedman $\chi^2(n = 6) = 9.33$, $p < .009$.

We would then need to report the results of further tests to determine which groups differed significantly and in what direction.

19.11 Kruskal–Wallis three or more unrelated samples test

The Kruskal–Wallis test is used in circumstances where there are *more than two* groups of independent or unrelated scores. All of the scores are *ranked* from lowest to highest irrespective of which group they belong to. The average rank in each group is examined. If the null hypothesis is true, then all groups should have more or less the same average rank.

Imagine that the reading abilities of children are compared under three conditions: (1) high motivation, (2) medium motivation, and (3) low motivation. The data might be as in Table 19.2. Different children are used in each condition so the data are unrelated. The scores on the dependent variable are on a standard reading test.

Table 19.2	Reading scores under three different levels of motivation	
High motivation	**Medium motivation**	**Low motivation**
17	10	3
14	11	9
19	8	2
16	12	5
18	9	1
20	11	7
23	8	6
21	12	
18	9	
	10	

19.12 Entering the data for the Kruskal–Wallis test

Step 1

In 'Variable View' of the 'Data Editor' label the first variable 'Conditions'. This defines the three types of condition.
Label the second variable 'Reading'. These are the reading scores.
Remove the two decimal places by changing the value here to 0 if necessary.

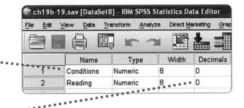

Step 2

In 'Data View' of the 'Data Editor' enter the values of the two variables in the first two columns.

	Conditions	Reading
1	1	17
2	1	14
3	1	19
4	1	16
5	1	18
6	1	20
7	1	23
8	1	21
9	1	18
10	2	10
11	2	11
12	2	8
13	2	12
14	2	9
15	2	11
16	2	8
17	2	12
18	2	9
19	2	10
20	3	3
21	3	9
22	3	2
23	3	5
24	3	1
25	3	7
26	3	6

19.13 Kruskal–Wallis test

Step 1

Select 'Analyze',
'Nonparametric Tests',
'Legacy Dialogs' and
'K Independent
Samples...'.

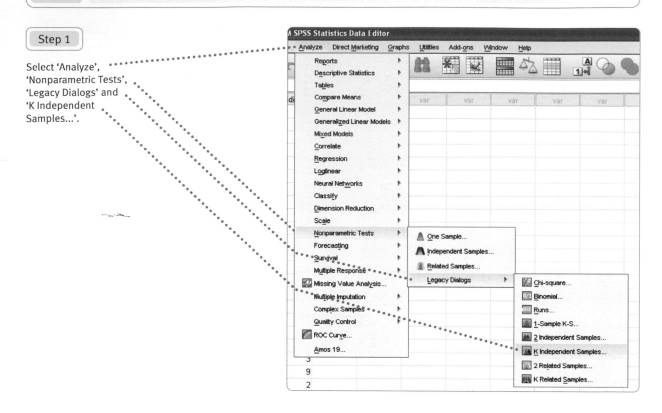

Step 2

Select 'Reading' and
the ➡ button to put
'Reading' in the 'Test
Variable List:' box.
Select 'Conditions' and
the ➡ button to put
'Conditions' in the
'Grouping Variable:'
box.
Select 'Define
Range...'.

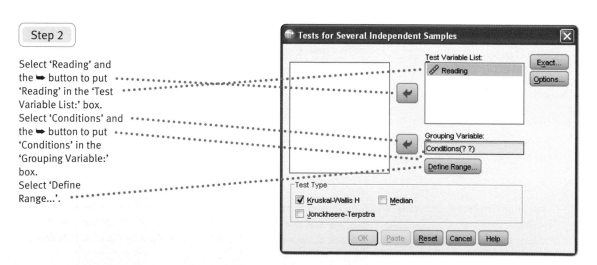

Step 3

Type '1' (for High) in the
box beside 'Minimum:'.
Type '3' (for Low) in the
box beside 'Maximum:'.
Select 'Continue'.
Select 'OK' in the
previous box.

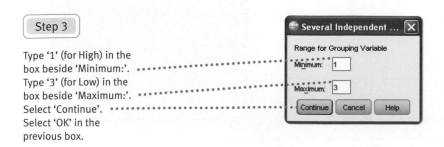

19.14 Interpreting the output for the Kruskal–Wallis test

The first of the two tables of the output can be ignored. It shows that the mean rank given to 'Reading' for the first group (i.e. High) is 22.00, the second group (i.e. Medium) is 12.20 and the third group (i.e. Low) is 4.43. This means that the scores for High tend to be larger than those for Medium which in turn tend to be higher than those for Low.

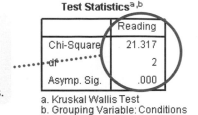

Ranks

	Conditions	N	Mean Rank
Reading	High	9	22.00
	Medium	10	12.20
	Low	7	4.43
	Total	26	

The second table shows the Kruskal–Wallis chi-square is 21.317, which is statistically significant at the .000 level. We would then need to determine which of these three groups differed significantly. We could use a Mann–Whitney U-test (Chapter 18) to do this.

Test Statistics[a,b]

	Reading
Chi-Square	21.317
df	2
Asymp. Sig.	.000

a. Kruskal Wallis Test
b. Grouping Variable: Conditions

REPORTING THE OUTPUT

We could report the results of this analysis as follows:

The Kruskal–Wallis test found that the reading scores in the three motivation conditions differed significantly, $\chi^2(2) = 21.31$, two-tailed $p = 0.001$.

We would then follow this with reporting the results of further tests to determine which groups differed significantly and in what direction.

Summary of SPSS Statistics steps for non-parametric tests for three or more groups

Data

- Name the variables in 'Variable View' of the 'Data Editor'.
- Enter the data under the appropriate variable names in 'Data View' of the 'Data Editor'.

Analysis

- For the Friedman test, select 'Analyze', 'Nonparametric Tests', 'Legacy Dialogs' and 'K Related Samples …'.
- Move the appropriate variables to the 'Test Variables:', check 'Friedman' is already selected and select 'OK'.
- For the Kruskal–Wallis test, select 'Analyze', 'Nonparametric Tests', 'Legacy Dialogs' and 'K Independent Samples …'.
- Move the dependent variable to the 'Test Variables List:' box and the grouping (independent) variable to the 'Grouping Variable' box.
- Select 'Define Groups …', define the two groups, select 'Continue' and then 'OK'.

Output

- Check to see if the p value is significant at .05 or less.
- If it is determine the direction of the difference.

For further resources including data sets and questions, please refer to the website accompanying this book.

PART 4

Analysis of variance

CHAPTER 20

The variance ratio test

Using the *F*-ratio to compare two variances

Overview

- The variance ratio test (*F*-test) indicates whether two unrelated sets of scores differ in the variability of the scores around the mean (i.e. are the variances significantly different?).

- This is clearly not the same question as asking whether two means are different, and one should remember that variances can be significantly different even though the means for the two sets of scores are the same. Consequently, examining the variances of the variables can be as important as comparing the means.

- Because few research questions are articulated in terms of differences in variance, researchers tend to overlook effects on variances and concentrate on differences between sample means. This should be avoided as far as possible.

- The *F*-test is probably more commonly found associated with the *t*-test and the analysis of variance.

20.1 What is the variance ratio test?

One of the conceptual clichés which is hard to break is the idea that the main function of statistical analysis in psychology is to compare groups in terms of the mean scores on a particular variable. Is it not possible to imagine circumstances in which two groups differ not in terms of their means but in terms of the spread of their scores? For example, what about eating behaviour when under emotional stress? We have probably all heard people say that they eat more when they are stressed (e.g. comfort eating) but others say that they can't eat when they are stressed. So if we compared the eating behaviour of a group of participants suffering from stress with that of a group not suffering stress then we would not necessarily expect the mean amounts of food eaten to be different between the two groups. What we might expect is that some stressed individuals eat a lot more and other stressed individuals eat a lot less. In contrast, those who are not stressed should show no such tendency. In other words, eating behaviour may be more extreme in both directions for the stressed group than for the non-stressed group. The mean amount of food eaten in both groups may average out the same. So it would be quite wrong to suggest that stress has no effect on eating behaviour simply because the means amounts eaten are the same.

What is the point of comparing variances?

- The effect of one variable on another is often construed as bringing about a change in the mean score.
- However, variables can have an influence on the spread of scores on a variable without making any difference to the mean score.

How important is it to understand?

- Analysis of variance (ANOVA) is a major branch of statistical analysis in psychology which is essential to the analysis of complex experimental designs among other things.

Can it be used more generally?

- One of the failings of psychologists is that they routinely apply statistical tests of significance without fully exploring what is happening in their data at the descriptive level.
- Regularly comparing the variances of aspects of your data may help you spot interesting features in your data but also problems.

FIGURE 20.1   About comparing variances

How can one detect this difference between the groups? The variance ratio (or *F*-ratio) is one way of identifying that the variability of scores in one group is greater than for the other group. The variance ratio is simply the ratio of the larger variance divided by the smaller variance. If the variances are significantly different, then this means that the variability of scores is different for the two samples. It is called the *F*-ratio because this statistic is distributed like the *F*-distribution, which is a statistical distribution like the *t*- and *z*-distributions.

The variance ratio is part of other statistical techniques. For example, it is a crucial statistic in analysis of variance (although SPSS actually calculates Levene's test which is slightly different). Levene's test is also calculated as part of the unrelated *t*-test procedure. The *t*-test assumes that the variances of the two samples are statistically similar. If they are not, then a different version of the *t*-test must be used from the SPSS output. Figure 20.1 highlights some of the points in comparing variances.

20.2   When to use the variance ratio test

It is legitimate to compare the variances of scores for different groups of participants. Routinely a researcher would obtain the variance (or standard deviation or standard error) for each sample of scores as part of the basic descriptive statistics reported in any study. Any large differences in variance could be tested using the variance ratio test.

It is good practice to check that the variances on a particular variable are similar for the different groups in the study. Normally variance (or the related measures of standard deviation or standard error) is routinely calculated for each variable or sample routinely as part of the basic descriptive statistics for the data. It would be appropriate to test for differences in variances if they seemed very different, say, for two different samples of participants. Most tests of differences between means routinely include variance ratios (*F*-ratios) in the SPSS output.

20.3   When not to use the variance ratio test

So long as the data are score data, use of the variance ratio test will be helpful in general.

20.4   Data requirements for the variance ratio test

The variance ratio needs to be computed on score data.

20.5   Problems in the use of the variance ratio test

There are no particular problems except that it is easily forgotten that variance is variation around the mean score of a sample. The value of the variance does not say anything about the mean score as such.

You can find out more about the variance ratio test in Chapter 19 of Howitt, D. and Cramer, D. (2011). *Introduction to Statistics in Psychology*, 5th edition. Harlow: Pearson.

20.6   The data to be analysed

To compute the variance ratio – or *F*-ratio – we divide the larger variance estimate by the smaller variance estimate. The variance estimate is produced by the Descriptives procedure first introduced in Chapter 7. The computation of the variance ratio is illustrated with the data in Table 20.1 (*ISP*, Table 19.2), which reports the emotional stability scores of patients who have had an electric current passed through either the left or the right hemisphere of the brain. In this chapter, the method involves a little manual calculation. However, it provides extra experience with SPSS.

An alternative way of achieving the same end is to follow the *t*-test procedures in Chapter 14. You may recall that the Levene *F*-ratio test is part of the output for that *t*-test. Although Levene's test is slightly different, it is a useful alternative to the conventional *F*-ratio test.

Table 20.1	Emotional stability scores from a study of ECT to different hemispheres of the brain	
Left hemisphere		**Right hemisphere**
20		36
14		28
18		4
22		18
13		2
15		22
9		1
Mean = 15.9		Mean = 15.9

20.7 Entering the data

Step 1

In 'Variable View' of the 'Data Editor' label the first variable 'Spheres'. Code left hemisphere as '1' and right hemisphere as '2'. Label these two values. Label the second variable 'Emotion'. These are the emotion scores.
Remove the two decimal places by changing the value here to 0 if necessary.

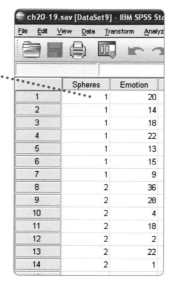

Step 2

In 'Data View' of the 'Data Editor' enter the values of the two variables in the first two columns.

	Spheres	Emotion
1	1	20
2	1	14
3	1	18
4	1	22
5	1	13
6	1	15
7	1	9
8	2	36
9	2	28
10	2	4
11	2	18
12	2	2
13	2	22
14	2	1

20.8 Variance estimate

Step 1

Select 'Analyze', 'Compare Means' and 'Means...'.

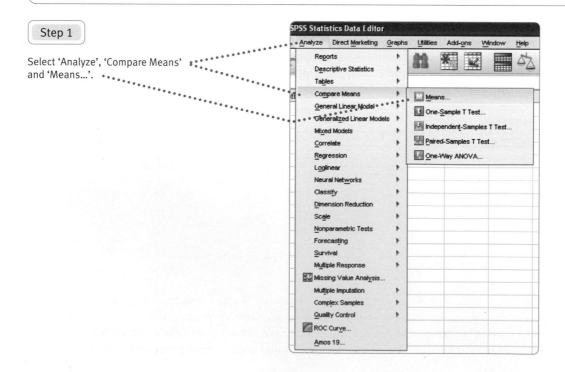

Step 2

Select 'Emotion' and the ➡ button beside the 'Dependent List:' box to put it there.
Select 'Spheres' and the ➡ button beside the 'Independent List:' box to put it there.
Select 'Options...'.

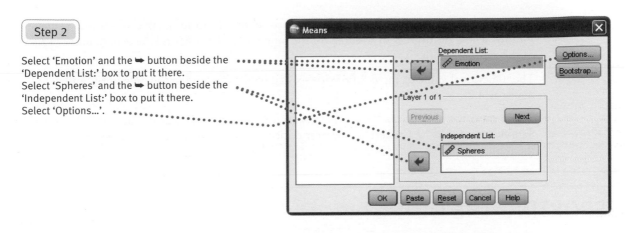

Step 3

Select 'Variance' and the ➡ button to put it in the 'Cell Statistics:' box.
Select 'Continue'.
Select 'OK' in the previous box.

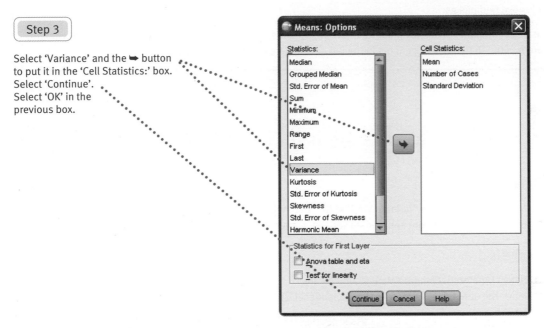

The variances for the two groups are in the last column of this table. It is 19.810 for the left hemisphere group.

Report

Emotion

Spheres	Mean	N	Std. Deviation	Variance
Left	15.86	7	4.451	19.810
Right	15.86	7	13.837	191.476
Total	15.86	14	9.875	97.516

20.9 Calculating the variance ratio from the output

- Divide the larger variance estimate in the output by the smaller variance estimate. The larger variance estimate is 191.476 (for 'Right'), which divided by the smaller one of 19.810 (for 'Left') gives a variance or F-ratio of 9.6656. This ratio is 9.66 when rounded *down* to two decimal places. We round down to provide a more conservative F-test and probability value.

- We need to look up the statistical significance of this ratio in a table of critical values of *F*-ratios where the degrees of freedom for the numerator (191.48) and the denominator (19.81) of the ratio are both six.

- The .05 critical value of the *F*-ratio with six degrees of freedom in the numerator and denominator is 4.28.

- The *F*-ratio we obtained is 9.66, which is larger than the .05 critical value of 4.28 (see *ISP* Significance Table 19.1 where the nearest critical value is 4.4 with five degrees of freedom in the numerator).

REPORTING THE OUTPUT

We would report these findings as:

The variance of emotionality scores of patients in the right hemisphere condition (191.48) was significantly larger, $F(6, 6) = 9.66$, $p < .05$, than those of patients in the left hemisphere condition (19.81).

Summary of SPSS Statistics steps for the variance ratio test

Data

- Name the variables in 'Variable View' of the 'Data Editor'.
- Enter the data under the appropriate variable names in 'Data View' of the 'Data Editor'.

Analysis

- Select 'Analyze', 'Compare Means' and 'Means ...'.
- Move the dependent variable to the 'Dependent List:' box and the independent variable to the 'Independent List:' box.
- Select 'Options ...' and move 'Variance' from the 'Statistics' to the 'Cell Statistics' box.

Output

- Divide the larger variance by the smaller variance and look up in an appropriate table such as Appendix J in *ISP* under the appropriate degrees of freedom whether the ratio is statistically significant.

For further resources including data sets and questions, please refer to the website accompanying this book.

CHAPTER 21

Analysis of variance (ANOVA)

Introduction to the one-way unrelated or uncorrelated ANOVA

Overview

- The unrelated/uncorrelated analysis of variance indicates whether several (two or more) groups of scores have very different means. It assumes that each of the sets of scores comes from different individuals. It is not essential to have equal numbers of scores for each set of scores.

- The different groups correspond to the independent variable. The scores correspond to the dependent variable.

- Basically the analysis of variance calculates the variation between scores and the variation between the sample means. Both of these can be used to estimate the variation in the population. If the between groups estimate is much bigger than the combined within groups estimate, it means that the variation due to the independent variable is greater than could be expected on the basis of the variation between scores. If this disparity is big enough, the difference in variability is statistically significant. This means that the independent variable is having an effect on the scores.

- The interpretation of the analysis of variance can be difficult when more than two groups are used. The overall analysis of variance may be statistically significant, but it is difficult to know which of the three or more groups is significantly different from the other groups.

- The solution is to break the analysis into several separate comparisons to assess which sets of scores are significantly different from other sets of scores. That is, which of the groups are significantly different from other groups.

- Ideally an adjustment should be made for the number of comparisons being made (see Chapter 24 on multiple comparisons for information on more sophisticated methods for doing this than those described in this chapter). This adjustment is necessary because the more statistical comparisons that are made, the more likely it is that some of the comparisons will be statistically significant.

21.1   What is one-way ANOVA?

The one-way ANOVA is essentially an extension of the unrelated *t*-test to cover the situation in which the researcher has three or more groups of participants. Indeed, one could use one-way ANOVA instead of the unrelated *t*-test where one has just two groups of participants but convention is against this. The sort of data which could be analysed using one-way ANOVA is illustrated in Table 21.1. This shows the data for a study which investigated the effect of different hormone treatments on depression scores. Do the mean scores for the groups differ significantly? It would appear from looking at Table 21.1 that the mean score for Group 1 is higher than those of the other two groups since if we calculate the means we find that they are: Group 1 mean = 9.67, Group 2 mean = 3.67 and Group 3 mean = 4.00.

Table 21.1	Data for a study of the effects of hormones		
	Group 1 **Hormone 1**	**Group 2** **Hormone 2**	**Group 3** **Placebo control**
	9	4	3
	12	2	6
	8	5	3

It is known as the one-way ANOVA since there is just one independent variable – that is, a single nominal variable forms the basis of the groups. In this case, that single variable is the drug treatment (Hormone 1, Hormone 2 and Placebo). It is possible to extend this design to have a second (or third, etc.) independent variable as discussed in Chapter 23 on the two-way ANOVA.

Essentially ANOVA works by comparing the variation in the group means with the variation within the groups using the variance ratio or *F*-ratio. The more variation there is between the group means compared to the variation within the groups then the more likely it is that the analysis will be statistically significant. The variation between the group means is the *between groups variance*; the variation within the groups is the *within groups variance* though it can be referred to as the *error variance* since it is the variation in scores which is unaccounted for. The bigger the error variance then the greater the possibility that we get variation between the cell means by chance. The calculation is a little more complex than this description implies but this will serve you well enough conceptually.

Confusingly, ANOVA does not use the terms between groups variance and error variance. Instead terms such as between groups mean square and error mean square are used. This is merely a convention as they are, in fact, the between groups variance estimate and the error variance estimate. They are variance estimates because one is trying to calculate the variance in the population of scores using information from a sample of scores.

The language used in ANOVA can be somewhat arcane – terms such as levels of treatment, split-plot designs, blocks and so forth reflect the origins of ANOVA in agricultural research. Gradually the terminology is changing but this takes a lot of time in psychological statistics.

ANOVA, like other statistical techniques, makes use of the concept of degrees of freedom. This is closely related to the number of participants in the study though it is not always apparent. This is because the total number of degrees of freedom (which is always one less than the number of scores) gets divided up in various ways. Fortunately, SPSS Statistics does all of this for you.

Finally, you need to understand that a significant ANOVA simply means that the variation in the group means is bigger than can be accounted for simply on the basis of the variation due to using samples. In other words, the independent variable (the groups) makes a difference to the dependent variable (the scores). It does not mean that all of the groups have means that are

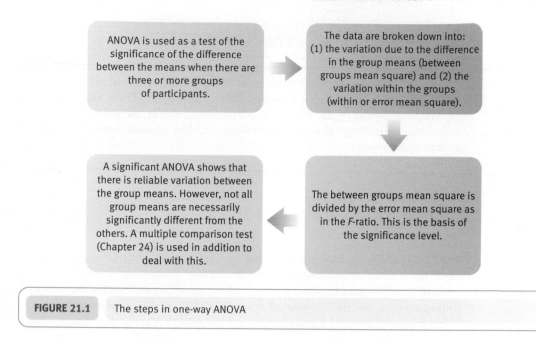

ANOVA is used as a test of the significance of the difference between the means when there are three or more groups of participants.

The data are broken down into: (1) the variation due to the difference in the group means (between groups mean square) and (2) the variation within the groups (within or error mean square).

A significant ANOVA shows that there is reliable variation between the group means. However, not all group means are necessarily significantly different from the others. A multiple comparison test (Chapter 24) is used in addition to deal with this.

The between groups mean square is divided by the error mean square as in the *F*-ratio. This is the basis of the significance level.

FIGURE 21.1   The steps in one-way ANOVA

significantly different from each other. In our example, the means are 9.67, 3.67 and 4.00. It is obvious that the last two means are not very different from each other whereas both are very different from the first mean. As part of ANOVA, it is usual to test which means are significantly different from each other. These tests are discussed in Chapter 24 on multiple comparisons. However, we do suggest simple procedures in this chapter which may be more readily understood by novices. Figure 21.1 shows the main steps in carrying out a one-way analysis of variance.

21.2   When to use one-way ANOVA

One-way ANOVA can be used when the means of three or more groups are to be compared. These must be independent groups in the sense that a participant can only be in one group and contribute one score to the data being analysed.

Conventionally, one-way ANOVA is not used when there are just two groups to be compared. A *t*-test is usually employed in these circumstances. However, it is perfectly proper to use one-way ANOVA in these circumstances. It will give exactly the same significance level as the *t*-test on the same data when the variances do not differ. When the variances differ it may be more appropriate to use an unrelated *t*-test where the variances are pooled together more appropriately.

21.3   When not to use one-way ANOVA

The main indication that there may be problems is when the test of the homogeneity of variances is statistically significant. This is discussed later in this chapter. This means that the variances of the different groups are very different, which makes it difficult to combine them. The extent to which this is a problem is not well documented but it is probably not so important where the group sizes are equal. If the variances do differ significantly, then some researchers might prefer to use the Kruskal–Wallis three or more unrelated conditions test which is available under 'Nonparametric Tests' on SPSS (see Chapter 19).

21.4   Data requirements for one-way ANOVA

One-way ANOVA requires three or more independent groups of participants, each of which contributes a single score to the analysis. Since one-way ANOVA essentially tests for differences between the means of the groups of participants, ideally the scores correspond to the theoretically ideal equal interval scale of measurement. The variances of the various groups need to be approximately equal.

21.5   Problems in the use of one-way ANOVA

Confusion arises because of the traditional advice that if the one-way ANOVA is not significant, then no further analysis should be carried out comparing the different group means with each other. This may have been sound advice once but it is somewhat outmoded. It is perfectly appropriate to use the Newman–Keuls test and Duncan's new multiple range test as multiple-comparisons tests (see Chapter 24) even if the ANOVA is not significant. This begs the question whether ANOVA adds anything in these circumstances – the answer is probably no. If ANOVA is significant, then any multiple-range test is appropriate.

You can find out more about one-way ANOVA in Chapter 20 of Howitt, D. and Cramer, D. (2011). *Introduction to Statistics in Psychology*, 5th edition. Harlow: Pearson.

21.6   The data to be analysed

The computation of a one-way unrelated analysis of variance is illustrated with the data in Table 21.1 (*ISP*, Table 20.2), which shows the scores of different participants in three conditions. It is a study of the effect of different hormone and placebo treatments on depression. So drug is the independent variable and depression the dependent variable.

21.7   Entering the data

Step 1

In 'Variable View' of the 'Data Editor', name the first row 'Condition'. Name conditions '1', '2' and '3' 'Hormone 1', 'Hormone 2' and 'Placebo Control' respectively.
Name the second row 'Depression'.
Change 'Decimal' places from '2' to '0'.

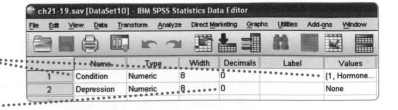

Step 2

In 'Data View' of the 'Data Editor' enter the data in the first two columns.
Save this file to use for Chapters 24 and 26.

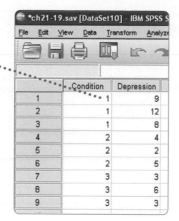

21.8   One-way unrelated ANOVA

Step 1

Select 'Analyze',
'Compare Means' and
'One-Way ANOVA...'.

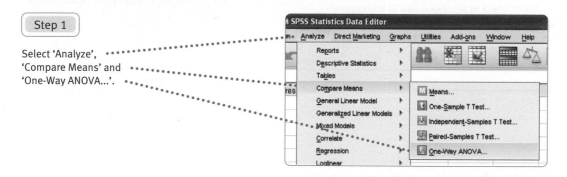

Step 2

Select 'Depression' and the ➡
button beside the 'Dependent
List:' box to put it there.
Select 'Condition' and the ➡
button beside the 'Factor:' box
to put it there.
Select 'Options...'.

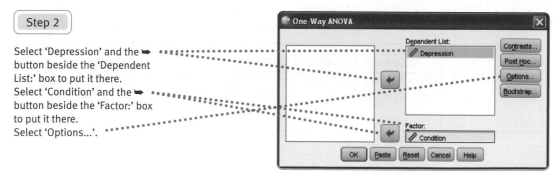

Step 3

Select 'Descriptive' and
'Homogeneity of variance
test'.
Select 'Continue'.
Select 'OK' from
previous box.

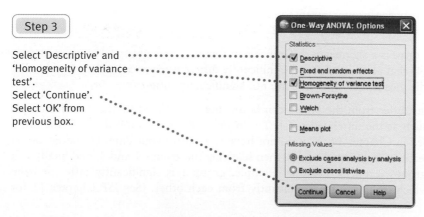

21.9 Interpreting the output

The first table provides various descriptive statistics such as the number (N) of cases, the mean and the standard deviation for the three conditions and the total sample.

Descriptives

Depression

	N	Mean	Std. Deviation	Std. Error	95% Confidence Interval for Mean		Minimum	Maximum
					Lower Bound	Upper Bound		
Hormone 1	3	9.67	2.082	1.202	4.50	14.84	8	12
Hormone 2	3	3.67	1.528	.882	-.13	7.46	2	5
Placebo control	3	4.00	1.732	1.000	-.30	8.30	3	6
Total	9	5.78	3.308	1.103	3.23	8.32	2	12

Test of Homogeneity of Variances

Depression

Levene Statistic	df1	df2	Sig.
.293	2	6	.756

The second table gives Levene's test of how similar the variances are. As this test is not significant (with a significance of .756), the variances are similar or homogeneous. If the variances were not similar, we should try to transform the scores to make them so. Otherwise there may be problems in interpreting the analysis of variance.

The third table shows the results of the analysis of variance. The *F*-ratio is significant at .011 as it is less than .05.

ANOVA

Depression

	Sum of Squares	df	Mean Square	F	Sig.
Between Groups	68.222	2	34.111	10.586	.011
Within Groups	19.333	6	3.222		
Total	87.556	8			

- The *F*-ratio is the between groups mean square divided by the within group mean square, which gives an *F*-ratio of 10.586 (34.111/3.222 = 10.5869).

- This indicates that there is a significant difference between the three groups. However, it does not necessarily imply that all the means are significantly different from each other. In this case, one suspects that the means 3.67 and 4.00 are not significantly different.

- Which of the means differ from the others can be further determined by the use of multiple comparison tests such as the unrelated *t*-test. To do this, follow the procedure for the unrelated *t*-test described in Chapter 14. You do not have to re-enter your data. However, do an unrelated *t*-test selecting the groups 1 and 2, then selecting the groups 1 and 3, and finally you would select the groups 2 and 3. For our example, group 1 is significantly different from groups 2 and 3, which do not differ significantly from each other. (See *ISP* Chapter 11 for more details.)

- Because we are doing three comparisons, the exact significance level of each t-test should be multiplied by 3 to obtain the Bonferroni significance level.

- It is useful to know how much variance the independent variable accounts for or explains. This is given by a statistic called eta squared. This statistic is not available with the one-way ANOVA SPSS procedure. It is available with the general linear model SPSS procedure, which is described in Chapters 22–27. If you use the univariate option of the general linear model you will see that eta squared for this analysis is about .78. This means that the three conditions account for about 80 per cent of the variance in the depression scores. Psychologists are encouraged to report this statistic as it gives an indication of the size of an effect.

REPORTING THE OUTPUT

We could report the results of the output as follows:

The effect of the drug treatment was significant overall, $F(2, 6) = 10.58$, $p = 0.011$, $\eta^2 = .78$. When a Bonferroni adjustment was made for the number of comparisons, the only significant difference was between the means of hormone treatment 1 and hormone treatment 2, $t(4) = 4.02$, two-tailed $p < .05$. The mean of hormone treatment 1 ($M = 9.67$, $SD = 2.08$) was significantly greater than that for hormone treatment 2 ($M = 3.67$, $SD = 1.53$). There was no significant difference between the mean of the placebo control and the mean of either hormone treatment 1 or hormone treatment 2.

Summary of SPSS Statistics steps for one-way ANOVA

Data

- Name the variables in 'Variable View' of the 'Data Editor'.
- Enter the data under the appropriate variable name in 'Data View' of the 'Data Editor'.

Analysis

- Select 'Analyze', 'Compare Means' and 'One-Way ANOVA …'.
- Move the dependent variable to the 'Dependent List:' box and the independent variable to the 'Factor:' box.
- Select 'Post Hoc …', the post hoc test you want and 'Continue'.
- Select 'Options', 'Descriptive', 'Homogeneity of variance test', 'Continue' and then 'OK'.

Output

- Check if the Sig(nificance) of the F ratio is significant at .05 or less.
- If so, check that the Sig(nificance) of the homogeneity of variances is not significant at more than .05.
- Check which means differ significantly with the post hoc test or further unrelated t-tests.

For further resources including data sets and questions, please refer to the website accompanying this book.

CHAPTER 22

Analysis of variance for correlated scores or repeated measures

Overview

- The correlated/related analysis of variance indicates whether several (two or more) sets of scores have very different means. However, it assumes that a single sample of individuals has contributed scores to each of the different sets of scores and that the correlation coefficients between sets of scores are large. It is also used in designs when matching has taken place.

- If your data do not meet these requirements then turn back to Chapter 21 on the unrelated analysis of variance.

- Changes in scores on a variable over time is a typical example of the sort of study which is appropriate for the correlated/related analysis of variance.

- If properly used, correlated/related designs can be extremely effective in that fewer participants are required to run the study. The reason is that once participants are measured more than once, it becomes possible to estimate the individual differences component of the variation in the data. In a study of memory, for example, some participants will tend to do well whatever the condition and others will tend to do poorly. These individual differences can be identified and adjusted for in the analysis. What would be classified as error variation in an unrelated analysis of variance is separated into two components – the individual differences component (within subjects error) and residual error. Effectively this means that the error term is reduced because the individual difference component has been removed. Since the error term is smaller, it is possible to get significant results with smaller numbers of participants than would be possible with an unrelated design.

22.1 What is repeated-measures ANOVA?

Just as the one-way ANOVA was an extension of the unrelated *t*-test, the repeated-measures ANOVA is an extension of the related *t*-test to cover three or more measures taken (usually) at different points in time. It is mainly used to investigate how scores change on a measure for a sample of cases over a period of investigation but can be used where participants in three groups are matched. It could be used as an alternative to the related *t*-test since exactly the same level of significance would be obtained. However, conventionally this is not done.

The whole point of repeated-measures, related-measures and correlated measures tests is that they can be very effective designs since they capitalise on the fact that participants are serving 'as their own controls'. That is to say, because individuals are measured at several different points in time, it is possible to estimate the tendency for individuals to give, say, particularly high scores or particularly low scores irrespective of the time that they are measured. In other words, it is possible to adjust the calculation for the consistencies in the responses of individuals measured at different points in time. This can be seen in the data in Table 22.1. As you can see, there are five participants who are given one of two drug treatments or an inert placebo. The scores are the amount of relief from pain felt in these three conditions. The greatest relief from pain is experienced in the Product-X condition where the average score is 8.00 compared to 6.00 in the Aspirin condition and 4.00 in the control Placebo condition. Thus the treatment condition does seem to have an effect.

It is equally important, however, to notice that Bob Robertson tends to get the highest relief from pain scores irrespective of the drug treatment condition. Bert Entwistle gets the lowest scores on average. The repeated-measures ANOVA simply takes this information and adjusts the scores to take account of these individual differences. In doing so, essentially it deals with some of the variance which would have been attributed to error in the unrelated measures one-way ANOVA. In other words, the error variance is essentially reduced. What this means is that the research findings have an increased chance of being significant simply because the error variance is smaller – that is, all things being equal.

Unfortunately, and this is largely overlooked by psychologists, all things are *not* equal. What happens is that the degrees of freedom associated with the error variance are reduced because of these individual difference adjustments. Depending on the size of the correlations between the scores in the different conditions, the analysis may be statistically significant (if the correlations are strong) but it may be statistically non-significant if the correlations are poor. But the researcher has to check on this.

The repeated measures ANOVA introduces yet more new terminology. This is because the error variance (error mean square) no longer contains the variance due to individual differences unlike the ANOVA in the previous chapter. This adjusted error variance is known as the residual variance although it is usually used as the measure of error. The individual difference variance is known as the between subjects or between people variance. Figure 22.1 shows the key steps in carrying out a repeated-measures or related samples ANOVA.

Table 22.1	Pain relief scores from a drugs experiment		
	Aspirin	**'Product X'**	**Placebo**
Bob Robertson	7	8	6
Mavis Fletcher	5	10	3
Bob Polansky	6	6	4
Ann Harrison	9	9	2
Bert Entwistle	3	7	5

Repeated measures ANOVA is used as a test of the significance of the difference between the means when there are three or more similar measures taken from a single group of participants or effective matching has been employed.

The data are broken down into: (1) the variation due to the differences in the group means (between groups mean square); (2) the variation due to the fact that the measures are correlated (individual differences); and (3) the variation due to the remaining error, which is known as residual mean square.

The repeated measures ANOVA is very efficient so long as the scores correlate substantially. If not, then the design is inefficient and it would be more appropriate to use an unrelated ANOVA though few, if any, psychologists would do so.

The between groups mean square is divided by the residual variance as in the F-ratio. The larger the between groups mean square relative to the residual mean square, the more likely are the findings to be statistically significant, all other things being equal.

FIGURE 22.1   Steps in the repeated measures ANOVA

22.2   When to use repeated-measures ANOVA

Correlated scores designs are relatively uncommon in psychology but do have some advantages in terms of their power in appropriate circumstances. They are, of course, essential if a group of participants is being studied over time on several occasions. Matching is harder to use though is perfectly appropriate as a way of achieving the requirement of correlated scores. It is essential that the dependent variable (the scores) is comparable at different points in time. Usually this is achieved by using the same measure on each occasion (or possibly alternative forms of the same measure).

22.3   When not to use repeated-measures ANOVA

We would argue that the repeated-measures ANOVA should be used only where the data show substantial correlations between the scores measured, for example, at different points in time. Otherwise, it is a form of analysis which has lower power compared with the unrelated samples ANOVA. That is to say, the findings may be non-significant for the repeated-measures ANOVA but significant if the same data are put through an unrelated ANOVA analysis. Unfortunately, except for carrying out the two analyses and comparing the outcomes, it is not possible to say just how much the scores need to correlate.

22.4   Data requirements for related-measures ANOVA

The repeated-measures ANOVA requires a complete set of scores from each participant. It cannot be done using partial data. The scores in each condition should ideally have similar variances. Overall, the scores should correlate between conditions.

22.5   Problems in the use of repeated-measures ANOVA

A major difficulty lies in the view that the research design rather than the statistical features of the data determine that a repeated-measures ANOVA should be used. Unfortunately, there is no guarantee that matching, for example, will result in strong correlations between the scores in different conditions – unless it is known that the matching variable correlates highly with the dependent variable. Furthermore, it is just an assumption that individuals will serve as their own controls – it can only be the case for measures which are fairly stable over time.

A more common problem is the lack of multiple comparisons measures for repeated-measures ANOVAs on SPSS Statistics. One solution to this is to carry out a number of related *t*-tests on selected pairs of conditions.

> You can find out more about the repeated-measures ANOVA in Chapter 21 of Howitt, D. and Cramer, D. (2011). *Introduction to Statistics in Psychology*, 5th edition. Harlow: Pearson.

22.6   The data to be analysed

The computation of a one-way repeated-measures analysis of variance is illustrated with the data in Table 22.1, which shows the scores of the same participants in three different conditions (*ISP*, Table 21.10).

22.7   Entering the data

Step 1

Name the first row 'Aspirin', the second row 'ProductX' and the third row 'Placebo'.
Change the 'Decimals' from '2' to '0'.

	Name	Type	Width	Decimals
1	Aspirin	Numeric	8	0
2	ProductX	Numeric	8	0
3	Placebo	Numeric	8	0

Step 2

In 'Data View' of the 'Data Editor' enter the data in the first three columns. As the data are related the values for the three conditions are in three columns.

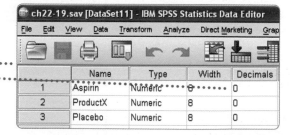

	Aspirin	ProductX	Placebo
1	7	8	6
2	5	10	3
3	6	6	4
4	9	9	2
5	3	7	5

22.8 One-way repeated-measures ANOVA

Step 1

Select 'Analyze', 'General Linear Model' and 'Repeated Measures…'.

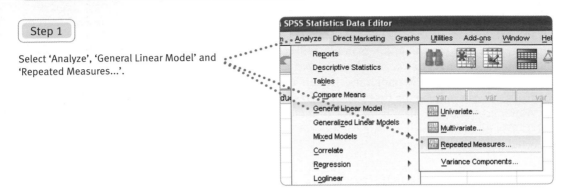

Step 2

Type '3' in the 'Number of Levels:' box.
Select 'Add'.
Select 'Define'.

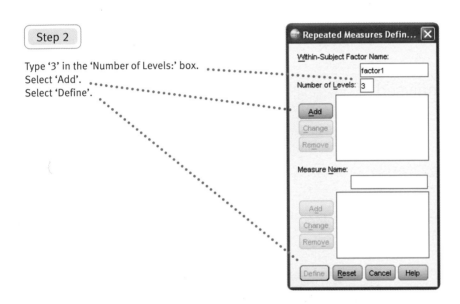

Step 3

Select each variable individually or all three together and the ➡ button beside the 'Within-Subjects Variables: (factor1)' box to put them there.
Select 'Options…'.

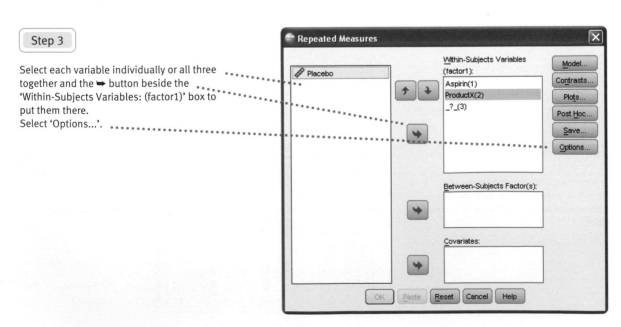

Step 4

Select 'Descriptive statistics'.
Select 'Estimates of effect size'.
Select 'Continue'.
Select 'OK' in the previous box.

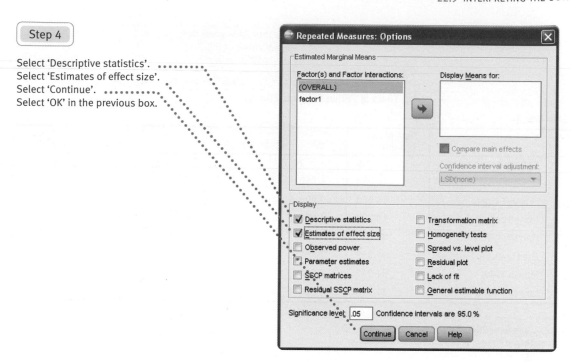

22.9 Interpreting the output

The output gives seven tables. Only the more important ones are shown below.

Descriptive Statistics

	Mean	Std. Deviation	N
Aspirin	6.00	2.236	5
ProductX	8.00	1.581	5
Placebo	4.00	1.581	5

The second table gives the mean and standard deviation for the three groups.

Mauchly's Test of Sphericity[b]

Measure:MEASURE_1

Within Subjects Effect	Mauchly's W	Approx. Chi-Square	df	Sig.	Epsilon[a]		
					Greenhouse-Geisser	Huynh-Feldt	Lower-bound
factor1	.862	.444	2	.801	.879	1.000	.500

Tests the null hypothesis that the error covariance matrix of the orthonormalized transformed dependent variables is proportional to an identity matrix.

a. May be used to adjust the degrees of freedom for the averaged tests of significance. Corrected tests are displayed in the Tests of Within-Subjects Effects table.
b. Design: Intercept
Within Subjects Design: factor1

The fourth table gives the results for Mauchly's test of sphericity. As this test is not significant, sphericity is assumed and we do not have to adjust the significance levels for the analysis. If it is significant, then one of the alternative tests in the next output box should be used.

The fifth table gives the significance of the *F*-ratio. It is .037 when sphericity is assumed.

Partial eta squared is .561.

Tests of Within-Subjects Effects

Measure:MEASURE_1

Source		Type III Sum of Squares	df	Mean Square	F	Sig.	Partial Eta Squared
factor1	Sphericity Assumed	40.000	2	20.000	5.106	.037	.561
	Greenhouse-Geisser	40.000	1.758	22.752	5.106	.045	.561
	Huynh-Feldt	40.000	2.000	20.000	5.106	.037	.561
	Lower-bound	40.000	1.000	40.000	5.106	.087	.561
Error(factor1)	Sphericity Assumed	31.333	8	3.917			
	Greenhouse-Geisser	31.333	7.032	4.456			
	Huynh-Feldt	31.333	8.000	3.917			
	Lower-bound	31.333	4.000	7.833			

Very little of the output is needed. Mostly it consists of similar analyses using slightly different tests. These are used when Mauchly's test is significant in the previous output table.

- The *F*-ratio is the Mean Square (MS) for 'factor1' (20.000) divided by Error(factor1) Mean Square (3.917). It is 5.106 (20.000/3.917 = 5.1059).

- The exact significance level of this *F*-ratio is 0.037. Since this value is smaller than .05, we would conclude that there is a significant difference in the mean scores of the three conditions overall.

- In order to interpret the meaning of the ANOVA as it applies to your data, you need to consider the means of each of the three groups of scores which are displayed in the second table. They are 6.00, 8.00 and 4.00. Which of these means are significantly different from the other means?

- You also need to remember that if you have three or more groups, you need to check where the significant differences lie between the pairs of groups. The related *t*-test procedure in Chapter 13 explains this. For the present example, only the difference between the means for 'Product X' and the 'Placebo' was significant. Because you are doing several *t*-tests, each exact probability for the *t*-tests should be multiplied by the number of *t*-tests being carried out. In our example, there are three comparisons, so each exact probability should be multiplied by 3. This is known as the Bonferroni adjustment or correction.

REPORTING THE OUTPUT

- We could describe the results of this analysis in the following way:

   A one-way repeated measures analysis of variance showed a significant treatment effect for the three conditions, $F(2, 8) = 5.10, p = .037$, partial $\eta^2 = .56$. The Aspirin mean was 6.00, the Product X mean 8.00, and the Placebo mean was 4.00. None of the three treatments differed from one another with related *t*-tests when a Bonferroni adjustment was made for the number of comparisons.

- This could be supplemented by an analysis of variance summary table such as Table 22.2. Drugs is factor1 in the output, and Residual Error is Error(factor1) from the fifth table in the output.
- Significant at 5% level.

Table 22.2	Analysis of variance summary table			
Source of variation	**Sum of squares**	**Degrees of freedom**	**Mean square**	**F-ratio**
Drugs	40.00	2	20.00	5.11*
Residual error	31.33	8	3.92	

Summary of SPSS Statistics steps for repeated-measures ANOVA

Data

- Name the variables in 'Variable View' of the 'Data Editor'.
- Enter the data under the appropriate variable name in 'Data View' of the 'Data Editor'.

Analysis

- Select 'Analyze', 'General Linear Model' and 'Repeated Measures …'.
- Enter the number of groups in the 'Number of Levels' and select 'Add' and 'Define'.
- Move the appropriate variables to the upper 'Within-Subjects Variables (factor1):' box and select 'Options'.
- Select 'Descriptive statistics', 'Estimates of effect size', 'Continue' and then 'OK'.

Output

- Check if Mauchly's test of sphericity is not significant with Sig. greater than .05.
- If it is not significant, check that the Sig(nificance) of the Factor for Sphericity Assumed in the Tests of Within-Subjects Effects table is .05 or less.
- If more than two means, check with related t-tests which means differ significantly.

For further resources including data sets and questions, please refer to the website accompanying this book.

CHAPTER 23

Two-way analysis of variance for unrelated/ uncorrelated scores

Overview

- Two-way analysis of variance allows you to compare the means of a dependent variable when there are *two* independent variables.

- If you have more than one *dependent* variable then you simply repeat the analysis for each dependent variable separately. On the other hand, if the several dependent variables are measuring much the same thing then they could be combined into a single overall measure using the summing procedures described in Chapter 43 or 44, or MANOVA could be used (Chapter 27).

- With SPSS Statistics, you do *not* need equal numbers of scores in each condition of the independent variable. If it is possible to have equal numbers in each condition, however, the analysis is optimal statistically.

- Although the two-way ANOVA can be regarded as an efficient design in so far as it allows two different independent variables to be incorporated into the study, its ability to identify interactions may be more important. An interaction is simply a situation in which the combined effect of two variables is greater than the sum of the effects of each of the two variables acting separately.

- The two-way ANOVA can be tricky to interpret. It is important to concentrate on the means in each condition and not simply on the complexities of the statistical output. It is important to note that ANOVAs proceed according to certain rules. The main effects are identified prior to the interactions. Sometimes, unless care is taken, the interaction is mistaken for the main effects – simply because variation is claimed for the main effects before it is claimed for the interaction. As with most statistical analyses, it is important to concentrate as much on the patterns of means in the data as the statistical probabilities.

23.1   What is two-way ANOVA?

ANOVA can be extended to include two or more independent variables. This is illustrated by the data in Table 23.1. This is a study of the effects of sleep deprivation on the number of mistakes that participants make on a task (the dependent variable). There are three groups of participants on the basis of the independent variable sleep deprivation. One group is deprived of sleep for 4 hours, one is deprived of sleep for 12 hours, and the third is deprived of sleep of 24 hours. However, there is a second independent variable – alcohol. Some of the participants are given alcohol as part of the study and others are not given alcohol. As can be seen, every combination of the alcohol variable and the sleep deprivation variable is used in the study. This is known as a two-way ANOVA because there are two independent variables. In each combination, in this case, data have been collected from three different participants – a total of 18 different participants. (If it were decided to consider male and female participants separately in addition, then this would be a three-way ANOVA because there are three independent variables – sleep deprivation, alcohol and gender.)

It is possible to regard the data in Table 23.1 as two separate studies:

- The effects of sleep deprivation on the number of mistakes made. In order to do this, one essentially ignores the alcohol condition so that there are six participants in the 4 hour sleep deprivation condition, six participants in the 12 hour sleep deprivation condition, and six participants in the 24 hour sleep deprivation condition. The average number of mistakes made in these three conditions would be compared.

- The effects of alcohol on the number of mistakes made. In this case, the sleep deprivation condition is ignored so that there are nine participants in the alcohol condition and another nine participants in the no-alcohol condition. The average number of mistakes made in the alcohol condition would be compared with the average number of mistakes made in the no-alcohol condition.

Thus really there are two separate studies contained within the one overall study. If it were two entirely separate studies, then one could compare the average errors in the sleep deprivation study using the one-way ANOVA discussed in Chapter 21 and the average errors for the alcohol study either using the unrelated *t*-test (Chapter 14) or the one-way ANOVA which is essentially the same thing for just two groups. This is more or less what happens in two-way ANOVA. Each of the separate studies, though, is called a *main effect*. So we would speak of the main effect of sleep deprivation and the main effect of alcohol. In the two-way ANOVA, the main effects are computed almost as if they were separate studies with one slight difference. That is, the degrees of freedom are slightly reduced for the error variance. The consequence of this is that the main effects are slightly less statistically significant than they would have been if the data had been

Table 23.1	Data for sleep deprivation experiment: number of mistakes on video test		
	Sleep deprivation		
	4 hours	**12 hours**	**24 hours**
Alcohol	16	18	22
	12	16	24
	17	25	32
No alcohol	11	13	12
	9	8	14
	12	11	12

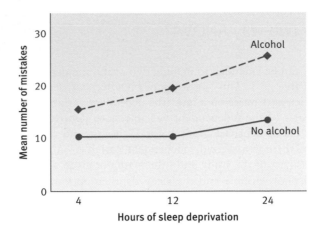

FIGURE 23.1   A chart illustrating the interaction of sleep deprivation with alcohol

analysed separately. In other words, there is a slight negative consequence involved in doing two studies essentially for the price of one.

However, there is nothing wrong in regarding each main effect as if it were a separate study.

There is an advantage in the use of two-way ANOVA. This is something known as the *interaction*. There is just one interaction for the two-way ANOVA but there would be four interactions for the three-way ANOVA. The number of interactions escalates with increasing numbers of independent variables in the design. Interaction is a difficult concept to understand. Looking at Table 23.1, it might be noticed that the number of errors made in the 24 hour sleep deprivation with alcohol condition seems to be disproportionately large. This is an indication that there may be an interaction since the scores in that condition are bigger than generally one would expect on the basis of the influence of alcohol acting alone and the effect of 24 hours of sleep deprivation acting alone. This becomes clearer if we put the means calculated from Table 23.1 on a chart (Figure 23.1). Notice that there are two lines on the chart – one for the alcohol conditions and the other for the no-alcohol condition. The vertical axis on the chart is the average number of mistakes.

If we look at the 4 hour sleep conditions first, it can be seen that on average a few more mistakes are made in the alcohol condition than the no alcohol condition. The 12 hour sleep deprivation conditions also indicate that even more mistakes are made on average in the alcohol condition than the no-alcohol condition. Finally, the 24 hour sleep deprivation conditions also show that more mistakes are made when alcohol is consumed but the difference is bigger than for any of the other sleep deprivation conditions. That is to say, the effect of being given alcohol is *not* consistently the same for each sleep deprivation condition. It is greater for some conditions than others. This is indicative of an interaction between the amount of sleep deprivation and whether or not alcohol had been consumed. The reason for this is that in statistics it is assumed that the effects of a variable are additive. In other words, there is an effect of being given alcohol which adds a certain amount to the effects of the different amounts of sleep deprivation. The crucial thing is that this effect of alcohol should be constant (the same amount) irrespective of the effect of the different amounts of sleep deprivation. This clearly is not the case in Figure 23.1 since more mistakes can be attributed to the influence of alcohol when participants have been sleep deprived for 24 hours than for those who had been sleep deprived, say, for 4 hours. This inconsistency of effect is the interaction. That is, alcohol has greater influence at some levels of sleep deprivation than others or, in other words, amount of sleep deprivation and alcohol interact to produce greater effects (greater numbers of mistakes) than the effects of alcohol alone plus the effects of the amount of sleep deprivation alone can account for.

Figure 23.2 illustrates one case where there is no interaction. Notice that the lines are parallel to each other. This means that alcohol has exactly the same effect on mistakes made irrespective of the amount of sleep deprivation.

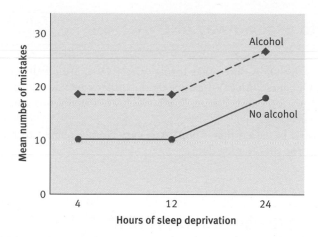

FIGURE 23.2   A chart illustrating a situation in which there is no interaction

In ANOVA, the main effects are calculated before the interaction is calculated. This means that priority is given to the main effects. Thus variation is attributed to the main effects as a priority and only the variation which is left over can potentially be attributed to the interaction. This is important when interpreting two-way ANOVA designs and others which involve potential interactions. Figure 23.3 highlights the main steps in carrying out a two-way ANOVA.

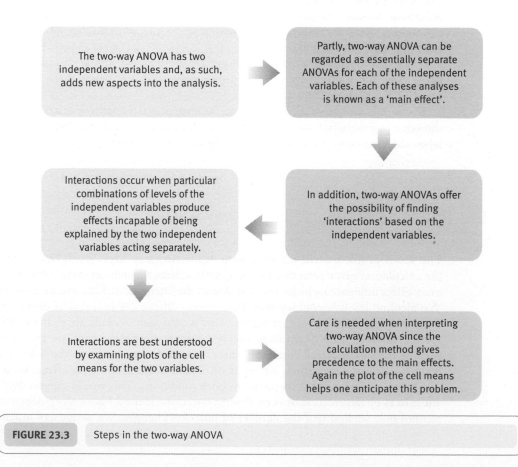

The two-way ANOVA has two independent variables and, as such, adds new aspects into the analysis.

Partly, two-way ANOVA can be regarded as essentially separate ANOVAs for each of the independent variables. Each of these analyses is known as a 'main effect'.

Interactions occur when particular combinations of levels of the independent variables produce effects incapable of being explained by the two independent variables acting separately.

In addition, two-way ANOVAs offer the possibility of finding 'interactions' based on the independent variables.

Interactions are best understood by examining plots of the cell means for the two variables.

Care is needed when interpreting two-way ANOVA since the calculation method gives precedence to the main effects. Again the plot of the cell means helps one anticipate this problem.

FIGURE 23.3   Steps in the two-way ANOVA

23.2 When to use two-way ANOVA

Two-way ANOVA, like other forms of ANOVA, are ideally suited to randomised experimental studies. Of course, ANOVA can be used in other contexts but care has to be taken to avoid making the analysis too cumbersome. Survey data are usually too complex to use ANOVA on, though not in every case.

It should be noted that ANOVA is very closely related to multiple regression, which may be preferred for survey-type data.

23.3 When not to use two-way ANOVA

While two-way ANOVA does have fairly restricted data requirements, it is actually difficult to avoid using it when interactions may be an important aspect of the analysis. This is because SPSS has no non-parametric alternatives to two-way ANOVA.

23.4 Data requirements for two-way ANOVA

The independent (grouping) variables should consist of relatively small numbers of categories otherwise the data analysis will be exceedingly cumbersome. This is not usually a problem for researchers designing randomised experiments as the practicalities of the research will place limits on what can be done.

It is best if the scores in each condition have more or less similar variances which SPSS can test is the case. As usual, many psychologists believe that it is better if the scores approximate an equal interval scale of measurement.

There does not have to be the same number of participants in each condition of the study though it is generally better if there are when the requirements of ANOVA are violated, such as when the variances of scores in each condition are different.

23.5 Problems in the use of two-way ANOVA

The major problem is the interpretation of the two-way ANOVA. It has to be appreciated that the calculation gives priority to finding main effects to such an extent that it can attribute to a main effect influences which really are due to the interaction. Take Figure 23.4. If you look at the 'No alcohol' line you will see that there are no differences in terms of mistakes between any of the three sleep deprivation conditions. This would seem to indicate that there is no main effect for the variable sleep deprivation simply because all of the means are the same. If we look at the line for the 'Alcohol' condition, there is no difference between the 4 hour and the 12 hour periods of sleep deprivation, which again would indicate no main effect for sleep deprivation. The number of mistakes is the same for both conditions. This would suggest that there is no main effect of sleep deprivation. It is only when we consider the 24 hour sleep deprivation group given alcohol that we find that the number of mistakes changes. We would say much the same of the

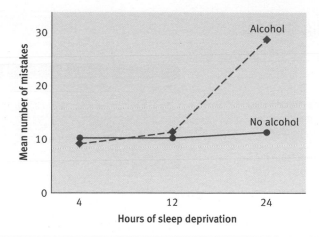

FIGURE 23.4   A case where an interaction may be confused with a main effect

alcohol variable. If we consider the 4 hour and the 12 hour periods of sleep deprivation, there is virtually no difference between the alcohol groups and the no-alcohol groups. This would be indicative of no main effect for alcohol. Only when we consider the 24 hour sleep deprivation groups do we get a big difference due to alcohol.

If the ANOVA tells us that there is a significant main effect for either or both alcohol and sleep deprivation, we need to be very cautious since this is at odds with what the pattern of the data in Figure 23.4 tells us. We suspect an interaction. However, it is possible that the interaction is not significant according to the SPSS output.

The message is that the ANOVA may be confusing things and we need to be very cautious when interpreting the output. This is a case where simple descriptive statistics are vital to the interpretation of our data.

Anyone contemplating more complex ANOVAs than this should be wary that three-way, four-way and so forth ANOVAs generate massive numbers of interactions which are extremely difficult to interpret. We would suggest that very complex ANOVAs should be avoided unless you have a very good reason for using them and until your level of statistical skill is well developed.

You can find out more about two-way ANOVA in Chapter 22 of Howitt, D. and Cramer, D. (2011). *Introduction to Statistics in Psychology*, 5th edition. Harlow: Pearson.

23.6   The data to be analysed

The computation of a two-way unrelated analysis of variance is illustrated with the data in Table 23.1. The table shows the scores of different participants in six conditions, reflecting the two factors of sleep deprivation and alcohol (*ISP*, Table 22.11). The purpose of the analysis is to evaluate whether the different combinations of alcohol and sleep deprivation differentially affect the mean number of mistakes made.

23.7 Entering the data

Step 1

In 'Variable View' of the 'Data Editor' name the first row 'Alcohol' and name its values of '1' and '2' 'Alcohol' and 'No alcohol' respectively.
Name the second row 'SleepDep' and name its values of '1', '2' and '3', '4 hrs' '12 hrs' and '24 hrs' respectively.
Name the third row 'Errors'.
Change 'Decimals' from '2' to '0'.

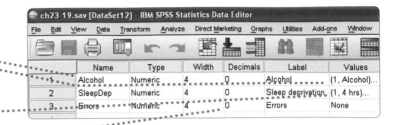

	Name	Type	Width	Decimals	Label	Values
1	Alcohol	Numeric	4	0	Alcohol	{1, Alcohol}...
2	SleepDep	Numeric	4	0	Sleep deprivation	{1, 4 hrs}...
3	Errors	Numeric	4	0	Errors	None

Step 2

In 'Data View' of the 'Data Editor' enter the data in the first three columns.

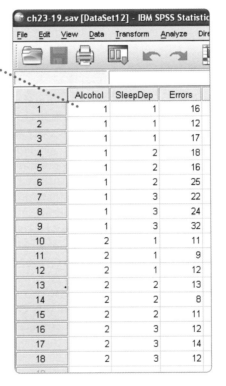

	Alcohol	SleepDep	Errors
1	1	1	16
2	1	1	12
3	1	1	17
4	1	2	18
5	1	2	16
6	1	2	25
7	1	3	22
8	1	3	24
9	1	3	32
10	2	1	11
11	2	1	9
12	2	1	12
13	2	2	13
14	2	2	8
15	2	2	11
16	2	3	12
17	2	3	14
18	2	3	12

23.8 Two-way unrelated ANOVA

Step 1

Select 'Analyze', 'General Linear Model' and 'Univariate...'.

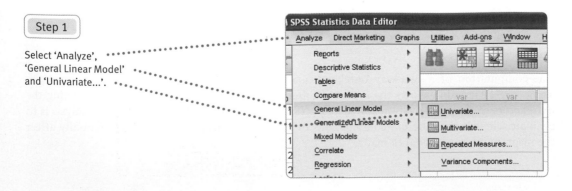

Step 2

Select 'Errors' and the ➡ button beside the
'Dependent Variable:' box to put it there.
Select 'Alcohol' and 'SleepDep' either
singly or together and the ➡ button beside
'Fixed Factor(s):' to put them there.
Select 'Options...'.

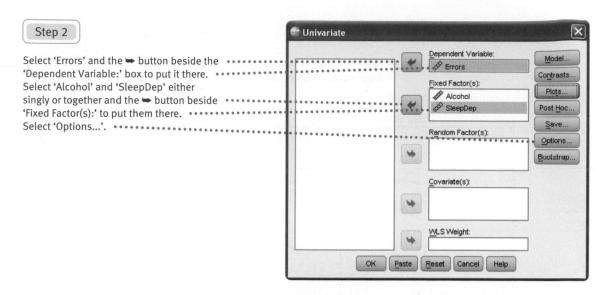

Step 3

Select 'Descriptive statistics', 'Estimates of
effect size' and 'Homogeneity tests'.
Select 'Continue'.
In the previous box which re-appears,
select 'Plots...'.

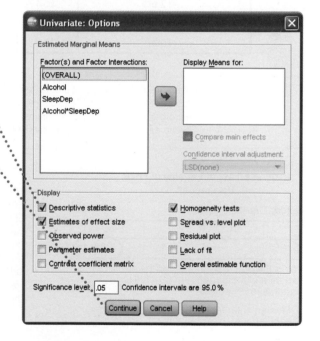

Step 4

Select 'Alcohol' and the
➡ button beside the
'Horizontal Axis:' box to
put it there.
Select 'SleepDep' and the
➡ button beside the
'Separate Lines:' box to
put it there.
Select 'Add'.
Select 'Continue'.
Select 'OK' in the previous
box which reappears.

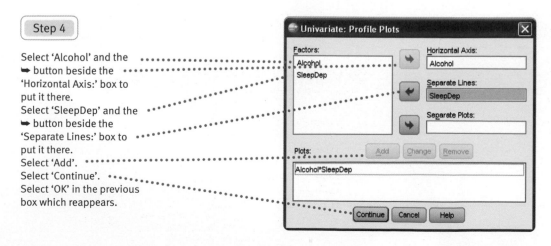

23.9 Interpreting the output

Descriptive Statistics

Dependent Variable:Errors

Alcohol	Sleep deprivation	Mean	Std. Deviation	N
Alcohol	4 hrs	15.00	2.646	3
	12 hrs	19.67	4.726	3
	24 hrs	26.00	5.292	3
	Total	20.22	6.099	9
No alcohol	4 hrs	10.67	1.528	3
	12 hrs	10.67	2.517	3
	24 hrs	12.67	1.155	3
	Total	11.33	1.871	9
Total	4 hrs	12.83	3.061	6
	12 hrs	15.17	5.981	6
	24 hrs	19.33	8.066	6
	Total	15.78	6.330	18

The second table provides the means, standard deviations and number (N) of cases for the two variables of 'Alcohol' and 'Sleep deprivation' separately and together. So the mean for the 'Alcohol' condition is given against the 'Total' (i.e. 20.22). The mean for the '4 hrs' 'Sleep deprivation' is given against the 'Total' (i.e. 12.83).

Levene's Test of Equality of Error Variances[a]

Dependent Variable:Errors

F	df1	df2	Sig.
2.786	5	12	.068

Tests the null hypothesis that the error variance of the dependent variable is equal across groups.

a. Design: Intercept + Alcohol + SleepDep + Alcohol * SleepDep

The third table gives Levene's test to see if the variances are similar. As the significance of this test is .068 (which is above .05) the variances are similar. If this test was significant, the scores should be transformed say by using a logarithmic scale to make the variances similar. This is a matter of trial and error – try different transformations until the variances become the same.

Tests of Between-Subjects Effects

Dependent Variable:Errors

Source	Type III Sum of Squares	df	Mean Square	F	Sig.	Partial Eta Squared
Corrected Model	546.444[a]	5	109.289	9.739	.001	.802
Intercept	4480.889	1	4480.889	399.287	.000	.971
Alcohol	355.556	1	355.556	31.683	.000	.725
SleepDep	130.111	2	65.056	5.797	.017	.491
Alcohol * SleepDep	60.778	2	30.389	2.708	.107	.311
Error	134.667	12	11.222			
Total	5162.000	18				
Corrected Total	681.111	17				

a. R Squared = .802 (Adjusted R Squared = .720)

The fourth table gives the significance levels for the two variables of 'Alcohol' and 'SleepDep' and their interaction as well as their partial etas squared. Partial eta squared is the sum of squares for an effect (e.g. 355.556 for alcohol) divided by the sum of squares for that effect added to the sum of squares for the error (355.556 + 134.667). In other word, it does not take account of the variance explained by the other effects.

- In the analysis of variance table, the *F*-ratio for the two main effects ('Alcohol' and 'SleepDep') is presented first.

- For the first variable of alcohol the *F*-ratio is 31.683, which is significant at less than the .0005 level. Since there are only two conditions for this effect we can conclude that the mean score for one condition is significantly higher than that for the other condition.

- For the second variable of sleep deprivation it is 5.797, which has an exact significance level of .017. In other words, this *F*-ratio is statistically significant at the .05 level, which means that the means of the three sleep conditions are dissimilar.

- Which of the means differ from the others can be further determined by the use of multiple comparison tests such as the unrelated *t*-test.

- The *F*-ratio for the two-way interaction between the two variables (Alcohol * SleepDep) is 2.708. As the exact significance level of this ratio is .107 we would conclude that there was no significant interaction. If this interaction was significant, we could determine which means of the six groups differed from each other by creating a new group variable from Alcohol and SleepDep (Chapter 45) and running a multiple comparison test (Chapter 24) in a one-way ANOVA (Chapter 21).

This plot is shown for the means of the six conditions. It has been edited with the 'Chart Editor'. The style of the different coloured lines has been changed so that they can be more readily distinguished.

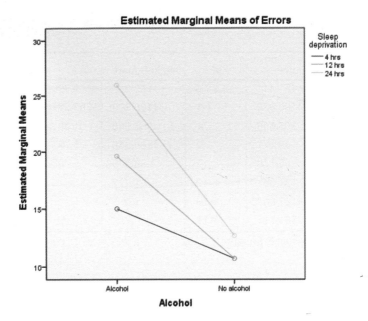

23.10   Editing the graph

To change the style of line, double click on the plot to select the 'Chart Editor'. Select the line in the legend to be changed.

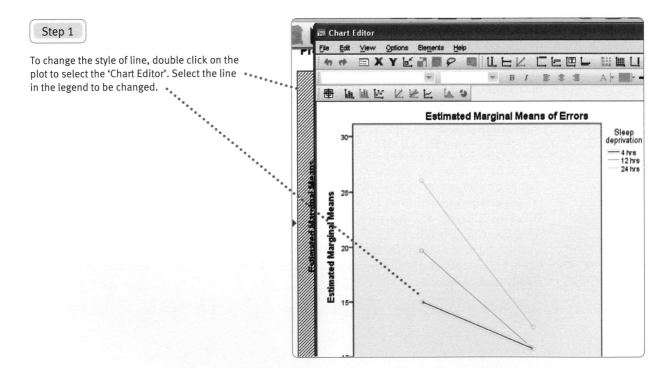

Step 2

Select the ▼ button next to 'Style' and select the style of line desired. Change the size of the line if desired.
Change the colour if desired. Select 'Apply' and then 'Close' (which changes from 'Cancel').
Select next line to be edited and repeat these steps.

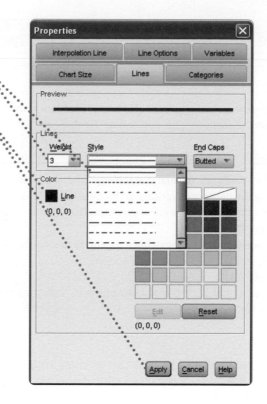

REPORTING THE OUTPUT

- We could report the results of the output as follows:

  A two-way unrelated ANOVA showed that significant effects were obtained for alcohol, $F(2, 12) = 31.68$, $p < 0.001$, partial $\eta^2 = .73$, and sleep deprivation, $F(2, 12) = 5.80$, $p = .017$, partial $\eta^2 = .49$, but not for their interaction, $F(2, 12) = 2.70$, $p = .107$, partial $\eta^2 = .31$.

- You may be required to give an analysis of variance summary table. A simple one, like that shown in Table 23.2, would leave out some of the information in the third table in the output, which is unnecessary.
- Because the 'SleepDep' factor has more than two conditions, we need to use an appropriate multiple comparison test to determine the means of which groups differ significantly (see Chapter 24).
- We also need to report the means and standard deviations of the groups which differ significantly. These descriptive statistics are given in the second table of the output.

Table 23.2	Analysis of variance summary table				
Source of variation	**Sums of squares**	**Degrees of freedom**	**Mean square**	**_F_-ratio**	**Probability**
Alcohol	355.56	1	355.56	31.68	< .001
Sleep deprivation	130.11	2	65.06	5.80	< .05
Alcohol with sleep deprivation	60.78	2	30.39	2.71	Not significant
Error	134.67	12	11.22		

Summary of SPSS Statistics steps for two-way ANOVA

Data

- Name the variables in 'Variable View' of the 'Data Editor'.
- Enter the data under the appropriate variable names in 'Data View' of the 'Data Editor'.

Analysis

- Select 'Analyze', 'General Linear Model' and 'Univariate ...'.
- Move the dependent variable to the 'Dependent Variable:' box.
- Move the independent variables to the 'Fixed Factor(s):' box.
- Select 'Plots ...'.
- Move one independent variable (Factor) to the 'Horizontal Axis:' box and the other to the 'Separate Lines:' box.
- Select 'Add' and 'Continue'.
- Select 'Options ...', 'Descriptive Statistics', 'Estimates of effect size', 'Homogeneity tests', 'Continue' and then 'OK'.

Output

- Check which *F*-ratios are significant by having a Sig(nificance) of .05 or less.
- Check that Levene's test shows the variances are homogeneous by having a Sig(nificance) level of more than .05.
- If a significant main effect has more than two groups, use further tests to determine which means differ.
- With a significant interaction use further tests to determine which means differ.

For further resources including data sets and questions, please refer to the website accompanying this book.

CHAPTER 24

Multiple comparisons in ANOVA

Overview

- This chapter extends the coverage of multiple *t*-tests from Chapters 21 and 22. It explains how to decide which particular pairs of means are significantly different from each other in the analysis of variance.

- The technique is used when you have more than two means. It adds no further information if there are only two means.

- ANOVAs with three or more means to compare are almost certain to benefit from employing multiple comparison methods.

- It is not possible to give definitive advice as to which multiple comparison test to use in different situations as there is no clear consensus in the literature.

24.1 What is multiple comparisons testing?

ANOVA brings problems of interpretation when the independent variable (the grouping variable) has three or more categories. The problem arises because if the ANOVA is significant, this does not indicate that all of the means are significantly different from each other – it merely implies that the pattern of means is unlikely to have occurred by chance as a consequence of random sampling from a population in which the groups do not differ. If the independent variable has just two categories, then a significant ANOVA indicates that the two means involved do differ significantly from each other.

Take the data in Table 24.1 which we have previously discussed in Chapter 21. We have added the group means on the dependent variable 'Depression' in each case. What is obvious is that the mean Depression score is higher in the Hormone 1 condition than in the other two conditions. Indeed, there seems to be little difference between the Hormone 2 group and the Placebo control group. If any comparison is statistically significant, it is likely to be the comparison between the

Table 24.1	Data for a study of the effects of hormones		
Group 1 **Hormone 1**	**Group 2** **Hormone 2**	**Group 3** **Placebo control**
9	4	3
12	2	6
8	5	3
Mean = 9.667	Mean = 3.667	Mean = 4.000

Hormone 1 group and the Hormone 2 group. The comparison between the Hormone 1 group and the Placebo control group is the next most likely significant difference. It is very unlikely that the comparison between the Hormone 2 group and the Placebo control group will be statistically significant.

It would seem to be obvious to compare the pairs of means using the *t*-test in these circumstances. There is just one problem with what is otherwise a good idea – that is, the more statistical analyses one carries out on any data the more likely that one is to get at least one statistically significant finding by chance. So if that method is to be used, then some adjustment should be made to deal with this problem. One way of doing this is the Bonferroni procedure which basically involves adjusting the significance level for the number of *t*-tests used. However, there is little point in doing this as SPSS Statistics has far better multiple comparisons tests available.

The problem is generally to decide which one(s) to use. It has to be said that there is little clarity in the literature on this matter. One way of dealing with this would be to use all of the available multiple comparisons tests on SPSS on the grounds that if they all lead to the same conclusion there is no problem – a problem only arises if they give different conclusions. However, we would recommend instead that you consider using the Newman–Keuls test (S-N-K in SPSS, the S standing for Student which the unrelated *t*-test is sometimes called) or Duncan's new multiple range test which are available on SPSS. The reasons are as follows:

● These are among the more powerful multiple comparison tests and are not prone to problems.

● They can be used when the ANOVA is not statistically significant overall. This is important because in the past it used to be claimed that ANOVA had to be significant before any paired comparisons could be tested for. This produced obviously anomalous situations in which certain means were clearly very different but had to be disregarded because of this unfortunate 'rule'. It should be added that some of the multiple comparisons tests were developed as alternatives to ANOVA anyway so it is legitimate to use them on their own.

These two tests assume that each condition has the same number of cases in them. A test which does not assume this is the Scheffé test. However, it is a more conservative test which means that differences are less likely to be significant.

Of course, the issue of multiple comparisons disappears when the independent variables in the ANOVA only have two different values. With a one-way ANOVA then it is easy to obtain multiple comparisons. However, with the two-way ANOVA and more complex versions one has to resort to breaking the analysis up into individual one-way ANOVAs in order to employ multiple comparisons. It probably is sound advice, wherever possible, then to plan the two-way ANOVA in a way in which each independent variable has only two values which circumvents the problem. However, this is not always desirable, of course. SPSS will calculate the multiple comparisons tests for two-way (and more complex) ANOVAs simply by doing two (or more) separate one-way ANOVAs.

24.2   When to use multiple comparisons tests

It is appropriate to carry out Newman–Keuls test (S-N-K in SPSS) or Duncan's new multiple range test whenever doing an unrelated samples one-way ANOVA irrespective of the outcome of the ANOVA. Generally, it is helpful to calculate multiple comparisons tests for one-way unrelated ANOVAs of the sort described in Chapter 21. If the variances of your groups depart markedly from being equal, it is possible to select a multiple comparisons test which does not assume equal variances. The Games–Howell would be a good choice in these circumstances. SPSS will calculate a homogeneity of variance test as one of the procedure's Options.

Multiple comparisons testing is available for two-way and more complex ANOVAs. However, be aware that SPSS will not attempt to calculate these where there are only two conditions of the independent (grouping) variable. Figure 24.1 outlines the main steps in carrying out multiple comparison tests.

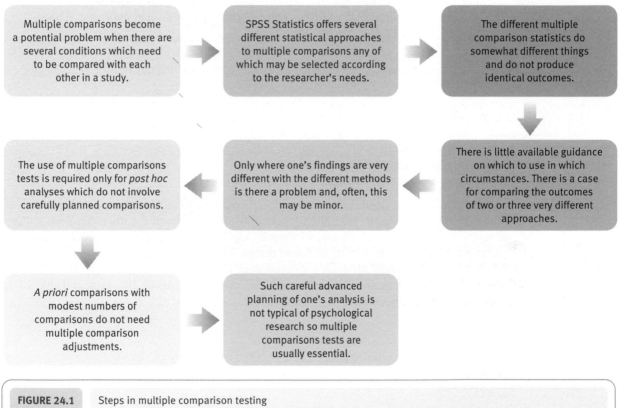

FIGURE 24.1   Steps in multiple comparison testing

24.3   When not to use multiple comparisons tests

Unfortunately, there is no point trying to do multiple comparisons tests on SPSS when dealing with correlated data as in the repeated-measures ANOVA (Chapter 22). It simply does not have any available. In these circumstances you could resort to multiple related *t*-tests between pairs of means.

24.4   Data requirements for multiple comparisons tests

There is a flexible choice of multiple comparisons tests in SPSS, Some of these do not require equality of variances. It is a simple matter to go into Options in SPSS for ANOVA analyses and select 'Homogeneity test'. If this is statistically significant then one should choose a multiple comparisons test which does not assume that the variances are equal.

24.5   Problems in the use of multiple comparisons tests

The output from SPSS for multiple comparisons tests is not entirely consistent for each of the tests. Choosing more than one test can make the output seem much more complicated so you may find it easier to check for homogeneity before running the multiple comparison test of your choice. This does mean running the analysis twice but this is just a small chore for SPSS.

> You can find out more about multiple comparison tests in Chapter 23 of Howitt, D. and Cramer, D. (2011). *Introduction to Statistics in Psychology*, 5th edition. Harlow: Pearson.

24.6   The data to be analysed

Knowing precisely where significant differences lie between different conditions of your study is important. The overall trend in the ANOVA may only tell you part of the story. SPSS has a number of *post hoc* procedures which are, of course, applied after the data are collected and not planned initially. They all do slightly different things. There is a thorough discussion of them in Howell, D. (2010). *Statistical Methods for Psychology* (7th edition). Belmont, CA: Duxbury. We will illustrate the use of these multiple comparison procedures (using the data in Table 24.1, which were previously discussed in Chapter 21).

24.7   Entering the data

Either select the saved file used in Chapter 21 or enter the data as follows:

Step 1

In 'Variable View' of the 'Data Editor', name the first row 'Condition'.
Name conditions '1', '2' and '3' 'Hormone 1', 'Hormone 2' and 'Placebo Control' respectively.
Name the second row 'Depression'.
Change 'Decimals' places from '2' to '0'.

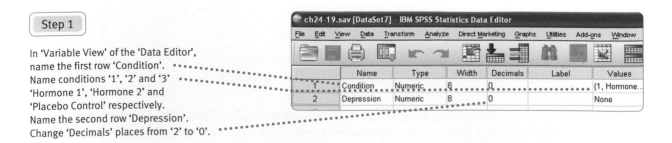

Step 2

In 'Data View' of the 'Data Editor' enter the data in the first two columns. Save this file to use for Chapter 26.

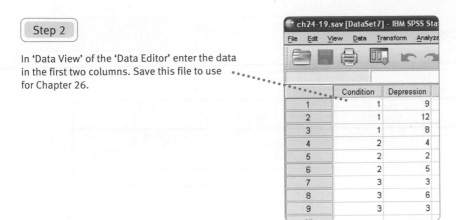

24.8 Multiple comparison tests

Step 1

Select 'Analyze', 'Compare Means' and 'One-Way ANOVA...'.

Step 2

Select 'Depression' and the ➡ button beside the 'Dependent List:' box to put it in the box.
Select 'Condition' and the ➡ button beside the 'Factor:' box to put it there.
Select 'Post Hoc...'.

Step 3

Select 'S-N-K',
'Duncan'
and 'Scheffe'.
Select 'Continue'.
Select 'OK' in the previous box.

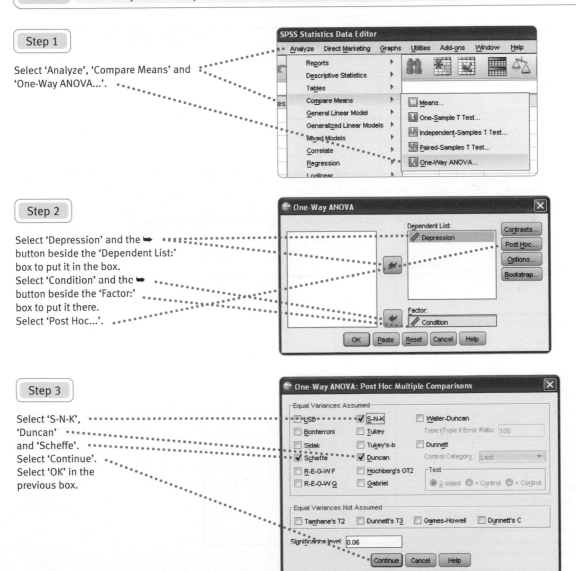

24.9   Interpreting the output

ANOVA

Depression

	Sum of Squares	df	Mean Square	F	Sig.
Between Groups	68.222	2	34.111	10.586	.011
Within Groups	19.333	6	3.222		
Total	87.556	8			

The first table shows the results for the analysis of variance. The F-ratio for the 'Between Groups' effect (i.e. the effects of hormones) is 10.586, which has an exact significance level of .011. In other words, the 'Between Groups' effect is significant. Overall the means for the three groups differ.

The second and last table gives the results for the three multiple comparison tests.

Multiple Comparisons

Dependent Variable:Depression

	(I) Condition	(J) Condition	Mean Difference (I-J)	Std. Error	Sig.	95% Confidence Interval Lower Bound	95% Confidence Interval Upper Bound
Scheffe	Hormone 1	Hormone 2	6.000*	1.466	.018	1.30	10.70
		Placebo control	5.667*	1.466	.023	.97	10.37
	Hormone 2	Hormone 1	-6.000*	1.466	.018	-10.70	-1.30
		Placebo control	-.333	1.466	.975	-5.03	4.37
	Placebo control	Hormone 1	-5.667*	1.466	.023	-10.37	-.97
		Hormone 2	.333	1.466	.975	-4.37	5.03

*. The mean difference is significant at the 0.05 level.

E.g. Using the Scheffe test, the Hormone 1 mean is significantly different from the Hormone 2 mean (sig. = .018) and the Placebo control mean (sig. = .023).

Depression

	Condition	N	Subset for alpha = 0.05 1	Subset for alpha = 0.05 2
Student-Newman-Keuls[a]	Hormone 2	3	3.67	
	Placebo control	3	4.00	
	Hormone 1	3		9.67
	Sig.		.828	1.000
Duncan[a]	Hormone 2	3	3.67	
	Placebo control	3	4.00	
	Hormone 1	3		9.67
	Sig.		.828	1.000
Scheffe[a]	Hormone 2	3	3.67	
	Placebo control	3	4.00	
	Hormone 1	3		9.67
	Sig.		.975	1.000

Hormone 2 and Placebo control are in the same subset – i.e. not significantly different.

Hormone 1 is the only group in this second subset. Consequently it is significantly different from the other two group means.

Means for groups in homogeneous subsets are displayed.

a. Uses Harmonic Mean Sample Size = 3.000.

- The last table, entitled 'Homogeneous Subsets', lists the sets of means which do not *differ* significantly from each other. So taking the section for the Student–Newman–Keuls test, there are two subsets of means. Subset 1 indicates that the Hormone 2 and Placebo control means of 3.67 and 4.00 do not differ significantly. Subset 2 contains just the Hormone 1 mean of 9.67. Thus the mean of Hormone 1 differs significantly from the means of both Hormone 2 and the Placebo control. However, the means of Hormone 2 and the Placebo control do not differ significantly. The pattern is identical for the Duncan and Scheffé tests in this case – it is not always so.

- Therefore the three multiple comparison tests all suggest the same thing: that there are significant differences between Hormone 1 and Hormone 2, and between Hormone 1 and the Placebo control. There are no other differences. So, for example, it is not possible to say that Hormone 1 and Hormone 2 are significantly different.

- The choice between the three tests is not a simple matter. Howell (2010) makes some recommendations.

REPORTING THE OUTPUT

We could report the results of the output as follows:

A one-way unrelated analysis of variance showed an overall significant effect for the type of drug treatment, $F(2, 6) = 10.59$, $p = 0.011$. Scheffé's test found that the Hormone 1 group differed from the Hormone 2 group, $p = .018$, and the Placebo Control, $p = .023$, but no other significant differences were found.

Summary of SPSS Statistics steps for multiple comparison tests

Data

- Name variables in 'Variable View' of the 'Data Editor'.
- Enter data under the appropriately named variables in 'Data View' of the 'Data Editor'.

Analysis

- Select 'Analyze' and appropriate analysis of variance option.
- Enter dependent and independent variables in boxes to the right.
- Select 'Post Hoc ...', 'Scheffe', 'Continue' and then 'OK'.

Output

- Check which means differ significantly by seeing if the Sig(nificance) is .05 or less.
- Note the direction of the difference.

For further resources including data sets and questions, please refer to the website accompanying this book.

CHAPTER 25

Two-way mixed analysis of variance (ANOVA)

Overview

- A mixed analysis of variance design is merely a two-way (or three-way, etc.) research design which contains *both* unrelated and related independent variables.

- Mixed designs generate rather more complex measures of error estimates compared with other forms of ANOVA. This means that special care needs to be taken when producing appropriate summary tables.

- Apart from that, concepts from related and unrelated ANOVAs apply.

25.1 What is two-way mixed ANOVA?

So long as you understand the two-way ANOVA (Chapter 23) and the concept of related data (Chapter 22), there is little new to learn for the mixed analysis of variance. It is simply a two-way ANOVA in which one of the two independent variables is based on repeated measures of the same variable such as when a group of participants is measured at several different points in time, for example. An example of a mixed-design is shown in Table 25.1. The study consists of two separate groups of children – the experimental and the control conditions or the experimental and control groups. These are unrelated groups since any child can be in either the experimental or control condition but not in both. That is, we have an unrelated independent variable. The other independent variable is a pre-test/post-test measure. Now, in this particular study, children are measured in both the pre-test and the post-test situations. In other words, pre-test/post-test is a related independent variable. What is happening in the study is that the self-esteem of children is being measured on two occasions – one before the experimental manipulation and the other after the experimental manipulation. The children in the experimental condition were praised whenever they exhibited good behaviour whereas there was no feedback for the children in the control condition. It should be fairly obvious that the self-esteem of the children in the experimental condition increases between the pre-test and the post-test.

Table 25.1	Pre- and post-test self-esteem scores in two conditions		
Conditions	Children	Pre-test	Post-test
Control	1	6	5
	2	4	6
	3	5	7
Experimental	4	7	10
	5	5	11
	6	5	12

Of course, all other things being equal, the fact that there is a related independent variable means that the error term can be adjusted for individual differences since participants have been measured more than once. If there is a strong individual difference component to the data then adjusting for this will make the ANOVA more powerful – that is, more able to detect significant differences if they exist.

There is little more that needs to be known since mixed ANOVA would be looking for the main effect of the pre-test/post-test variable and the main effect of the control/experimental variable together with any interaction between the two. Actually the interaction is crucial in this study since it is this which tests the hypothesis that praise raises self-esteem. The main effects are of relatively little interest in this example, though this will not always be the case. Figure 25.1 presents the essential steps in carrying out a two-way mixed analysis of variance.

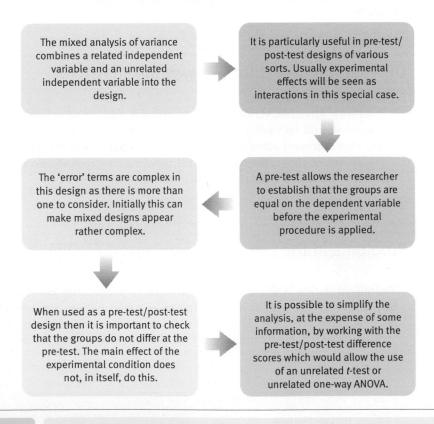

FIGURE 25.1   Essential steps in two-way mixed analysis of variance

25.2   When to use two-way mixed ANOVA

A two-way mixed design ANOVA will be particularly appropriate for studies which investigate change over a period of time. So it will be used when a single group of participants are studied at different time-points. It can also be used where a group of participants serves in a number of different conditions of an experiment (though counterbalancing of the order of serving in the different conditions would be a design improvement).

Such a design is readily turned into a mixed design simply by, for example, breaking the sample down into two separate samples – males and females. Of course, there may be other unrelated independent variables which could be used.

25.3   When not to use two-way mixed ANOVA

The usual strictures about related measures apply to the mixed analysis of variance. That is, where the data are supposed to be related scores, then the scores should correlate. In our example, the pre-test scores should correlate with the post-test scores. If they do not, then it is not advantageous to use the mixed analysis of variance and it can be counterproductive.

You may find it useful to try both the two-way ANOVA and the mixed ANOVA on your data since this can be informative. If you get more significant results with the two-way ANOVA than the mixed ANOVA then it may be that your related measure is not in fact a related measure. In these circumstances, it would be acceptable to use the results of the two-way ANOVA, though we know of no cases where researchers have done that.

25.4   Data requirements for two-way mixed ANOVA

A two-way mixed ANOVA requires a related and an unrelated independent variable. The variances should be similar for all conditions though the test used (the Box test of equality of covariance matrices) actually checks that the covariances in the data are similar as well.

As discussed above, the scores on the repeated measures should correlate with each other for this form of analysis to be maximally effective.

25.5   Problems in the use of two-way mixed ANOVA

The mixed design ANOVA in SPSS produces a particularly bewildering array of output. It is helpful to dispense with as much of this as possible and just concentrate on the essentials if you are new to the procedure. Our step-by-step instructions should help you with this.

What is particularly confusing is that different error terms are used in different parts of the analysis.

You can find out more about mixed ANOVA in Chapter 24 of Howitt, D. and Cramer, D. (2011). *Introduction to Statistics in Psychology*, 5th edition. Harlow: Pearson.

25.6 The data to be analysed

A two-way mixed analysis of variance has one unrelated factor and one related factor. Factors are independent variables. We will illustrate this analysis with the data in Table 25.1 (*ISP*, Table 24.1), which consists of the self-esteem scores of children measured before and after an experimental manipulation in which half the children (chosen at random) were praised for good behaviour (experimental condition) while the other half were given no feedback (control condition).

25.7 Entering the data

Step 1

In 'Variable View' of the 'Data Editor', name the first row 'Condition'.
Name conditions '1' and '2' 'Control' and 'Experimental' respectively.
Name the second and third rows 'Pretest' and 'Posttest' respectively.
Change 'Decimals' places from '2' to '0'.

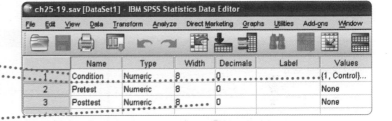

Step 2

In 'Data View' of the 'Data Editor' enter the data in the first three columns.

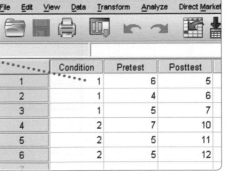

25.8 Two-way mixed ANOVA

Step 1

Select 'Analyze',
'General Linear Model' and
'Repeated Measures...'.

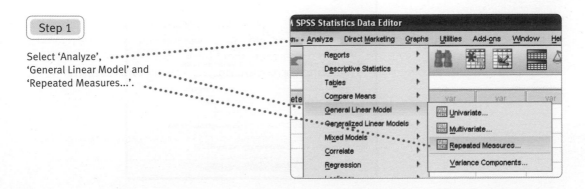

Step 2

Type '2' in the 'Number of
Levels:' box.
Select 'Add'.
Select 'Define'.

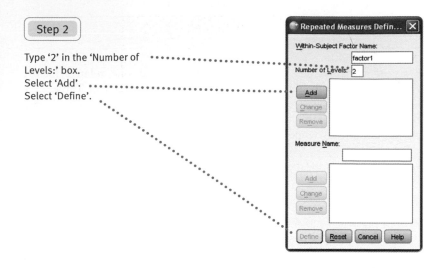

Step 3

Select 'Pretest' and
'Posttest' either alone or
together and the ➡ button
beside the 'Within-Subjects
Variables(factor1):' box to
put them there.
Select 'Condition' and the
➡ button beside the
'Between-Subjects
Factor(s):' box to put it
there.
Select 'Options...'.

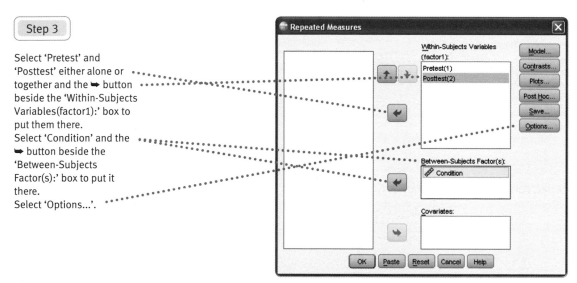

Step 4

Select 'Descriptive statistics',
'Estimates of effect size' and
'Homogeneity tests'.
Select 'Continue'.
In previous box which
reappears, select 'Plots...'.

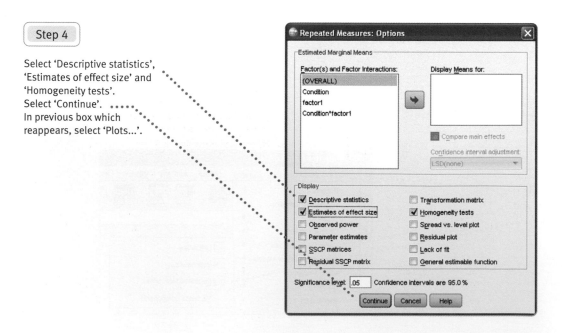

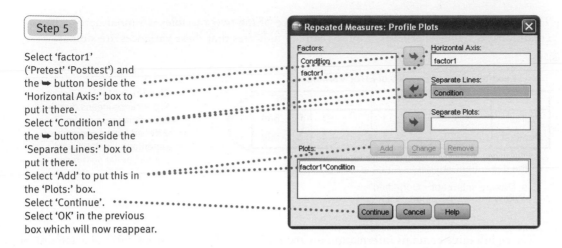

Step 5

Select 'factor1'
('Pretest' 'Posttest') and
the ➡ button beside the
'Horizontal Axis:' box to
put it there.
Select 'Condition' and
the ➡ button beside the
'Separate Lines:' box to
put it there.
Select 'Add' to put this in
the 'Plots:' box.
Select 'Continue'.
Select 'OK' in the previous
box which will now reappear.

25.9 Interpreting the output

The output gives 10 tables and a plot. Only the more important tables are shown here.
The second table gives the mean and standard deviation for the two groups.

Descriptive Statistics

	Condition	Mean	Std. Deviation	N
Pretest	Control	5.00	1.000	3
	Experimental	5.67	1.155	3
	Total	5.33	1.033	6
Posttest	Control	6.00	1.000	3
	Experimental	11.00	1.000	3
	Total	8.50	2.881	6

The third table shows whether the covariances matrices of the post-test are equal across the
two conditions. This analysis of variance assumes that they are. As the significance level of .951
is greater than .05, the matrices are similar and this assumption is met.

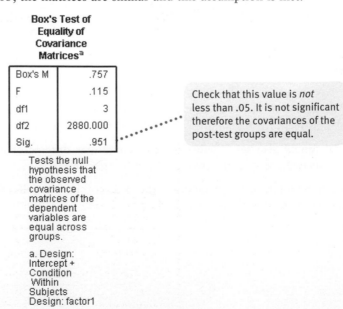

**Box's Test of
Equality of
Covariance
Matrices[a]**

Box's M	.757
F	.115
df1	3
df2	2880.000
Sig.	.951

Tests the null
hypothesis that
the observed
covariance
matrices of the
dependent
variables are
equal across
groups.

a. Design:
Intercept +
Condition
Within
Subjects
Design: factor1

Check that this value is *not*
less than .05. It is not significant
therefore the covariances of the
post-test groups are equal.

The ninth table shows whether the error variance of the two variables is similar across the two conditions. A significance level of more than .05 indicates that these variances are similar.

Levene's Test of Equality of Error Variances[a]

	F	df1	df2	Sig.
Pretest	.308	1	4	.609
Posttest	.000	1	4	1.000

Check these for significance. Significance means that the error variances are significantly different for the two or more conditions – either for the pre-test or the post-test.

Tests the null hypothesis that the error variance of the dependent variable is equal across groups.

a. Design: Intercept + Condition
Within Subjects Design: factor1

The eighth table contains information for the *F*-test. The *F*-test that is of particular interest to us is that for the interaction between the within-subjects and between-subjects factor (factor1 * Condition). This *F*-ratio is 7.682 and has a probability value of .05. In other words, this interaction is just significant. If we look at the means for the four groups we can see that while the mean for the control condition increases little from pre-test (5.00) to post-test (6.00), the mean for the experimental condition shows a larger increase from pre-test (5.67) to post-test (11.00).

Tests of Within-Subjects Contrasts

Measure:MEASURE_1

Source	factor1	Type III Sum of Squares	df	Mean Square	F	Sig.	Partial Eta Squared
factor1	Linear	30.083	1	30.083	16.409	.015	.804
factor1 * Condition	Linear	14.083	1	14.083	7.682	.050	.658
Error(factor1)	Linear	7.333	4	1.833			

This indicates a just significant interaction between pre-test/post-test and experimental condition. In other words, there are differences between the cells which cannot be explained by the pre-existing differences between the groups of participants or simply by changes in all conditions between the pretest and the posttest. Take a look at Table 25.3 which will help clarify the trends in the means.

These are the partial etas squared.

- To determine whether these increases were statistically significant we could run a related *t*-test between the pre- and post-test scores for the two conditions separately (with Bonferroni adjustment for the number of comparisons carried out).

- We could also see whether the two conditions differed at pre-test and at post-test with an unrelated *t*-test for the two test periods separately.

The graph on the facing page shows the means of the four cells which may help you to grasp more quickly the relationships between them.

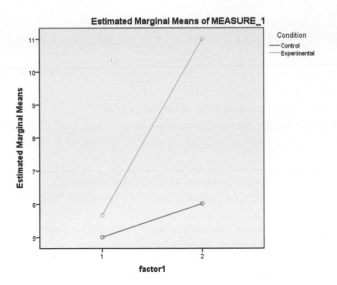

Estimated Marginal Means of MEASURE_1

REPORTING THE OUTPUT

● We could report the results of the output as follows:

> The interaction between the two conditions and the change over time was statistically significant, $F(1, 4) = 7.68$, $p = .05$, partial $\eta^2 = .66$. While the pre-test means did not differ significantly, $t(4) = 0.76$, two-tailed $p = .492$, the post-test mean for the experimental condition ($M = 11.00$, $SD = 1.00$) was significantly higher, $t(4) = 6.12$, two-tailed $p = .004$, than that for the control condition ($M = 6.00$, $SD = 1.00$). The increase from pre-test ($M = 5.67$, $SD = 1.15$) to post-test ($M = 11.00$, $SD = 1.00$) was significant for the experimental condition, $t(2) = 4.44$, two-tailed $p = .047$, but not for the control condition, $t(2) = 1.00$, two-tailed $p = .423$.

● An analysis of variance table for this analysis is presented in Table 25.2.
● It is also useful to include a table of means (M) and standard deviations (SD) as shown in Table 25.3.

Table 25.2	ANOVA summary table for a two-way mixed design			
Source of variance	**Sums of squares**	**Degrees of freedom**	**Mean square**	**F-ratio**
Between subjects factor	24.08	1	24.08	72.25*
Between subjects error	1.33	1	1.33	
Within subjects factor	30.08	1	30.08	16.41*
Within subjects error	7.33	4	1.83	
Interaction	14.08	1	14.08	7.68*

*Significant at .05 level.

Table 25.3	Means and standard deviations of the pre- and post-tests for the control and experimental conditions			
	Pre-test		**Post-test**	
Conditions	*M*	*SD*	*M*	*SD*
Control	5.00	1.00	6.00	1.00
Experimental	5.67	1.15	11.00	1.00

Summary of SPSS Statistics steps for mixed ANOVA

Data

- Name the variables in 'Variable View' of the 'Data Editor'.
- Enter the data under the appropriate variable names in 'Data View' of the 'Data Editor'.

Analysis

- Select 'Analyze', 'General Linear Model' and 'Repeated Measures …'.
- Enter number of conditions for related independent variable in 'Number of levels:' box. Select 'Add' and 'Define'.
- Move the related or repeated measures variables to the 'Within-Subjects Variables:' box.
- Move the unrelated factor to the 'Between-Subjects Variables:' box.
- Select 'Plots …'. Move related variable to 'Horizontal Axis:' box and unrelated variable to 'Separate Lines:' box. Select 'Add' and 'Continue'.
- Select 'Options', 'Descriptive statistics', 'Estimates of effect size', 'Homogeneity tests', 'Continue' and 'OK'.

Output

- The effect of most interest in a pre-test–post-test design is the interaction between the related and the unrelated variable.
- Check whether this interaction is significant in the 'Tests of Within-Subjects Contrasts' table by seeing if the Sig(nificance) is .05 or less.
- If it is significant, conduct further tests to determine which means differ significantly.
- In a repeated measures design where the order of conditions is controlled, it is important to determine if there is a carryover effect.

For further resources including data sets and questions, please refer to the website accompanying this book.

CHAPTER 26

Analysis of covariance (ANCOVA)

Overview

- The analysis of covariance (ANCOVA) is a variant of ANOVA. ANCOVA allows the researcher to control or adjust for variables which correlate with the dependent variable before comparing the means on the dependent variable. These variables are known as covariates of the dependent variable.

- To the extent that the levels of the covariates are different for the different research conditions, unless you adjust your dependent variable for the covariates you will confuse the effects of your independent variables with the influence of the pre-existing differences between the conditions caused by different levels of the covariates.

- By controlling for the covariates, essentially you are taking their effect away from your scores on the dependent variable. Thus having adjusted for the covariates, the remaining variation between conditions cannot be due to the covariates.

- One common use of ANCOVA is in pre-test/post-test designs. Assume that the pre-test suggests that the different conditions of the experiment have different means prior to testing (e.g. the experimental and control groups are different), ANCOVA may be used to adjust for these pre-test differences.

26.1 What is analysis of covariance (ANCOVA)?

Analysis of covariance (ANCOVA) basically allows the researcher to control (adjust for) variables which are not part of the main ANOVA design but may be correlated with both the independent and the dependent variables and so produce a spurious relationship in the ANOVA analysis. In randomised experiments, it is assumed that random assignment of participants to various conditions of the study equates the conditions in terms of all influences other than the independent variables. Of course, randomisation only works in the long run – it may not fully equate the participants in each of the conditions prior to the experimental treatment.

Furthermore, ANCOVA can also be used to analyse non-experimental studies. In this case, it is perfectly possible that participants in the different conditions of the experiment are different in some respect related to the dependent variable. ANCOVA allows you to explore possible influential variables which are not part of the basic ANOVA design.

Table 26.1	Data for a study of the effects of hormones (analysis of covariance)					
Group 1 **Hormone 1**		**Group 2** **Hormone 2**		**Group 3** **Placebo control**	
Pre	Post	Pre	Post	Pre	Post
5	9	3	4	2	3
4	12	2	2	3	6
6	8	1	5	2	3

Table 26.1 contains data which can be used to illustrate the use of ANCOVA. The study is the one which we used in Chapter 21 though it has been extended. We have three groups of participants who have been given one of three hormone treatments (Hormone 1, Hormone 2, or Placebo control). The dependent variable is a measure of depression. The study has been changed such that it now includes a pre-test measure of depression as well as a post-treatment measure. Ideally, the pre-test measures should have the same mean irrespective of the condition. But you will notice that the mean pre-test score for the Hormone 1 group is higher than for the other two groups. It is possible that the higher post-test scores of the Hormone 1 group is simply a consequence of their starting with higher levels of depression and not an effect of the hormone treatment.

It is possible to compute an ANOVA for these data but treating the pre-test measure of depression as a covariate. Two important things can be achieved:

- ANCOVA on SPSS Statistics will give you the means on the dependent variable adjusted for the covariate. In other words, it will equate all the participants on the covariate and make adjustments to the scores of the dependent variable to reflect this equality.

- ANCOVA adjusts the error variance to take into account the tendency for individual differences to affect the dependent variable. This gives a new significance level which can be more statistically significant, though, of course, this does not have to be so.

Figure 26.1 highlights the main steps in carrying out an analysis of covariance.

26.2 When to use ANCOVA

The analysis of covariance needs to be used with restraint. It is undesirable to have more than a few covariates for example. These covariates should be selected because it is known that they correlate with the dependent variable – if not then there is no point in using it. Covariates should correlate poorly with each other to be effective in ANCOVA.

26.3 When not to use ANCOVA

There is a requirement that the relationship between the covariate and the dependent variable should be the same throughout the data. It should not be the case that the regression slope is different in the different conditions of the study because ANCOVA works with the average slope which may not be appropriate within a particular condition if the regression is not consistent. SPSS offers a test of the homogeneity of regression. This simply means that the regression slope is constant for each condition of the independent variable.

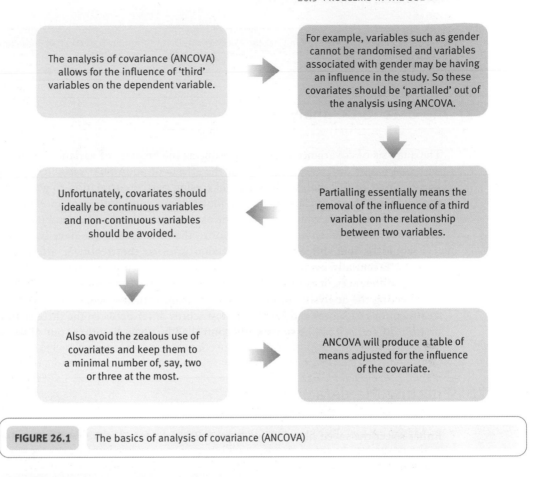

FIGURE 26.1   The basics of analysis of covariance (ANCOVA)

Covariates which are unreliable can cause difficulty because ANCOVA assumes that all covariates are reliable. Of course, few psychological variables are reliable in the same sense that, say, a measure of age is. Estimates of the adjusted mean may be over-inflated if covariates are unreliable and it is likely that the test of significance is reduced in power.

26.4   Data requirements for ANCOVA

Generally any ANOVA model can be used as an ANCOVA. The covariates in ANCOVA should have a clear *linear* relationship with the dependent variable otherwise ANCOVA will not be helpful. Curvilinear relationships are inappropriate because of the way in which ANCOVA works. Avoid the use of non-continuous variables such as marital status as covariates in ANCOVA. These could, however, potentially be used as an additional independent variable.

26.5   Problems in the use of ANCOVA

Conceptually be very wary when speaking of causality in relation to the analysis of covariance. Controlling for a covariate has no strong implication in relation to the issue of causality.

In general, it is fairly difficult to meet in full the requirements of data for ANCOVA which should encourage extra caution where the results of an ANCOVA are marginal in terms of significance.

> You can find out more about mixed ANCOVA in Chapter 25 of Howitt, D. and Cramer, D. (2011). *Introduction to Statistics in Psychology*, 5th edition. Harlow: Pearson.

26.6   The data to be analysed

The analysis of covariance is much the same as the analysis of variance dealt with elsewhere but with one major difference. This is that the effects of additional variables (covariates) are taken away as part of the analysis. It is a bit like using partial correlation to get rid of the effects of a third variable on a correlation. We will illustrate the computation of an ANCOVA with the data shown in Table 26.1, which are the same as those presented in Tables 21.1 and 24.1 except that depression scores taken immediately prior to the three treatments have been included.

It could be that differences in depression prior to the treatment affect the outcome of the analysis. Essentially by adjusting the scores on the dependent variable to 'get rid' of these pre-existing differences, it is possible to disregard the possibility that these pre-existing differences are affecting the analysis. So, if (a) the pre-treatment or test scores are correlated with the post-treatment or test scores and (b) the pre-test scores differ between the three treatments, then these pre-test differences can be statistically controlled by covarying them out of the analysis.

26.7   Entering the data

Either select the saved file used in Chapters 21 and 24 and make the changes below or enter the data as follows:

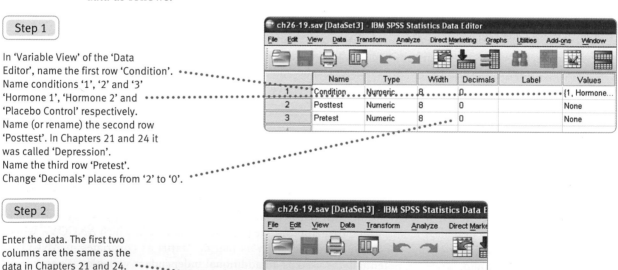

Step 1

In 'Variable View' of the 'Data Editor', name the first row 'Condition'. Name conditions '1', '2' and '3' 'Hormone 1', 'Hormone 2' and 'Placebo Control' respectively. Name (or rename) the second row 'Posttest'. In Chapters 21 and 24 it was called 'Depression'. Name the third row 'Pretest'. Change 'Decimals' places from '2' to '0'.

Step 2

Enter the data. The first two columns are the same as the data in Chapters 21 and 24. The data for the third column are new.

26.8   One-way ANCOVA

Select 'Analyze', 'General
Linear Model' and
'Univariate...'.

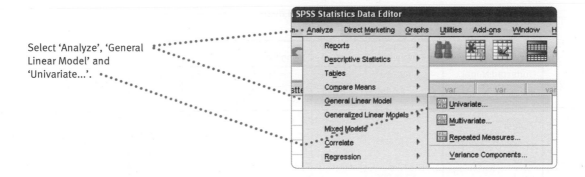

26.9   Testing that the slope of the regression line within cells is similar

Step 1

Select 'Posttest' and the
➡ button beside the
'Dependent Variable:'
box to put it there.
Select 'Condition' and
the ➡ button beside the
'Fixed Factor(s):' box to
put it there.
Select 'Pretest' and
the ➡ button beside the
'Covariate(s):' box to put
it there.
Select 'Model...'.

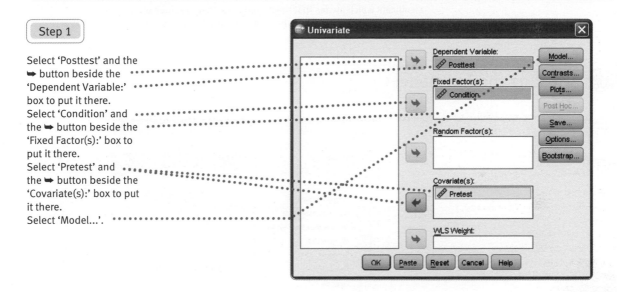

Step 2

Select 'Custom'.
Select 'Condition' and
the ➡ button to put it
in the 'Model:' box.
Do the same for 'Pretest'.
Select both variables and
the ➡ button to put the
interaction between them
in the 'Model:' box.
Select 'Continue'.
Select 'OK' in the previous box.

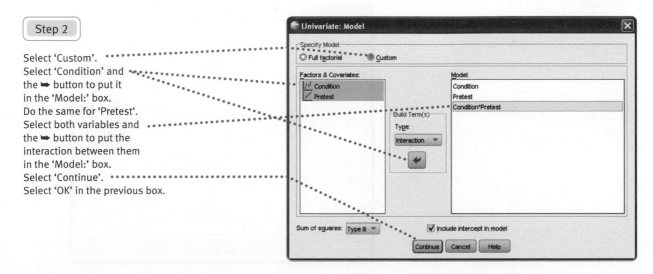

26.10 Interpreting the output

Tests of Between-Subjects Effects

Dependent Variable:Posttest

Source	Type III Sum of Squares	df	Mean Square	F	Sig.
Corrected Model	82.722[a]	5	16.544	10.269	.042
Intercept	19.230	1	19.230	11.936	.041
Condition	24.333	2	12.167	7.552	.067
Pretest	.100	1	.100	.062	.819
Condition * Pretest	12.571	2	6.286	3.901	.146
Error	4.833	3	1.611		
Total	388.000	9			
Corrected Total	87.556	8			

a. R Squared = .945 (Adjusted R Squared = .853)

The interaction between the 'Condition' and the covariate is not significant which means that the prerequisite that the slope of the regression line within the three conditions is similar is met.

26.11 Testing the full model

Step 1

Select 'Analyze', 'General Linear Model' and 'Univariate...'.

Step 2

Select 'Model...'.

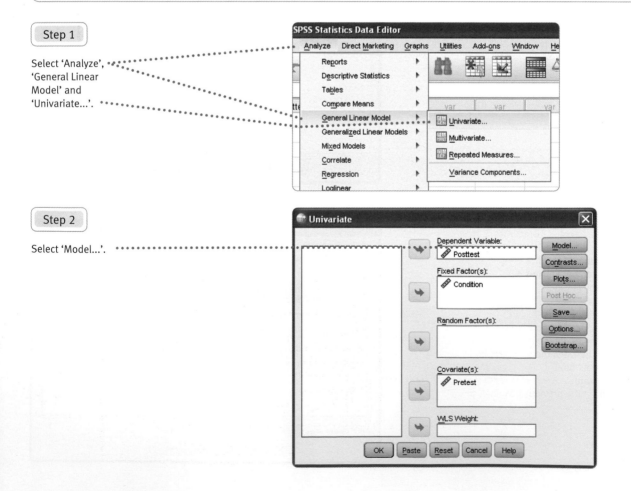

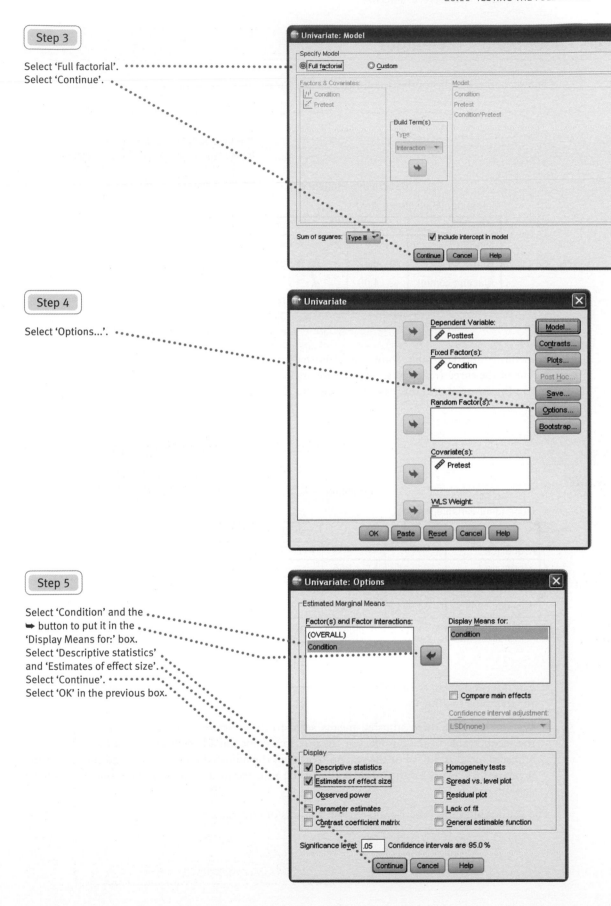

Step 3

Select 'Full factorial'.
Select 'Continue'.

Univariate: Model

Specify Model
● Full factorial   ○ Custom

Factors & Covariates:
Condition
Pretest

Build Term(s)
Type:
Interaction

Model:
Condition
Pretest
Condition*Pretest

Sum of squares: Type III   ☑ Include intercept in model

Continue   Cancel   Help

Step 4

Select 'Options...'.

Univariate

Dependent Variable:
Posttest

Fixed Factor(s):
Condition

Random Factor(s):

Covariate(s):
Pretest

WLS Weight:

Model...
Contrasts...
Plots...
Post Hoc...
Save...
Options...
Bootstrap...

OK   Paste   Reset   Cancel   Help

Step 5

Select 'Condition' and the
➡ button to put it in the
'Display Means for:' box.
Select 'Descriptive statistics'
and 'Estimates of effect size'.
Select 'Continue'.
Select 'OK' in the previous box.

Univariate: Options

Estimated Marginal Means

Factor(s) and Factor Interactions:
(OVERALL)
Condition

Display Means for:
Condition

☐ Compare main effects

Confidence interval adjustment:
LSD(none)

Display
☑ Descriptive statistics   ☐ Homogeneity tests
☑ Estimates of effect size   ☐ Spread vs. level plot
☐ Observed power   ☐ Residual plot
● Parameter estimates   ☐ Lack of fit
☐ Contrast coefficient matrix   ☐ General estimable function

Significance level: .05   Confidence intervals are 95.0 %

Continue   Cancel   Help

26.12 | Interpreting the output

The second table shows the unadjusted means for the three conditions.

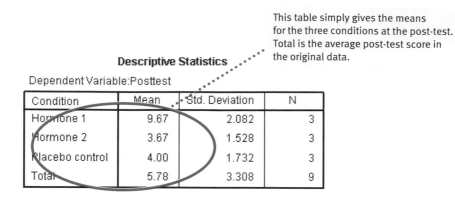

This table simply gives the means for the three conditions at the post-test. Total is the average post-test score in the original data.

Descriptive Statistics

Dependent Variable:Posttest

Condition	Mean	Std. Deviation	N
Hormone 1	9.67	2.082	3
Hormone 2	3.67	1.528	3
Placebo control	4.00	1.732	3
Total	5.78	3.308	9

The fourth and last table shows the adjusted means for these three conditions. The adjusted means of the three treatments are what the means are when all groups are adjusted to be identical on the covariate (in this case pre-treatment depression scores).

The means for the post-test given in this output table have been adjusted for the effect of the covariate on the three conditions. The effect of the covariate has effectively been removed from the data.

Condition

Dependent Variable:Posttest

Condition	Mean	Std. Error	95% Confidence Interval	
			Lower Bound	Upper Bound
Hormone 1	10.881[a]	1.955	5.856	15.906
Hormone 2	2.952[a]	1.443	-.756	6.661
Placebo control	3.500[a]	1.269	.237	6.763

a. Covariates appearing in the model are evaluated at the following values: Pretest = 3.11.

- The adjusted mean is 10.881 for the first treatment, 2.952 for the second treatment and 3.500 for the third treatment.

- We can see that these adjusted means seem to differ from the unadjusted means shown in the second table of the output. For the first treatment the adjusted mean is 10.88 and the unadjusted mean is 9.67. For the second treatment the adjusted mean is 2.95 and the unadjusted mean is 3.67, while for the third treatment the adjusted mean is 3.50 and the unadjusted mean is 4.00.

The third table shows the F-ratio for the analysis of covariance.

Tests of Between-Subjects Effects

Dependent Variable:Posttest

Source	Type III Sum of Squares	df	Mean Square	F	Sig.	Partial Eta Squared
Corrected Model	70.151[a]	3	23.384	6.718	.033	.801
Intercept	27.684	1	27.684	7.953	.037	.614
Pretest	1.929	1	1.929	.554	.490	.100
Condition	26.425	2	13.213	3.796	.099	.603
Error	17.405	5	3.481			
Total	388.000	9				
Corrected Total	87.556	8				

a. R Squared = .801 (Adjusted R Squared = .682)

Following removal of the effects of the covariate, there is not a significant difference between the means of the three conditions as the significance is .099 which is not statistically significant.

Partial eta squared is .603.

- The F-ratio for the main effect is 3.796 (13.213/3.481 = 3.796).
- The probability of this F-ratio is .099. In other words, it is greater than the .05 critical value and so is not statistically significant.

REPORTING THE OUTPUT

- We could report the results of the output as follows:

    A one-way ANCOVA showed that when pre-test depression was covaried out, the main effect of treatment on post-test depression was not significant, $F(2, 5) = 3.80$, $p = .099$, partial $\eta^2 = .60$.

    You would normally also report the changes to the means once the covariate has been removed.
- If necessary, we could give an ANCOVA summary table as in Table 26.2.

Table 26.2	ANCOVA summary table for effects of treatments on depression controlling for pre-treatment depression			
Source of variance	**Sums of squares**	**Degrees of freedom**	**Mean square**	**F-ratio**
Covariate (pre-treatment depression scores)	1.93	1	1.93	0.55
Main effect (treatment)	26.43	2	13.21	3.80*
Residual error	17.41	5	3.48	

*Significant at .05 level.

Summary of SPSS Statistics steps for ANCOVA

Data

- Name the variables in 'Variable View' of the 'Data Editor'.
- Enter the data under the appropriate variable names in 'Data View' of the 'Data Editor'.

Analysis

- If the potential covariate is reasonably strongly related to the dependent variable, proceed with ANCOVA.
- Select 'Analyze', 'General Linear Model' and 'Univariate …'.
- Move dependent variable to 'Dependent Variable:' box, independent variable to 'Fixed Factor:' box and covariate variable to 'Covariate:' box.
- To check that the slope of the regression line is similar in each condition, select 'Model:' and 'Custom:'. Move the independent variable, the covariate and the interaction between the independent variable and covariate box to the box on the right.
- If the interaction between the independent variable and the covariate is not significant by having a Sig(nificance) of greater than .05, proceed with the ANCOVA.
- Select 'Analyze', 'General Linear Model', 'Univariate …', 'Model', 'Full factorial' and 'Continue'.
- Select 'Options …', 'Descriptive statistics' and 'Estimates of effect size'. Move the independent variable to the 'Display Means for:' box.

Output

- Check if the independent variable is significant by seeing if its Sig(nificance) is .05 or less.
- If it is and there are more than two conditions, conduct further tests to determine which adjusted means differ significantly.

For further resources including data sets and questions, please refer to the website accompanying this book.

CHAPTER 27

Multivariate analysis of variance (MANOVA)

- The multivariate analysis of variance (MANOVA) is a variant of ANOVA which is applied when you wish to analyse several dependent variables at the same time. These dependent variables should be *conceptually* related to the hypothesis and each other. They need to be scores rather than categorical variables.

- Essentially MANOVA computes a composite variable based on several dependent variables and then tests to see whether the means of the groups on the combined dependent variables differ significantly.

- MANOVA helps deal with the problems created by multiple significance tests being applied to the same data.

- If MANOVA is statistically significant then it is appropriate to test the significance of the individual dependent variables using ANOVAs and also to explore the combination of variables used in the MANOVA using discriminant function analysis (see Chapter 28).

27.1 What is multivariate analysis of variance (MANOVA)?

Multivariate analysis of variance (MANOVA) is a test of the statistical significance of differences between several groups. It is different from ANOVA in that the means which are compared are a composite variable based on combining several score variables statistically. It is most commonly used by researchers using randomised experimental designs but, like ANOVA, not exclusively so. Read Chapter 21 on one-way ANOVA if you haven't already done so as this forms the starting point for the present chapter.

Table 27.1 summarises a study for which MANOVA may be the preferred test of significance. It is taken from *ISP*, Table 26.2. The study is of the effectiveness of different team building methods in sport. There are three groups (or conditions) – team building with a sports psychologist, team building with a sports coach, and a no team building control. The hypothesis is that team building has a positive influence on team cohesiveness. Three dependent variables were measured

Table 27.1	Data suitable for a MANOVA analysis							
Group (independent variable)								
Team building with sports psychologist			**Team building with sports coach**			**No team building controls**		
Dependent variables			**Dependent variables**			**Dependent variables**		
Like*	Gym	Game	Like	Gym	Game	Like	Gym	Game
9**	12	14	4	6	15	9	6	10
5	9	14	5	4	12	1	2	5
8	11	12	4	9	15	6	10	12
4	6	5	3	8	8	2	5	6
9	12	3	4	9	9	3	6	7
9	11	14	5	3	8	4	7	8
6	13	14	2	8	12	1	6	13
6	11	18	6	9	11	4	9	12
8	11	22	4	7	15	3	8	15
8	13	22	4	8	28	3	2	14
9	15	18	5	7	10	2	8	11
7	12	18	4	9	9	6	9	10
8	10	13	5	18	18	3	8	13
6	11	22	7	12	24	6	14	22

*Like = Difference between ratings of most and least liked team members, Gym = Number of gym sessions voluntarily attended, and Game = Number of games played.
**The scores are the scores on the three dependent variables.

following this experimental manipulation. These are (1) the difference between ratings of most and least liked team members, (2) the number of gym sessions voluntarily attended, and (3) the number of games played. The researchers considered that these were measures which would indicate the success of the team building sessions. There are other possibilities too, of course, and it is difficult to choose which one of the three should be used. After all, they are all measures of different aspects which might be expected of successful team building, so why not use them all?

One could analyse the data in Table 27.1 by (a) using separate ANOVA analyses for each of the three different dependent variables measured or (b) by combining the dependent variables in some way to give a composite variable mean score which is then analysed in a single analysis. The latter is the approach used in MANOVA. In fact, in the MANOVA procedure in SPSS Statistics both these approaches are taken though the ANOVAs are not considered unless the MANOVA is statistically significant. Why is this?

Researchers do not use MANOVA simply because it uses a composite measure of several dependent variables in combination. It has the major advantage that it deals with the problem of multiple tests of significance. If one carries out multiple tests of significance on any data (such as several ANOVAs in this case) there is an increased likelihood that at least one of the tests of significance will be statistically significant. All things being equal, the more tests of significance that are performed on any data the more likely it is that at least one test will be statistically significant by chance in any data analysis. By using a single combined dependent variable, MANOVA avoids the problem of multiple tests of significance.

Things are a little more complicated than this in practice:

● If the MANOVA is statistically significant then it is legitimate to examine the results of ANOVAs performed on each of the several dependent variables which were combined for the

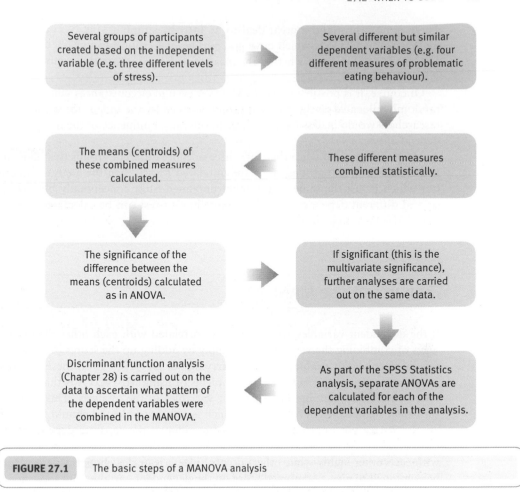

Several groups of participants created based on the independent variable (e.g. three different levels of stress).	Several different but similar dependent variables (e.g. four different measures of problematic eating behaviour).

The means (centroids) of these combined measures calculated.	These different measures combined statistically.

The significance of the difference between the means (centroids) calculated as in ANOVA.	If significant (this is the multivariate significance), further analyses are carried out on the same data.

Discriminant function analysis (Chapter 28) is carried out on the data to ascertain what pattern of the dependent variables were combined in the MANOVA.	As part of the SPSS Statistics analysis, separate ANOVAs are calculated for each of the dependent variables in the analysis.

FIGURE 27.1   The basic steps of a MANOVA analysis

MANOVA. (You will probably find that there is more MANOVA output on SPSS which refers to these individual ANOVAs than refers to MANOVA.)

● You will notice that so far we have said nothing about how the dependent variables are combined in MANOVA. Just how are the dependent variables combined to give the combined variable in MANOVA? Without this information it is not possible to interpret the meaning of the MANOVA precisely. The complication is that this requires a separate and distinct analysis using discriminant function analysis, which is described in the next chapter (Chapter 28). So, if you have a significant MANOVA, then you will probably wish to do a discriminant function analysis. (The separate ANOVAs do not tell you anything about the combined dependent variable for the simple reason that the several dependent variables will correlate to different extents with each other.)

All of this means that a MANOVA analysis involves several steps. These steps and some other essential features of a MANOVA analysis are summarised in Figure 27.1. This is a flow-diagram for the important elements of a data analysis using MANOVA.

27.2   When to use MANOVA

MANOVA can be used when the researcher has employed a randomised experimental design in which there are several groups of participants but where several different variables have been measured to reflect different aspects of the dependent variable described in the hypothesis.

There is no point (and a great deal to be lost) by simply trying to combine different variables together in MANOVA which have no *conceptual* relationship to each other and the hypothesis. MANOVA will indicate whether the means for the groups on the combined dependent variables differ significantly.

Of course, it is possible to use MANOVA even in circumstances where participants where not randomly allocated to the different groups formed by the independent variable. In such a study, researchers would be less likely to refer to the causal influence of the independent variable on the dependent variable.

MANOVA is used when there are three or more different groups (just like one-way ANOVA). If there are just two groups (as in the unrelated *t*-test), then the analysis becomes Hotelling's *t*-test which is just the same as any other unrelated *t*-test conceptually but uses a combination of several different dependent variables. Hotelling's *t*-test can be calculated simply by carrying out the MANOVA procedure on SPSS.

27.3   When not to use MANOVA

If the dependent variables are very highly correlated with each other then there may be simpler ways of combining the variables, e.g. simply by adding up the scores of each participant over the several dependent variables and then carrying out an ANOVA on these summated scores. What is the advantage of this? One of the problems of using MANOVA is that degrees of freedom are lost owing to the number of dependent variables in the analysis. This, effectively, reduces the statistical significance of MANOVA which is not desirable and should be avoided. So if there is an alternative but unproblematic way of combining the separate dependent variables then this can be employed to good effect. This will not work where some dependent variables correlate with each other highly while others do not since it is unclear how the dependent variables should be combined to give a combined score on the dependent variable. But these are the circumstances in which MANOVA is likely to be at its most effective and so would be the preferred approach.

We would not recommend the use of MANOVA if you have many dependent variables to combine. For example, if one has a lengthy questionnaire (say, ten or more questions) then it would not be helpful to treat answers to each question as a separate dependent variable as part of MANOVA as there are more informative ways of combining items to yield a small number of combined variables. It would be better, for example, to factor analyse (Chapter 30) the set of dependent variables. In this way, the structure of the items will be clarified. It is likely that a very small number of significant factors will emerge. Individuals may be given a score on each factor using the factor score procedure of SPSS which puts a score on each factor at the end of the SPSS spreadsheet (see page 287). These factor scores may be used as variables in ANOVA.

27.4   Data requirements for MANOVA

The independent variable (grouping variable) consists of three or more different categories for MANOVA (do not use score variables as the independent variable unless they have been recoded into a small number of categories).

Always try to ensure that the sample sizes of each group in the MANOVA are the same. This helps avoid problems in the analysis if assumptions related to something called the equality of the covariance matrices are not met.

The two sample equivalent to the *t*-test is known as the Hotelling's *t*-test but it would be calculated on SPSS using the MANOVA procedure. There need to be several dependent variables which take the form of numerical scores.

27.5   Problems in the use of MANOVA

It is possible to run before you can walk with MANOVA. Many of the features of SPSS MANOVA output require understanding of ANOVA – *post hoc* multiple comparisons tests and the like are discussed in the chapters on ANOVA and are not dealt with in this chapter for reasons of clarity. So it is best to have a grounding in ANOVA before moving on to MANOVA. If you need to cut corners then try a crash-course in ANOVA by studying Chapter 21 on one-way ANOVA and then trying out what you have learnt on your data by using just one dependent variable – also study *post hoc* tests in Chapter 24.

> You can find out more about MANOVA in Chapter 26 of Howitt, D. and Cramer, D. (2011). *Introduction to Statistics in Psychology*, 5th edition. Harlow: Pearson.

27.6   The data to be analysed

The data in Table 27.1 is used as the example for SPSS analysis for MANOVA.

27.7   Entering the data

Step 1

In 'Variable View' of the 'Data Editor', name the first row 'group'. Name conditions '1', '2' and '3' 'teampsychologist', 'teamcoach' and 'control' respectively.
Name the second, third and fourth rows 'leastliked', 'gymsession' and 'gamesplayed' respectively.
Change 'Decimals' places from '2' to '0'.

	Name	Type	Width	Decimals	Label	Values
1	group	Numeric	8	0		(1, teampsy...
2	leastliked	Numeric	8	0	difference betw...	None
3	gymsessions	Numeric	8	0	number of gym...	None
4	gamesplayed	Numeric	8	0	number of gam...	None

Step 2

Enter the data.
'Group' is the independent variable.
These are the dependent variables.

	group	leastliked	gymsessions	gamesplayed
1	1	9	12	14
2	2	4	6	15
3	3	9	6	10
4	1	5	9	14
5	2	5	4	12
6	3	1	2	5
7	1	8	11	12
8	2	4	9	15
9	3	6	10	12
10	1	4	6	5
11	2	3	8	8
12	3	2	5	6
13	1	9	12	3

27.8   MANOVA

Step 1

Select 'Analyze',
'General Linear Model' and
'Multivariate...'.

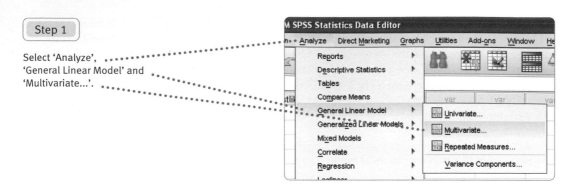

Step 2

Move the dependent variables from the large
left-hand panel to the small 'Dependent
Variables:' box.
Move the independent variable from the left to
the 'Fixed Factor (s):' box.
Select 'options'.

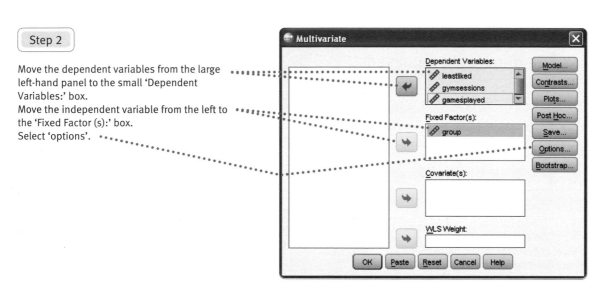

Step 3

Move 'group' over to the right-hand 'Display
Means for:' box.
Select 'Descriptive statistics' and
'Estimates of effect size'.
Select 'Continue'.
Select 'OK' from the
previous box, which reappears.

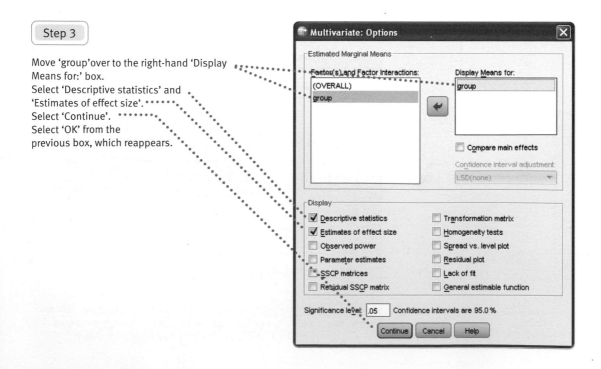

27.9 | Interpreting the output

The interpretation of the SPSS output from MANOVA is complicated when options are selected but there is only a small core of the SPSS output which contains the results of the MANOVA analysis. If this is significant then one would normally go on to examine other aspects of the output which deal with the individual dependent variables – and probably go on to do a discriminant function analysis as detailed in Chapter 28.

The SPSS output for MANOVA contains the results of separate ANOVA analyses for each separate dependent variable. These are considered when the overall MANOVA is significant. However, the individual ANOVAs actually do not tell us about the combined dependent variable since we need to know the contribution of each dependent variable to the combined scores before we know this. This is why it is usually recommended that a discriminant function analysis is carried out in addition.

Between-Subjects Factors

		Value Label	N
group	1	teampsychologist	14
	2	teamcoach	14
	3	control	14

The first table merely identifies the groups based on the independent variable and gives the sample size involved. It is good that equal sample sizes were used as this helps avoid technical problems.

Descriptive Statistics

	group	Mean	Std. Deviation	N
difference between ratings of most and least liked team members	teampsychologist	7.29	1.637	14
	teamcoach	4.43	1.222	14
	control	3.79	2.259	14
	Total	5.17	2.305	42
number of gym sessions voluntarily attended	teampsychologist	11.21	2.082	14
	teamcoach	8.36	3.565	14
	control	7.14	3.110	14
	Total	8.90	3.384	42
number of games played	teampsychologist	14.93	5.784	14
	teamcoach	13.86	6.011	14
	control	11.29	4.322	14
	Total	13.36	5.512	42

The second table gives the mean scores on the three dependent variables for each grouping variable.

Multivariate Tests[c]

Effect		Value	F	Hypothesis df	Error df	Sig.	Partial Eta Squared
Intercept	Pillai's Trace	.937	184.710[a]	3.000	37.000	.000	.937
	Wilks' Lambda	.063	184.710[a]	3.000	37.000	.000	.937
	Hotelling's Trace	14.976	184.710[a]	3.000	37.000	.000	.937
	Roy's Largest Root	14.976	184.710[a]	3.000	37.000	.000	.937
group	Pillai's Trace	.490	4.109	6.000	76.000	.001	.245
	Wilks' Lambda	.522	4.733[a]	6.000	74.000	.000	.277
	Hotelling's Trace	.892	5.349	6.000	72.000	.000	.308
	Roy's Largest Root	.865	10.953[b]	3.000	38.000	.000	.464

a. Exact statistic
b. The statistic is an upper bound on F that yields a lower bound on the significance level.
c. Design: Intercept + group

The third table contains the MANOVA analysis. There are several Multivariate tests which are very similar. Use the Pillai's trace as using all four is cumbersome. The MANOVA is significant at the .001 level. Thus the group means on the composite dependent variable differ significantly at the .001 level. Thus the hypothesis is supported.

Tests of Between-Subjects Effects

Source	Dependent Variable	Type III Sum of Squares	df	Mean Square	F	Sig.	Partial Eta Squared
Corrected Model	difference between ratings of most and least liked team members	97.190[a]	2	48.595	15.709	.000	.446
	number of gymn sessions volutarily attended	122.333[b]	2	61.167	6.869	.003	.260
	number of games played	98.143[c]	2	49.071	1.668	.202	.079
Intercept	difference between ratings of most and least liked team members	1121.167	1	1121.167	362.438	.000	.903
	number of gymn sessions volutarily attended	3330.381	1	3330.381	374.000	.000	.906
	number of games played	7493.357	1	7493.357	254.676	.000	.867
group	difference between ratings of most and least liked team members	97.190	2	48.595	15.709	.000	.446
	number of gymn sessions volutarily attended	122.333	2	61.167	6.869	.003	.260
	number of games played	98.143	2	49.071	1.668	.202	.079
Error	difference between ratings of most and least liked team members	120.643	39	3.093			
	number of gymn sessions volutarily attended	347.286	39	8.905			
	number of games played	1147.500	39	29.423			
Total	difference between ratings of most and least liked team members	1339.000	42				
	number of gymn sessions volutarily attended	3800.000	42				
	number of games played	8739.000	42				
Corrected Total	difference between ratings of most and least liked team members	217.833	41				
	number of gymn sessions volutarily attended	469.619	41				
	number of games played	1245.643	41				

a. R Squared = .446 (Adjusted R Squared = .418)
b. R Squared = .260 (Adjusted R Squared = .223)
c. R Squared = .079 (Adjusted R Squared = .032)

The fourth table gives the significance of the individual ANOVAs applied to the three dependent variables separately. As can be seen, two of the dependent variables differ significantly according to the group. However, the dependent variable 'number of games played' does not produce significant results.

Estimated Marginal Means

group

Dependent Variable	group	Mean	Std. Error	95% Confidence Interval	
				Lower Bound	Upper Bound
difference between ratings of most and least liked team members	teampsychologist	7.286	.470	6.335	8.237
	teamcoach	4.429	.470	3.478	5.379
	control	3.786	.470	2.835	4.737
number of gymn sessions volutarily attended	teampsychologist	11.214	.798	9.601	12.827
	teamcoach	8.357	.798	6.744	9.970
	control	7.143	.798	5.530	8.756
number of games played	teampsychologist	14.929	1.450	11.996	17.861
	teamcoach	13.857	1.450	10.925	16.789
	control	11.286	1.450	8.353	14.218

The fifth table shows the means of the three dependent variables for the three groups. Which means differ for each dependent variable needs to be determined with further SPSS analyses (see Chapter 20).

REPORTING THE OUTPUT

In circumstances in which the MANOVA is not significant then you could briefly note this as follows:

The hypothesis that team work training improves team training was tested using MANOVA. However, the null hypothesis was supported by the data, Pillai's trace $F() =$, ns, partial $\eta^2 =$.

In the example, the findings were significant and you might write:

MANOVA showed there was a significant multivariate effect of team building on the three dependent variables, Pillai's trace $F(6, 76) = 4.12$, $p < .01$, partial $\eta^2 = .25$. Each dependent variable was subjected to a further ANOVA in order to show whether this trend is the same for each of the separate dependent variables. For the measure of the difference between favourite and least favourite team member, an ANOVA showed there was an overall significant difference between the means, $F(2, 39) = 15.71$, $p = .001$, partial $\eta^2 = .45$. Similarly, the mean number of gym sessions attended was different according to group, $F(2, 39) = 61.17$, $p = .03$, partial $\eta^2 = .26$. However, the number of games played failed to reach statistical significance, $F(2, 39) = 49.07$, ns, partial $\eta^2 = .08$.

Note that had we selected *post hoc* tests during the analysis, it would be possible to specify exactly which groups differed significantly from each other on each dependent variable. We would recommend that you routinely do this. It is explained in Chapter 24.

Furthermore, if the findings for MANOVA are significant, you are likely to wish to carry out a discriminant function analysis as described in the next chapter.

Summary of SPSS Statistics steps for MANOVA

Data

● Name the variables in 'Variable View' of the 'Data Editor'.
● Enter the data under the appropriate variable names in 'Data View' of the 'Data Editor'.

Analysis

● Select 'Analyze', 'General Linear Model' and 'Multivariate ...'.
● Move the dependent variables to the 'Dependent Variables:' box and the independent variable(s) to the 'Independent Variable(s):' box.
● Select 'Options ...' and move the independent variable to the 'Display Means for:' box. Select 'Descriptive statistics', 'Estimates of effect size', 'Continue' and 'OK'.

Output

● Check in the Multivariate Tests table if Pillai's trace F for the independent variable is significant with a Sig(nificance) of .05 or less.
● If it is significant, check in the 'Tests of Between-Subjects Effects' table which of the dependent variables the independent variable has a significant effect on with a Sig(nificance) of .05 or less.
● If there are more than two groups use further tests to determine which means differ significantly from each other.

For further resources including data sets and questions, please refer to the website accompanying this book.

CHAPTER 28

Discriminant function analysis (for MANOVA)

Overview

- Discriminant function analysis has its major application in understanding better the meaning of a significant MANOVA (see Chapter 27). It can be used alone as a way of assessing whether a set of predictor (independent) score variables effectively distinguishes different categories or groups of participants. However, logistic regression (Chapters 37 and 38) does a far better job of this.

- A discriminant function is basically a computed variable which combines several score variables. It is computed in such a way that the differences between the groups of participants on the means (centroids) of the discriminant function are maximised.

- There may be several discriminant functions obtained in an analysis. The number depends on the characteristics of the data, especially the number of independent variables. Each discriminant function is uncorrelated with the others.

- Discriminant function analysis and MANOVA are mathematically similar. Discriminant function analysis can help understand the meaning of a significant MANOVA analysis since it identifies the combination of score variables which led to the significant MANOVA.

- Discriminant function analysis produces a classification table which indicates the accuracy of the group membership predictions based on the independent variables.

- Confusingly, the independent variables and dependent variables are reversed for discriminant function analysis compared with MANOVA. That is, the independent variable in MANOVA is the dependent variable in discriminant function analysis.

28.1 What is discriminant function analysis?

Discriminant function analysis is a way of assessing whether members of different groups can be identified on the basis of their scores on a set of variables. It is sometimes used as a final stage of MANOVA (see Chapter 27) where it indicates which of a set of scores had been used by MANOVA as the 'combined' dependent variable. In discriminant function analysis a mathematical

Table 28.1	Data suitable for a discriminant function analysis							
Group (dependent variable)								
Team building with sports psychologist			**Team building with sports coach**			**No team building controls**		
Independent variables			Independent variables			Independent variables		
Like*	Gym	Game	Like	Gym	Game	Like	Gym	Game
9**	12	14	4	6	15	9	6	10
5	9	14	5	4	12	1	2	5
8	11	12	4	9	15	6	10	12
4	6	5	3	8	8	2	5	6
9	12	3	4	9	9	3	6	7
9	11	14	5	3	8	4	7	8
6	13	14	2	8	12	1	6	13
6	11	18	6	9	11	4	9	12
8	11	22	4	7	15	3	8	15
8	13	221	4	8	28	3	2	14
9	15	18	5	7	10	2	8	11
7	12	18	4	9	9	6	9	10
8	10	13	5	18	18	3	8	13
6	11	22	7	12	24	6	14	22

*Like = Difference between ratings of most and least liked team members, Gym = Number of gym sessions voluntarily attended, and Game = Number of games played.
**The scores are the scores on the three independent variables.

function (or functions) is calculated based on weighted combination of variables. This function is calculated in such a way that the group differences on the function are maximised. This is probably best understood in terms of an example and, don't worry, all of the hard work of analysis is done by SPSS Statistics – you just need to understand what it all means.

Table 28.1 summarises a study for which MANOVA may be the preferred test of significance. It should be familiar to you as it is more or less the same as that used to illustrate MANOVA in Chapter 27. Eagle-eyed readers may notice that there is a difference. The grouping variable (team building) has been labelled the dependent variable whereas the score variables have been labelled the independent variables. This is exactly the reverse of how the variables were labelled for MANOVA. This reversal of labelling is the consequence of MANOVA being a test of differences between means and discriminant function analysis being a form of regression. Essentially, what we are trying to do when using discriminant function analysis is to use the scores on the three score variables (the difference between ratings of most and least liked team members, the number of gym sessions voluntarily attended and the number of games played) to discriminate participants in terms of the group that they belong to (the team building with a sports psychologist group or the team building with a sports coach group or the no team building control group).

A discriminant function is a weighted combination of variables. In this example, the discriminant function would be based on a weighted combination of the independent variables like Gym and Games that are the independent variables in the table. Just to illustrate the general idea, the formula for the discriminant function might include the following weights applied to scores on each variable:

> Choose a set of independent variables which differentiate between different groups of participants.

> What weighted combination of these independent variables best differentiates the different groups of participants? This is known as a discriminant function.

> The means for each group on the discriminant function are known as the centroids.

FIGURE 28.1   The steps of a discriminant function analysis

Discriminant function = $(2 \times \text{Like}) + (1 \times \text{Gym}) + (3 \times \text{Games})$

For example, 'Like' is weighted twice as much (2×) as 'Gym'. To calculate this discriminant function we would apply the above formula for the scores of every participant in the study. Then we can calculate the means of each of the groups of participants on the resulting discriminant function. A group mean on a discriminant function is called a centroid.

A discriminant function can be regarded as a derived variable based on a particular combination of other variables. Of course, the problem is that there are many different discriminant functions possible. Each would have its own pattern of weightings. So which discriminant function or pattern of weightings do we choose? Since we wish to discriminate between the groups of participants in our research we choose the discriminant function which has the maximum difference between the groups in terms of the group means (or centroids). There is only one discriminant function that meets this requirement and that is the one we use. Fortunately the SPSS calculation of the discriminant function analysis does all of this work of identifying this one discriminant function for us.

Actually, it is a little misleading to talk of a single discriminant function emerging from an analysis since there may be several discriminant functions depending on the data – and especially the number of variables. Discriminant functions are independent of each other – that is, they are not correlated. This ensures that the discriminant functions have the maximum possible power to differentiate between the groups of participants. Figure 28.1 outlines the main steps in a discriminant function analysis.

28.2   When to use discriminant function analysis

If you carry out a MANOVA and find that it is statistically significant (Chapter 27), a discriminant function analysis may be used to explore the nature of the weighted combination of variables which was calculated as part of this analysis. Mathematically MANOVA and discriminant function analysis have much in common. The discriminant functions identified in discriminant function analysis are the basis of the combination of variables used in MANOVA.

It is possible to use discriminant function analysis in any circumstance in which you wish to use a set of score variables to predict membership of groups of participants. It can thus be regarded as a form of regression in which score variables are used to predict membership of groups (as opposed to scores on a dependent variable). This use of discriminant function analysis is suggested in a number of popular SPSS texts but should be regarded as superseded by logistic regression (Chapters 38 and 39).

28.3   When not to use discriminant function analysis

As indicated above, discriminant function analysis has no advantages over the more powerful and flexible logistic regression methods (Chapters 38 and 39). It is built on a less adequate mathematical model than logistic regression and does not have the latter's flexibility. For example, logistic regression can be used with both score variables and nominal (category) variables as predictors. Binary logistic regression is capable of producing quite complex models of one's data. Basically, this is a case of no-contest and discriminant function analysis should be used only in support of MANOVA analyses. Why not use logistic regression in relation to MANOVA? Unfortunately this is not appropriate as the underlying statistical model is different in logistic regression whereas the statistical models underlying MANOVA and discriminant function analysis are essentially the same.

28.4   Data requirements for discriminant function analysis

Discriminant function analysis requires a set of score variables and a single nominal (category) variable which defines the groups used. Cases can only be in one of the groups.

There is no point in deriving the groups from a score variable such as by classifying high, medium and low scoring groups on the basis of that variable. Multiple regression would handle this situation better by using the grouping variable which has been measured as a score variable (the dependent variable) in its original form.

28.5   Problems in the use of discriminant function analysis

The main problem in using discriminant function analysis is its use of the concept of 'centroids', which is not a self-evident concept. However, a centroid is merely a group mean on a discriminant function so is easily understood as a mean of a particular combination of variables.

Another problem is that the meaning of a discriminant function lies in the variables which were used to derive the discriminant function. Some of the variables will be strongly related to the discriminant function and others will relate to it poorly. So the variables which correlate with (or load on) a discriminant function are the ones which help you to identify what the discriminant function is.

You can find out more about discriminant function analysis in Chapter 27 of Howitt, D. and Cramer, D. (2011). *Introduction to Statistics in Psychology*, 5th edition. Harlow: Pearson.

28.6   The data to be analysed

The data in Table 28.1 is used as the example for the SPSS analysis for discriminant function analysis.

28.7 Entering the data

As you will be using discriminant function analysis as an adjunct to MANOVA you will already have entered the data into SPSS. Section 27.7 describes this. If you have not entered or saved the data, carry out the following steps.

Step 1

In 'Variable View' of the 'Data Editor', name the first row 'group'.
Name conditions '1', '2' and '3' 'teampsychologist', 'teamcoach' and 'control' respectively.
Name the second, third and fourth rows 'leastliked', 'gymsession' and 'gamesplayed' respectively.
Change 'Decimals' places from '2' to '0'.

Step 2

Enter the data.
'Group' is the independent variable.
These are the dependent variables.

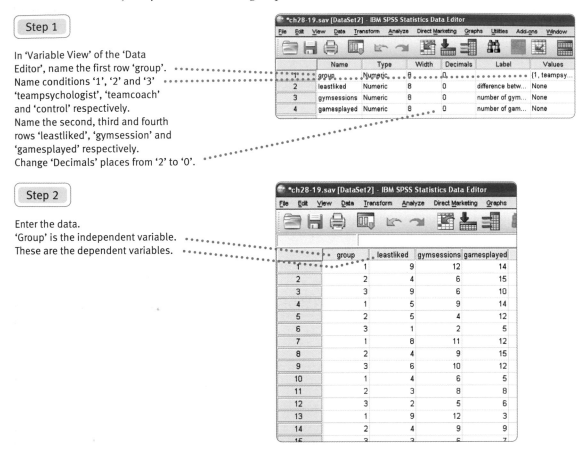

28.8 Discriminant function analysis

Step 1

Select 'Analyze', 'Classify' and 'Discriminant...'.

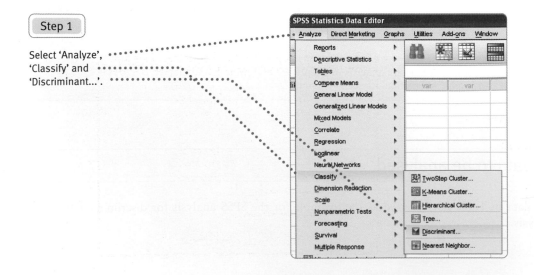

Step 2

Move the independent variables (the score variables) from the left-hand box to the 'Independents:' box.
Move 'group' from the left-hand box to the 'Grouping Varlable:' box.
Select 'Define Range'.

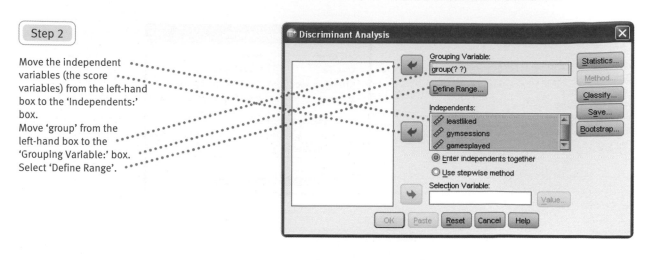

Step 3

Enter the smallest numeric code given to the grouping variable.
Enter the largest numeric code given to the grouping variable.
Select 'Continue'.

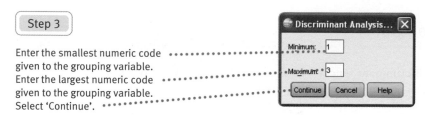

Step 4

Select 'Classify...'.

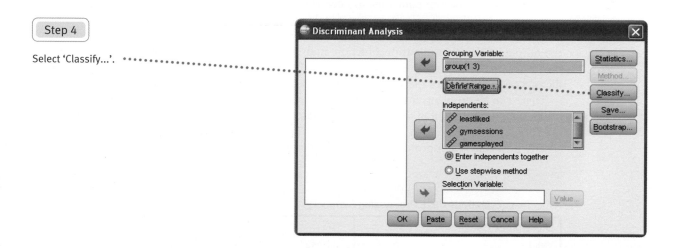

Step 5

Select 'Compute from group sizes' and 'Summary table'.
Select 'Continue' which returns the previous dialog box.
Select 'Statistics...' in this box.

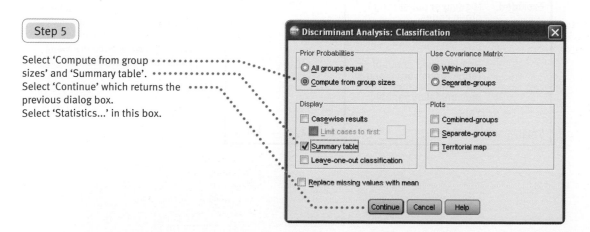

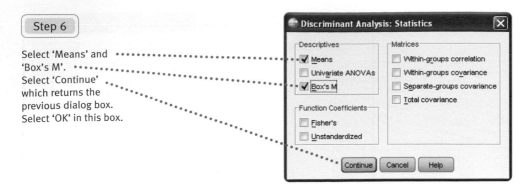

Step 6

Select 'Means' and 'Box's M'.
Select 'Continue' which returns the previous dialog box.
Select 'OK' in this box.

28.9 Interpreting the output

The interpretation of discriminant function analysis on SPSS is reasonably straightforward if the appropriate options are chosen. The key thing to remember is that a discriminant function is a sort of variable that is based on several measured variables. As such, in order to understand what the discriminant function measures it is necessary to know how each of the score variables which contribute to the discriminant function correlate with that discriminant function. This information is to be found in the structure matrix table.

This, taken in combination with the classification results table, gives a basic understanding of the output of the discrimination function analysis. If the classification results table is examined then it is apparent what the extent is to which the predicted classification on the basis of the predictor variables matches the known, actual classification of participants into the various groups.

There can be problems if the Box test is statistically significant but generally speaking, so long as the sample sizes of all of the groups are equal, this can be ignored. However, if the sample sizes are very different from each other then be very cautious about your analysis. It is not possible to specify how much of a problem a significant Box test is in these circumstances.

Analysis Case Processing Summary

Unweighted Cases		N	Percent
Valid		42	100.0
Excluded	Missing or out-of-range group codes	0	.0
	At least one missing discriminating variable	0	.0
	Both missing or out-of-range group codes and at least one missing discriminating variable	0	.0
	Total	0	.0
Total		42	100.0

It is useful to check this first table to make sure that the data analysed are what you expect. For example, you may have entered the wrong codes for the groups earlier.

Group Statistics

group		Mean	Std. Deviation	Valid N (listwise) Unweighted	Weighted
teampsychologist	difference between ratings of most and least liked team members	7.29	1.637	14	14.000
	number of gymn sessions volutarily attended	11.21	2.082	14	14.000
	number of games played	14.93	5.784	14	14.000
teamcoach	difference between ratings of most and least liked team members	4.43	1.222	14	14.000
	number of gymn sessions volutarily attended	8.36	3.565	14	14.000
	number of games played	13.86	6.011	14	14.000
control	difference between ratings of most and least liked team members	3.79	2.259	14	14.000
	number of gymn sessions volutarily attended	7.14	3.110	14	14.000
	number of games played	11.29	4.322	14	14.000
Total	difference between ratings of most and least liked team members	5.17	2.305	42	42.000
	number of gymn sessions volutarily attended	8.90	3.384	42	42.000
	number of games played	13.36	5.512	42	42.000

This table gives basic descriptive statistics for each group on the dependent variables. It is clear in this case that the means for the team psychologist group are consistently higher than for the team coach group whose mean scores are consistently higher than for the control condition when each variable is considered in turn. This is in perfect agreement with the hypothesis.

Test Results

Box's M		18.700
F	Approx.	1.383
	df1	12
	df2	7371.000
	Sig.	.166

Tests null hypothesis of equal population covariance matrices.

The Box test basically assesses whether the variances of the different dependent variables are similar. If it is significant then there may be problems. In this case it is not significant so there is no problem.

However, if it is significant and the group sizes are equal then once again there is no major problem. Only when it is significant and the group sizes are very different should you be cautious about proceeding.

■ Summary of canonical discriminant functions

Eigenvalues

Function	Eigenvalue	% of Variance	Cumulative %	Canonical Correlation
1	.865[a]	97.0	97.0	.681
2	.027[a]	3.0	100.0	.162

a. First 2 canonical discriminant functions were used in the analysis.

Two discriminant functions have been identified. One is substantial as it accounts for 97% of the reliable variance whereas the other is quite small in comparison as it explains only 3.0%.

Wilks' Lambda

Test of Function(s)	Wilks' Lambda	Chi-square	df	Sig.
1 through 2	.522	24.685	6	.000
2	.974	1.007	2	.605

Wilks' Lambda indicates that the two discriminant functions are statistically significant though discriminant function 2 is not statistically significant alone.

Standardized Canonical Discriminant Function Coefficients

	Function 1	Function 2
difference between ratings of most and least liked team members	.848	-.305
number of gymn sessions volutarily attended	.302	.047
number of games played	-.037	.996

This is not a very useful table when independent variables are measured on dissimilar scales so this table is best ignored.

Structure Matrix

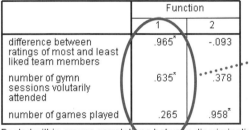

	Function 1	Function 2
difference between ratings of most and least liked team members	.965[*]	-.093
number of gymn sessions volutarily attended	.635[*]	.378
number of games played	.265	.958[*]

Pooled within-groups correlations between discriminating variables and standardized canonical discriminant functions Variables ordered by absolute size of correlation within function.

*. Largest absolute correlation between each variable and any discriminant function

These numbers are the correlations of each independent variable with the first discriminant function.
It is clear that the difference between ratings of most and least liked team members is the biggest contributor to the first discriminant function and number of games played is the biggest contributor to Function 2.

Classification Results[a]

		group	Predicted Group Membership teampsychologist	teamcoach	control	Total
Original	Count	teampsychologist	11	2	1	14
		teamcoach	3	4	7	14
		control	4	2	8	14
	%	teampsychologist	78.6	14.3	7.1	100.0
		teamcoach	21.4	28.6	50.0	100.0
		control	28.6	14.3	57.1	100.0

a. 54.8% of original grouped cases correctly classified.

This table gives the accuracy of the classification based on the independent variables. The team building by the psychologist is the most accurately predicted (78.6%) whereas team building by the coach is the least accurately predicted (28.6%) which is less than the chance level of 33.3%.

REPORTING THE OUTPUT

The results of this discriminant function analysis could be written up in the following way:

A discriminant analysis was carried out using the three predictors (1) the difference between the most and least liked team member, (2) the number of gym sessions voluntarily attended and (3) the number of games played. The grouping variable was team building, which consisted of three different conditions – team building with a sports psychologist, team building with a sports coach and the no team building controls. Two discriminant functions differentiated membership of these conditions effectively and accounted for 97 per cent and 3 per cent of the variance respectively. Wilks' lambda was statistically significant for the combined functions, $X^2(6) = 24.69$, $p < .001$, but was not significant when the first function was removed, $X^2(2) = 1.01$, *p ns*. The first discriminant function differentiated the psychologist teamwork building group from the other two groups. The variable 'the difference between the most and least liked team member' correlated with the discriminant function at .96. The variable 'number of gym sessions attended' also correlated but at the lower level of .64. The second discriminant function maximally distinguished the sports coach's team work group from the other two groups and loaded most strongly with the number of games played (.96). Classification based on the discriminant functions was good with about 80 per cent of the cases correctly classified compared with the 33 per cent chance accuracy. Approximately 79 per cent of the psychologist's team group were correctly identified, with 14 per cent misclassified as belonging to the coach's team building group. Of the control group 57 per cent were correctly identified, with 29 per cent misclassified in the psychologist's team building group. Of the coach's team building group 29 per cent were correctly identified, with 50 per cent misclassified as the control team members.

Summary of SPSS Statistics steps for discriminant function analysis

Data

- Name the variables in 'Variable View' of the 'Data Editor'.
- Enter the data under the appropriate variable names in 'Data View' of the 'Data Editor'.

Analysis

- Select 'Analyze', 'Classify' and 'Discriminant ...'.
- Move grouping variable to 'Grouping Variable:' box.
- Select 'Define Range'. Put the lowest group number code in the 'Minimum' box and the highest in the 'Maximum' box. Select 'Continue'.
- Move predictor variables to the 'Independents' box and select 'Statistics'.
- Select 'Means', 'Box's M' and 'Continue'.
- Select 'Classify', 'Compute from group size', 'Summary table', 'Continue' and 'OK'.

Output

- Check whether any of the Wilks' lambda for the discriminant functions are significant with a Sig(nificance) of .05 or less.
- If so check in the 'Structure Matrix' table which predictors are most highly correlated as these best discriminate the groups.
- Determine the percentage of correct classification for the groups.

For further resources including data sets and questions, please refer to the website accompanying this book.

PART 5

More advanced correlational statistics

CHAPTER 29

Partial correlation

Overview

- If you suspect that a correlation between two variables is affected by their correlations with yet another variable, it is possible to adjust for the effects of this additional variable by using the partial correlation procedure.

- The correlation between two variables (before partialling) is known as the zero-order correlation.

- Using SPSS Statistics, it is also possible to simultaneously control for several variables which may be affecting the correlation coefficient.

- If the variable to be controlled for consists of a small number of nominal categories, it is useful to explore the correlations for separate sets of cases based on the control variable. For example, if gender is to be controlled, then separate your sample into a male subsample and then a female subsample. Explore what the correlations between your main variables are for these two groups. Often this clarifies the effect of partialling in unanticipated ways.

29.1 What is partial correlation?

We know that variables correlate with each other. Height correlates with weight, age correlates with general knowledge, experiencing violence in childhood correlates with being violent as an adult, and so forth. But it is not simply that pairs of variables correlate but that pairs of variables correlate with other variables. The patterns are, of course, complex. Partial correlation takes a correlation between two variables and adjusts that correlation for the fact that they correlate with a third variable. Essentially it calculates what the correlation would be if the two variables did not correlate with this third variable.

Table 29.1 gives some data which illustrate what is meant. The basic data consist of two variables – one is scores on a measure of numerical intelligence and the other variable is scores on a measure of verbal intelligence. These two variables correlate highly with a Pearson correlation of .97. But the table also contains a third variable – age. There is a correlation between verbal intelligence and age of .85 and a correlation between numerical intelligence and age of .80. In other words, part of the correlation between verbal intelligence and numerical intelligence is associated with age. Partial correlation tells us what the correlation between verbal intelligence and numerical intelligence would be *if* their association with age were removed. That is, what the correlation would be if the correlations with age were zero. Usually researchers would refer to

Table 29.1	Numerical and verbal intelligence test scores and age	
Numerical scores	**Verbal scores**	**Age**
90	90	13
100	95	15
95	95	15
105	105	16
100	100	17

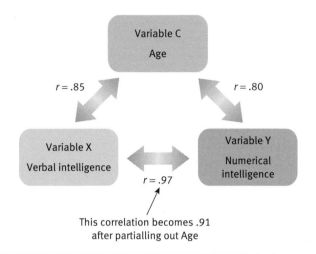

FIGURE 29.1   The correlations between the three variables before partialling

age (the variable being removed) as the third variable. As can be seen in Figure 29.1, the correlation becomes .91. Figure 29.2 highlights the main steps in a partial correlation analysis.

This is a relatively simple calculation but SPSS also allows you to control for several third variables at the same time with scarcely any additional effort. A partial correlation in which one variable is partialled out is called a first order partial correlation, if two variables are partialled out it is called a second order partial correlation, and so forth. You will see the main correlation (before partialling) referred to as the zero order partial correlation.

The change in the size of the correlation between verbal intelligence and numerical intelligence following partialling is not at all great in this case. There is an important lesson here since it is all too easy to believe that a third variable, if it has strong correlations with the two main variables, will produce a massive change in interpretation when partialled out. This is not always the case as can be seen. Also be aware that it is possible to have any pattern of change following partialling – a positive correlation can become a negative one, a zero correlation can become substantial, and a large correlation can become smaller – and quite what will happen is not easily predictable especially by inexperienced researchers.

29.2   When to use partial correlation

Partial correlation is built on the assumption that the relationships between the variables in question are linear. That is, if plotted on a scattergram the points are best represented by a straight line. Where the relationships depart from linearity then partial correlation may be misleading.

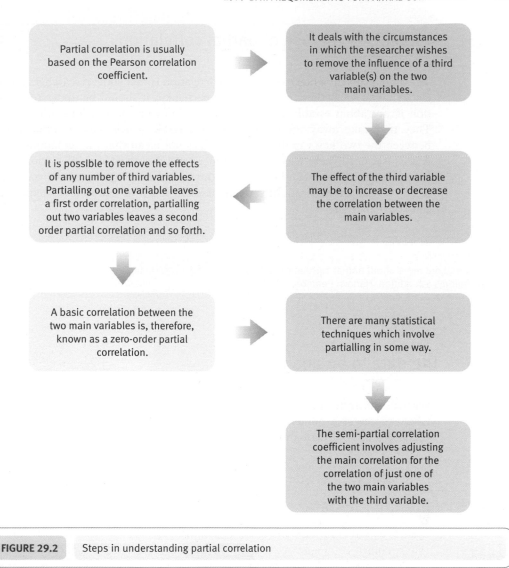

FIGURE 29.2   Steps in understanding partial correlation

This aside, the partial correlation can be a useful technique for estimating the influence of a third variable on any correlation coefficient.

29.3   When not to use partial correlation

Many techniques in statistics can do a similar job to partial correlation – that is, control for the influences of third variables, for example, analysis of covariance and multiple regression. If you are using these, then it is sensible to partial out influences of third variables using these techniques rather than carry out an additional partial correlation analysis.

29.4   Data requirements for partial correlation

At a minimum, partial correlation requires three score variables, each of which should ideally be normally distributed. There can be any number of third variables partialled at the same time though it is recommended that the ones to include are carefully planned to be the minimum needed.

29.5   Problems in the use of partial correlation

The main problem when using partial correlation is that some researchers try to make too strong claims about what they can achieve. In particular, it should be understood that partial correlation is not about establishing causality by removing the spurious influence of third variables. They may take away the influence of a variable which may be influencing the relationship between the two key variables but that does not mean that the remaining partialled correlation is a causal one. There may be any number of other third variables which need to be removed before any confidence can be gained about the relationship being a causal one. It is the research design which enables causality to be assessed (such as randomised experiments) and not the statistical analysis.

You can find out more about partial correlation in Chapter 29 of Howitt, D. and Cramer, D. (2011). *Introduction to Statistics in Psychology*, 5th edition. Harlow: Pearson.

29.6   The data to be analysed

We will illustrate the computation of partial correlations with the raw scores in Table 29.1, which represent a numerical intelligence test score, a verbal intelligence test score and age in years. We will correlate the two test scores partialling out age.

29.7   Entering the data

Step 1

In 'Variable View' of the 'Data Editor' name the three variables 'Num_IQ', 'Verb_IQ' and 'Age'.
Remove the two decimal places by setting the number to 0.

	Name	Type	Width	Decimals
1	Num_IQ	Numeric	8	0
2	Verb_IQ	Numeric	8	0
3	Age	Numeric	8	0

Step 2

In 'Data View' of the 'Data Editor' enter the numerical IQ scores in the first column, the verbal IQ scores in the second column and age in the third column.

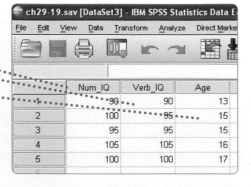

	Num_IQ	Verb_IQ	Age
1	90	90	13
2	100	95	15
3	95	95	15
4	105	105	16
5	100	100	17

29.8   Partial correlation

Step 1

Select 'Analyze', 'Correlate' and 'Partial...'.

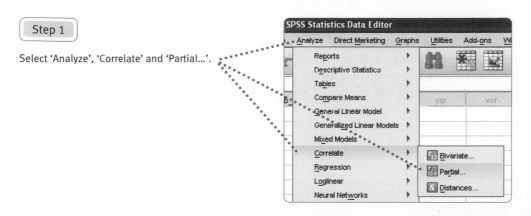

Step 2

Select 'Num_IQ' and 'Verb_IQ' and the
➡ button to put these two variables into
the 'Variables:' box.
Select 'Age' and the ➡button to put it
into the 'Controlling for:' box.
Select 'OK'.

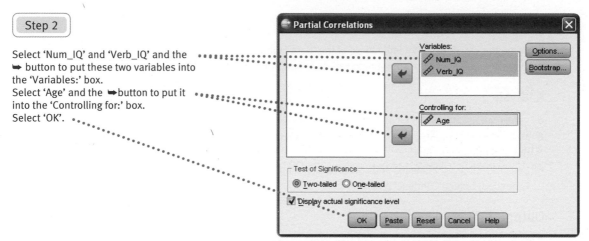

29.9   Interpreting the output

The partial correlation between
numerical and verbal intelligence
controlling for age is .776.

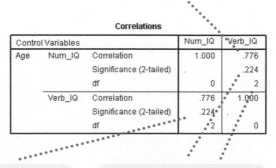

Correlations

Control Variables			Num_IQ	Verb_IQ
Age	Num_IQ	Correlation	1.000	.776
		Significance (2-tailed)	.	.224
		df	0	2
	Verb_IQ	Correlation	.776	1.000
		Significance (2-tailed)	.224	.
		df	2	0

This cell gives the
same information.

The two-tailed significance
is .224 which is not significant.
There are two degrees of freedom.

REPORTING THE OUTPUT

We need to mention the original correlation between numerical and verbal intelligence which is .92. So we could report the results as follows:

The correlation between numerical and verbal intelligence is .92, *df* = 3, two-tailed *p* = .025. However, the correlation between numerical and verbal intelligence controlling for age declines to 0.78, which is not significant, two-tailed *p* = .224. In other words, there is no significant relationship between numerical and verbal intelligence when age is controlled.

Summary of SPSS Statistics steps for partial correlation

Data

- Name the variables in 'Variable View' of the 'Data Editor'.
- Enter the data under the appropriate variable names in 'Data View' of the 'Data Editor'.

Analysis

- Select 'Analyze', 'Correlate' and 'Partial ...'.
- Move the two variables to be correlated to the 'Variables:' box and move the variables to be controlled to the 'Controlled for:' box.
- Select 'Options ...' if 'Means and standard deviations', 'Zero-order correlations' or 'Exclude cases pairwise' is needed. Select 'Continue' and 'OK'.

Output

- Check for any difference in size and sign between the zero-order correlation and the partial correlation.
- Note whether the partial correlation is significant with a Significance of .05 or less.

For further resources including data sets and questions, please refer to the website accompanying this book.

CHAPTER 30

Factor analysis

Overview

- There are two types of factor analysis: exploratory and confirmatory factor analysis. SPSS Statistics does not compute confirmatory factor analysis directly. Exploratory factor analysis is probably the more important and SPSS has a very extensive package of options for this.

- (Exploratory) factor analysis allows one to make sense of a complex set of variables by reducing them to a smaller number of factors (or supervariables) which account for many of the original variables. Although it is possible to obtain valuable insights from a matrix of correlations between several variables, the sheer size of the matrix may make this difficult even with a relatively small number of variables.

- Factor analysis is commonly used when trying to understand the pattern of responses of people completing closed-ended questionnaires. The items measuring similar things can be identified through factor analysis and, consequently, the structure of replies to the questionnaire.

- Factor analysis, however, includes a variety of techniques and approaches which can be bewildering. We provide a 'standard' approach which will serve the purposes of most researchers well.

- Factor analysis requires a degree of judgement, especially on the matter of the number of factors to extract. The speed of computer factor analyses means that more than one approach may be tried even within quite a short analysis session. It is a useful exercise to explore the effects of varying the method of analysis in order to assess the effect of this on one's conclusions.

30.1 What is factor analysis?

Factor analysis is a different sort of statistical technique from those in previous chapters. It is one of a number of statistical procedures that can be used to try to identify patterns in fairly large sets of data with substantial numbers of variables. There are other techniques such as cluster analysis which do a similar thing. One of the commonest uses of factor analysis is where one has written a number of questions (items) to measure such things as attitudes, personality and so forth. Such questionnaires tend to be fairly lengthy and there is little to be done with them using the statistical techniques so far covered in this book. One could, perhaps, produce a correlation matrix of answers to all of the items on the measure but, say, if you have 20 items then this results in a 20 × 20 correlation matrix of every item correlated with every other item. This is a

Table 30.1	Scores of nine individuals on six variables					
Individual	**Batting**	**Crosswords**	**Darts**	**Scrabble**	**Juggling**	**Spelling**
1	10	15	8	26	15	8
2	6	16	5	25	12	9
3	2	11	1	22	7	6
4	5	16	3	28	11	9
5	7	15	4	24	12	7
6	8	13	4	23	14	6
7	6	17	3	29	10	9
8	2	18	1	28	8	8
9	5	14	2	25	10	6

Table 30.2	Correlation matrix of the six variables					
	Batting	**Crosswords**	**Darts**	**Scrabble**	**Juggling**	**Spelling**
Batting	1	.00	.91	.05	.96	.10
Crosswords		1	.08	.88	.02	.80
Darts			1	.01	.90	.29
Scrabble				1	.08	.79
Juggling					1	.11
Spelling						1

table containing 400 correlation coefficients which will not even fit on a page of SPSS output. Consequently it is not simply mentally difficult to handle but it is physically virtually unmanageable too.

Factor analysis is usually based on such a correlation matrix. The matrix is produced by SPSS as part of doing the factor analysis so it does not have to be produced separately. Table 30.1 contains a very simplified set of data which in many ways would be suitable for a factor analysis. It contains performance scores on six variables (batting, crosswords, darts, Scrabble, juggling and spelling) from nine participants for purposes of illustration. In truth, one would not normally do a factor analysis on data from so few cases but we should emphasise that this is for purposes of illustration not to be emulated directly.

A correlation matrix for all of these variables is shown in Table 30.2. Because a correlation matrix is symmetrical around the top left to bottom right diagonal, we have left out some of the correlations for clarity's sake. The diagonal contains correlations of 1 because the correlation of a set of scores with itself is a perfect correlation. What can be made of this correlation matrix? Well it is quite a simple one and deliberately highly structured so you may be able to work out patterns, but this would be difficult in many cases.

If we subject this correlation matrix to a factor analysis we get a factor loading matrix such as that in Table 30.3. Each of the six variables is listed and there are columns headed 'Factor 1' and 'Factor 2'. There are also a lot of numbers. The numbers are known as 'factor loadings' but they are nothing other than correlation coefficients. The new idea in this is the factors. Factors are empirically derived variables based on the data in the correlation matrix. They are made up of

Table 30.3	Orthogonal factor loading matrix for six skills	
Variable	**Factor 1**	**Factor 2**
Skill at batting	.98	−.01
Skill at crosswords	.01	.95
Skill at darts	.96	.10
Skill at Scrabble	−.08	.95
Skill at juggling	.98	−.01
Skill at spelling	.15	.91

combinations of variables which tend to correlate. So the factor loadings are the correlations of variables with an empirically derived 'combined variable' or factor.

The researcher's task is to make sense of each of these factors. This is done by looking at which set of variables correlates highly with the factor. For Factor 1 these are batting, darts and juggling. What do these variables have in common? They seem to reflect sensory-motor skills and Factor 1 could be labelled with this phrase. Factor 2 has its high correlations with crosswords, Scrabble and spelling. These all seem to reflect verbal skills and Factor 2 could be given that name. Of course, the naming of the factors requires some creativity and insight.

Well that is basically what factor analysis produces. However, there are a number of other issues that need to be considered:

- What sort of factor analysis? There are many sorts of factor analysis but for most psychological data we would recommend principal components analysis or, failing that, principal axes analysis which is more suitable if the typical correlations are low. With some data, principal axes analysis simply fails to work properly on SPSS. This is not a fault in SPSS.

- How many factors? A factor analysis will produce as many factors as variables in the analysis. This is not helpful as many of the factors are simply nonsensical error. At the very minimum, a factor should have an eigenvalue of 1.00 or greater. The eigenvalue is simply the sum of the squared factor loadings on a particular factor. As such it is a measure of the amount of variance associated with that factor. You will also see it referred to as the latent root. However, in order to estimate the number of factors to use it is more helpful to use the scree test. This produces a graph and where the graph of the eigenvalue of each of the factors levels off (i.e. becomes virtually a straight line). The number before this break is usually the number of factors that should be used.

- Rotation. If you choose rotation in factor analysis then you are more likely to get a factor matrix which is readily interpretable. We would recommend that you use some method of rotation though this very occasionally only changes things a little.

- It is often desirable to get factor scores for each factor. These are essentially a score for each participant on each factor so factors can be treated just like any other variable. SPSS saves them as an additional column(s) on the SPSS data spreadsheet. You will see a button on the SPSS Factor Analysis window labelled 'Scores' which will allow you to save the factors as variables. Factor scores look odd on the spreadsheet because they are standardised around a mean of zero just like z-scores of which they are a variant.

SPSS describes factor analysis as a dimension reduction procedure. This is because it takes a number of variables and condenses them into a number of factors which are still variables and can be used like any other score variable. Figure 30.1 gives a schematic presentation of a factor analysis.

Factor analysis requires a substantial number of scores or binary variables intended to measure a particular thing or concept. Typical data would be questions measuring personality or attitudes.

Ideally a big sample is required: at a bare minimum about 50 cases but some authorities suggest 10 or more participants for each variable.

Probably the best choices of types of factor analysis is principal components (or principal axes especially if the typical correlations are small).

It may be necessary to run the factor analysis at least twice. The first time choose the scree test as this helps decide the number of factors to use. Then run the analysis again stipulating this number of factors.

It is usually helpful to rotate the factors. Doing so usually makes the factors more easily interpreted.

Factors can be saved as new variables in the form of factor scores. These can then be used in any other statistical procedure that uses score variables.

FIGURE 30.1   Steps in factor analysis

30.2   When to use factor analysis

Factor analysis can be used whenever you have a set of score variables (or it can be used with binary (yes–no) variables). It identifies patterns in the correlations between these variables. However, it is not desirable to put every variable that meets these criteria into your factor analysis since the output will be somewhat meaningless. It is better to confine your factor analysis to a set of variables which is relatively coherent in the first place. So if you have a number of items which were intended to measure 'strength of religious feeling', say, then do not throw into the analysis other variables such as gender, age and number of marathons run in a year as what you get may not make too much sense at all. But this would be advice that should generally be applied to statistical analysis. Aim to have a coherent set of data which addresses the questions that you are interested in and avoid simply collecting data because you can and avoid statistically analysing data simply because the data are there.

30.3   When not to use factor analysis

Factor analysis merely identifies patterns in data and it is not, for example, a test of significance. Neither does it have independent and dependent variables. So if either of these is in your mind then you do not need factor analysis.

Also it makes little sense to carry out a factor analysis if one has less than a handful of variables. So it is unlikely that you would wish to do a factor analysis where, say, you had five or fewer variables.

30.4   Data requirements for factor analysis

There is some controversy about the number of participants required for a study using factor analysis. The suggestions vary widely according to different sources. Generally speaking, you would not carry out a factor analysis with fewer than about 50 participants. Of course, for student work then a smaller sample than this would be common and in this context probably acceptable as part of learning and assessment. However, one also finds suggestions that for a factor analysis to be stable then one should have a minimum of about 10 participants per variable. Sometimes the figure given is greater than this. Basically, the more participants the better, though factor analysis may be illuminating even if one is at the lower end of acceptable numbers of participants.

30.5   Problems in the use of factor analysis

Our experience is that beginners have the greatest difficulty with the least statistical of the aspects of factor analysis – that is the interpretation of factors. Partly this is because many statistical techniques seem to require little special input from the user. Factor analysis, however, requires some inductive reasoning and is not merely a rule-following task. However, the task is really relatively straightforward since it merely involves explaining why a number of variables should be closely related. Usually, and this is a rule-of-thumb, we recommend that the variables which have a factor loading of about .5 on a particular factor should be considered when trying to name the factor. Of course, –.5 and bigger also counts.

Of course, factor analysis is a somewhat abstract procedure so it can take a little while to grasp the core ideas. But persevere since factor analysis is the first step towards a variety of procedures which help you explore complex data rather than simply test significance.

You can find out more about factor analysis in Chapter 30 of Howitt, D. and Cramer, D. (2011). *Introduction to Statistics in Psychology*, 5th edition. Harlow: Pearson.

30.6   The data to be analysed

The computation of a principal components analysis is illustrated with the data in Table 30.1, which consist of scores on six variables for nine individuals. This is only for illustrative purposes. It would be considered a ludicrously small number of cases to do a factor analysis on. Normally, you should think of having at least two or three times as many cases as you have variables. The following is a standard factor analysis which is adequate for most situations. However, SPSS has many options for factor analysis.

30.7 Entering the data

Step 1

In 'Variable View' of the 'Data Editor' name the six variables 'Batting', 'Crosswords' and so on. Remove the two decimal places by setting the number to 0.

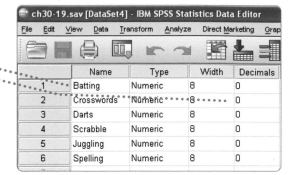

Step 2

In 'Data View' of the 'Data Editor' enter the values.

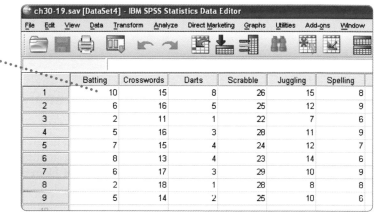

30.8 Principal components analysis with orthogonal rotation

Step 1

Select 'Analyze', 'Dimension Reduction' and 'Factor...'.

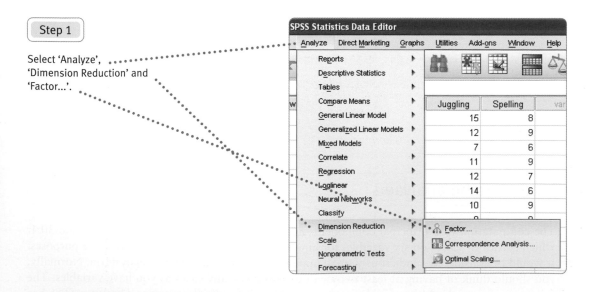

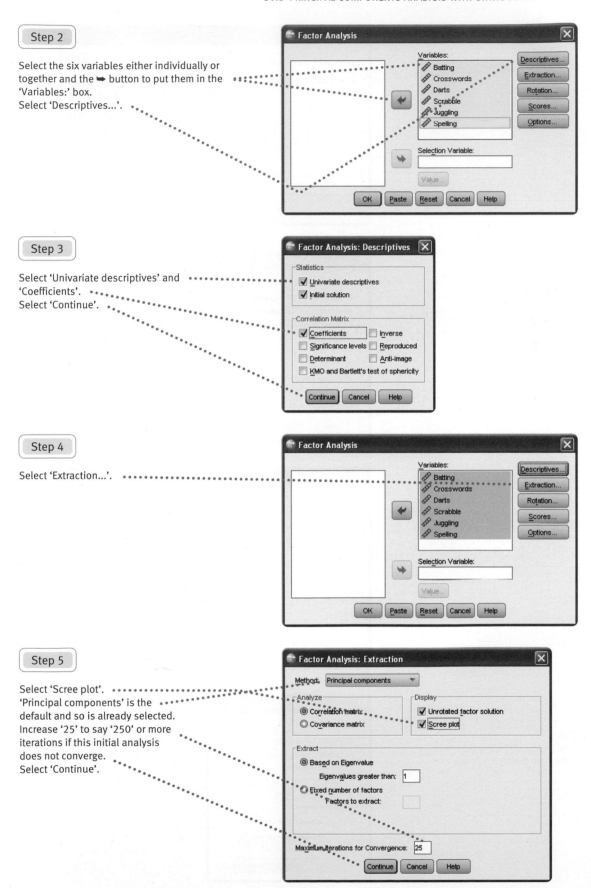

Step 2

Select the six variables either individually or together and the ➡ button to put them in the 'Variables:' box.
Select 'Descriptives...'.

Step 3

Select 'Univariate descriptives' and 'Coefficients'.
Select 'Continue'.

Step 4

Select 'Extraction...'.

Step 5

Select 'Scree plot'.
'Principal components' is the default and so is already selected.
Increase '25' to say '250' or more iterations if this initial analysis does not converge.
Select 'Continue'.

Step 6

Select 'Rotation...'.

Step 7

Select 'Varimax' for orthogonal or
unrelated factor or 'Direct Oblimin' for
oblique or related factors.
Increase '25' to say '250' or more iterations
if this rotated analysis does not converge.
Select 'Continue'.

Step 8

Select 'Options...'.

Step 9

Select 'Sorted by size'.
Select 'Continue'.
Select 'OK' in the main box.

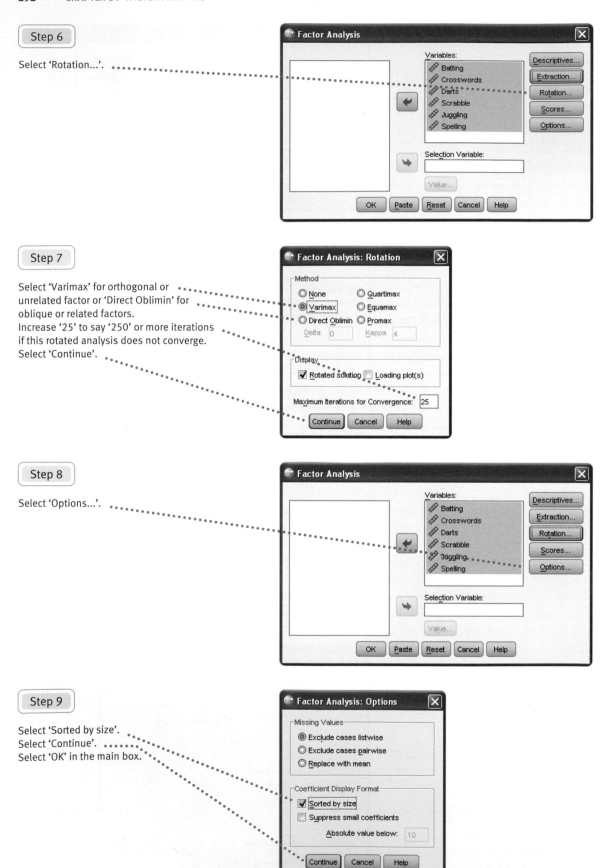

30.9   Interpreting the output

- The first table gives the mean, standard deviation and number of cases for each variable.

Descriptive Statistics

	Mean	Std. Deviation	Analysis N
Batting	5.67	2.598	9
Crosswords	15.00	2.121	9
Darts	3.44	2.186	9
Scrabble	25.56	2.404	9
Juggling	11.00	2.598	9
Spelling	7.56	1.333	9

N shows the number of cases on which your analysis is based. This information is essential particularly if you have missing data.

Note that if one variable has a .00 standard deviation, a factor analysis will not be conducted. The variables having .00 standard deviation need to be excluded for the analysis to run.

- The second table presents the correlation matrix. From this it appears that there are two groups of variables that are strongly intercorrelated. One consists of batting, juggling and darts, and the other of crosswords, Scrabble and spelling. These have been indicated – but remember that as a correlation matrix is symmetrical that only the lower half below the diagonal has been marked. Normally in factor analysis the correlation matrix is much more difficult to decipher than this. Our data are highly stylised.

Correlation Matrix

		Batting	Crosswords	Darts	Scrabble	Juggling	Spelling
Correlation	Batting	1.000	.000	.910	-.047	.963	.096
	Crosswords	.000	1.000	.081	.883	.023	.795
	Darts	.910	.081	1.000	-.005	.902	.291
	Scrabble	-.047	.883	-.005	1.000	-.080	.789
	Juggling	.963	.023	.902	-.080	1.000	.108
	Spelling	.096	.795	.291	.789	.108	1.000

The largest correlations have been enclosed. Because factor analysis usually involves a lot of variables and there is a limit to what can be got on a computer screen, normally the correlation matrix is difficult to see in its entirety.

- The fourth table shows that two principal components were initially extracted in this case. The computer ignores factors with an eigenvalue of less than 1.00. This is because such factors consist of uninterpretable error variation. Of course, your analysis may have even more (or fewer) factors. Note that 'Rotation Sums of Squared Loadings' are not displayed if related factors are requested.

Total Variance Explained

Component	Initial Eigenvalues			Extraction Sums of Squared Loadings			Rotation Sums of Squared Loadings		
	Total	% of Variance	Cumulative %	Total	% of Variance	Cumulative %	Total	% of Variance	Cumulative %
1	2.951	49.186	49.186	2.951	49.186	49.186	2.876	47.931	47.931
2	2.579	42.981	92.167	2.579	42.981	92.167	2.654	44.236	92.167
3	.264	4.401	96.567						
4	.124	2.062	98.630						
5	.058	.974	99.604						
6	.024	.396	100.000						

Extraction Method: Principal Component Analysis.

The first two factors (components) will be analysed further by the computer as their eigenvalues are larger than 1.00. If an analysis fails to converge, increase the number of iterations in the 'Extraction' sub-dialog box (Step 5).

- The Scree Plot also shows that a break in the size of eigenvalues for the factors occurs after the second factor: the curve is fairly flat after the second factor. Since it is important in factor analysis to ensure that you do not have too many factors, you may wish to do your factor analysis and rotation stipulating the number of factors once you have the results of the Scree test. (This can be done by inserting the number in the 'Number of factors:' in the 'Factor Analysis: Extraction' sub-dialog box.) In the case of our data this does not need to be done since the computer has used the first two factors and ignored the others because of the minimum eigen-value requirement of 1.00. It is not unusual for a component analysis to be recomputed in the light of the pattern which emerges.

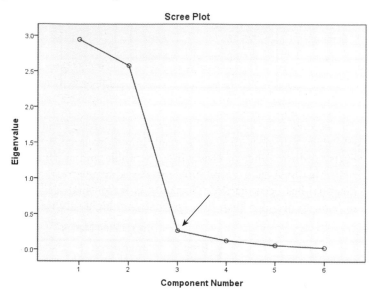

- These two components are then orthogonally rotated and the loadings of the six variables on these two factors are shown in the fifth table entitled 'Rotated Component Matrix'.

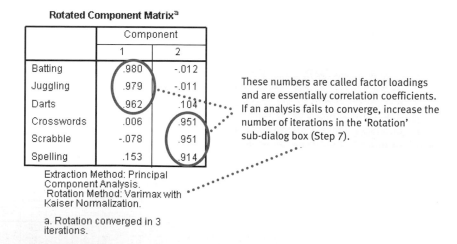

Rotated Component Matrix[a]

	Component	
	1	2
Batting	.980	-.012
Juggling	.979	-.011
Darts	.962	.104
Crosswords	.006	.951
Scrabble	-.078	.951
Spelling	.153	.914

Extraction Method: Principal Component Analysis.
Rotation Method: Varimax with Kaiser Normalization.

a. Rotation converged in 3 iterations.

These numbers are called factor loadings and are essentially correlation coefficients. If an analysis fails to converge, increase the number of iterations in the 'Rotation' sub-dialog box (Step 7).

- The variables are ordered or sorted according to their loading on the first factor from those with the highest loadings to those with the lowest loadings. This helps interpretation of the factor since the high loading items are the ones which primarily help you decide what the factor is.

- On the first factor, 'batting' has the highest loading (.980) followed by 'juggling' (.979) and 'darts' (.962).

- On the second factor, 'crosswords' and 'Scrabble' have the highest loading (.951) followed by 'spelling' (.914). The apparent lack of difference in size of loading of 'crosswords' and

'Scrabble' is due to rounding. This can be seen if you double click on the rotated component matrix table and then double click on these two loadings in turn to show more decimal places.

- We would interpret the meaning of these factors in terms of the content of the variables that loaded most highly on them.
- The percentage of variance that each of the orthogonally rotated factors accounts for is given in the third table under '% of variance' in the 'Rotation Sums of Squared Loadings' section. It is 47.931 for the first factor and 44.236 for the second factor.

REPORTING THE OUTPUT

- It would be usual to tabulate the factors and variables, space permitting. Since the data in our example are on various tests of skill, the factor analysis table might be as in Table 30.3. The figures have been given to two decimal places.
- The exact way of reporting the results of a factor analysis will depend on the purpose of the analysis. One way of describing the results would be as follows:

A principal components analysis was conducted on the correlations of the six variables. Two factors were initially extracted with eigenvalues equal to or greater than 1.00. Orthogonal rotation of the factors yielded the factor structure given in Table 30.3. The first factor accounted for 48 per cent of the variance and the second factor 44 per cent. The first factor seems to be hand–eye coordination and the second factor seems to be verbal flexibility.

Since the factors have to be interpreted, differences in interpretation may occur.

Summary of SPSS Statistics steps for factor analysis

Data

- Name the variables in 'Variable View' of the 'Data Editor'.
- Enter the data under the appropriate variable names in 'Data View' of the 'Data Editor'.

Analysis

- Select 'Analyze', 'Dimension Reduction' and 'Factor ...'.
- Move the variables to be analysed to the 'Variables:' box.
- Select 'Descriptives ...', 'Univariate descriptives' and 'Continue'.
- Select 'Extraction ...', 'Method' of extraction if different from principal components, 'Scree test' and 'Continue'.
- Select 'Rotation ...', method of rotation and 'Continue'.
- Select 'Options ...', 'Sorted by size', 'Continue' and 'OK'.

Output

- The analysis may not complete if any variable has no variance and if the number of iterations needed is greater than the default of 25. Adjust analysis accordingly.
- Check the sample size in the 'Descriptive Statistics' table.
- Check the meaning of the variables correlating most highly on the rotated factors to determine the meaning of the factors.

For further resources including data sets and questions, please refer to the website accompanying this book.

CHAPTER 31

Item reliability and inter-rater agreement

Overview

- Reliability is a complex matter, as the term refers to a range of very different concepts and measures. It is easy to confuse them.

- Item alpha reliability and split-half reliability assess the internal consistency of the items in a questionnaire – that is, do the items tend to be measuring much the same thing?

- Split-half reliability on SPSS Statistics refers to the correlation between scores based on the first half of items you list for inclusion and the second half of the items. This correlation can be adjusted statistically to maintain the original questionnaire length.

- Coefficient alpha is merely the average of all possible split-half reliabilities for the questionnaire and so is generally preferred, as it is not dependent on how the items are ordered. Coefficient alpha can be used as a means of shortening a questionnaire while maintaining or improving its internal reliability.

- Inter-rater reliability (here assessed by kappa) is essentially a measure of agreement between the ratings of two different raters. Thus it is particularly useful for assessing codings or ratings by 'experts' of aspects of open-ended data; in other words, the quantification of qualitative data. It involves the extent of exact agreement between raters on their ratings compared with what agreement would be expected by chance. Note then that it is different from the correlation between raters, which does not require *exact* agreement to achieve high correlations but merely that the ratings agree relatively for both raters.

- Other forms of reliability, such as the consistency of a measure taken at two different points in time (test–retest reliability), could be assessed simply using the correlation coefficient (Chapter 10).

31.1　What are item reliability and inter-rater agreement?

■ Alpha and split-half reliability

The concept of reliability is one of the most important in psychological methodology alongside that of validity. Both of these concepts are applied to psychological test scores and other measures. The concept of reliability covered in this chapter is primarily to do with the construction of psychological tests and scales. There are two important and very different meanings of reliability:

- External reliability which considers the stability of scores on psychological tests and other measures over time. This is typically known as test–retest reliability and essentially consists of the correlation between measures taken at different points in time.

- Internal reliability which considers the internal consistency of a test or scale. Alpha and split-half reliabilities are both examples of measures of internal consistency.

Psychological tests and scales consist of a number of items which are intended to measure a psychological concept such as self-esteem, locus of control, personality characteristics and so forth. A lot of items are used simply because it is believed that a single item is not powerful enough to adequately measure the concept. So the assumption is that all of the items on the scale are measuring aspects of the psychological concept in question. This has several implications – one of which is that the items will all correlate with each other over a sample of participants. (Factor analysis could be used to explore this pattern, of course.) Another implication is that any subset or subgroup of items on the test or scale should show a good correlation with any other subset of items on the test or scale. This is essentially because all of the items are assumed to be measuring the same thing.

Table 31.1 contains data illustrative of the situation. There are data from 10 participants for a four item questionnaire. Normally, to score such a scale one would simply sum the scores on the four individual item to give a total score for each participant on the four-item scale. So do these items form an internally consistent (i.e. reliable) scale? The approach of split-half reliability is simply to split the items into two halves and sum the total for each half. So Item 1 + Item 2 would be the score of each participant on the first half of the items and Item 3 + Item 4 would be the score of each participant on the second half of the items.

Essentially, the split-half reliability for these data would be the correlation between the sum of the first half of items with the sum of the second half of these items. One would expect a good

Table 31.1	Data for 10 cases from a four-item questionnaire			
Cases	Item 1	Item 2	Item 3	Item 4
1	1	3	5	6
2	2	1	1	2
3	1	1	1	1
4	5	2	4	2
5	6	4	3	2
6	5	4	5	6
7	4	5	3	2
8	2	1	2	1
9	1	2	1	1
10	1	1	2	2

correlation between the two if the items on the scale are internally consistent. There is one slight problem with this. We have so far calculated the reliability of just halves of the scale and not the total scale. So an adjustment is made to the split-half reliability to estimate the reliability of the full-length scale. This is an easy computation but SPSS does it for you automatically so no further action is needed on your part.

Alpha reliability (or Cronbach's coefficient alpha) is conceptually much the same as split-half reliability but deals with an obvious problem with split-half-reliability. How one splits a scale into two halves will affect the size of the reliability. If Item 1 and Item 3 are summed and Item 2 and Item 4 summed then this will give a different answer from summing the first two and last two items. This is a little less than satisfactory. Alpha reliability gets around this problem by calculating the average of all possible split-half reliabilities based on a scale. That is, it works out all possible ways of choosing two different halves of the set of items. In other words, alpha reliability is based on more information in the data and, consequently, is a better estimate of the internal reliability.

Inter-rater agreement

This is used in a quite different set of circumstances. Self-evidently, it would be used when the data are from raters who are making assessments of the characteristics of individuals. For example, imagine that the study is comparing the ratings of a forensic psychologist and a psychiatrist of the risk posed to the public by sex offenders, a rating of 3 being the highest risk level. This situation is illustrated in Table 31.2. At first sight, one might assume that the reliability of these ratings could be assessed by correlating the ratings of the forensic psychologist with those of the psychiatrist. There is a problem with this which is due to the fact that correlation coefficients actually assess covariation in data and do not reflect agreement on the actual scores. So, for example, if a number such as 7 were added to each of the ratings made by the psychiatrist, this would make no difference to the size of the correlation coefficient between the two sets of ratings but would make a big difference to the mean score.

The kappa coefficient is a measure of the extent to which the actual ratings of the forensic psychologist and the psychiatrist are identical. We could take the data in Table 31.2 and recast it into a table which shows the frequency of agreements and disagreements between the two (see Table 31.3). It is probably obvious that the agreement is largely due to the high number of high ratings (extremely dangerous) made since there is little agreement at the lower rating levels. The

Table 31.2	Ratings of risk by two professionals of 12 offenders	
Sex offenders	**Forensic psychologist**	**Psychiatrist**
1	3	3
2	3	3
3	3	3
4	1	1
5	1	2
6	3	3
7	2	3
8	3	3
9	2	3
10	3	3
11	3	3
12	3	3

Table 31.3	Agreements and disagreements between the forensic psychologist and the psychiatrist on ratings of sex offenders with marginal totals added			
		Psychiatrist		
		1	2	3
Psychologist	1	1	1	0
	2	0	0	2
	3	0	0	8

kappa coefficient makes allowance for this sort of tendency by adjusting for the chance level of agreement based on the data.

Figure 31.1 highlights some of the main ideas in item reliability and inter-rater agreement.

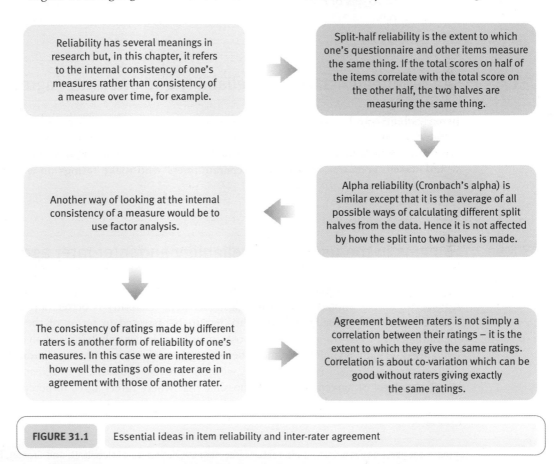

Reliability has several meanings in research but, in this chapter, it refers to the internal consistency of one's measures rather than consistency of a measure over time, for example.

Split-half reliability is the extent to which one's questionnaire and other items measure the same thing. If the total scores on half of the items correlate with the total score on the other half, the two halves are measuring the same thing.

Another way of looking at the internal consistency of a measure would be to use factor analysis.

Alpha reliability (Cronbach's alpha) is similar except that it is the average of all possible ways of calculating different split halves from the data. Hence it is not affected by how the split into two halves is made.

The consistency of ratings made by different raters is another form of reliability of one's measures. In this case we are interested in how well the ratings of one rater are in agreement with those of another rater.

Agreement between raters is not simply a correlation between their ratings – it is the extent to which they give the same ratings. Correlation is about co-variation which can be good without raters giving exactly the same ratings.

FIGURE 31.1   Essential ideas in item reliability and inter-rater agreement

31.2   When to use item reliability and inter-rater agreement

Item reliability should be used whenever you have a group of items which you intend to combine into a total score. You may wish to employ factor analysis too in these circumstances. If you are using a ready-made scale (e.g. a commercial test) then it may be appropriate to assess its reliability on the sample you are researching as reliability is not totally a feature of a test but may be affected by the nature of the sample and the circumstances of the research.

Inter-rater agreement in the form of coefficient kappa can be used in any circumstances where independent ratings are being used. Do not forget, however, that the chances of agreement are reduced if there are many rating categories. So consider the extent to which you might wish to combine rating categories for the purposes of the analysis should you have more than a few categories.

31.3   When not to use item reliability and inter-rater agreement

Not all test and scale construction methods are built on the assumption that scores on individual items are intercorrelated and can be summed to give a total score. Be on the lookout for such circumstances and consider carefully the applicability of item reliability methods. These exceptional circumstances are unlikely, however.

Inter-rater agreement is important in some circumstances but not in all circumstances. So low inter-rater agreement may be unimportant for research studies and having a good correlation between the two sets of ratings sufficient to indicate that the measure is reliable for the purposes of the research.

31.4   Data requirements for item reliability and inter-rater agreement

Item reliability can be used for response scales using Likert ratings as well as those using binary (yes–no) answer formats. It cannot be used for responses which are on a three or more category nominal scale or where the responses are not entered into SPSS in the form of scores.

Inter-rater agreement requires two separate raters who make ratings on a small number of categories.

31.5   Problems in the use of item reliability and inter-rater agreement

Item analysis does not generally cause any problems. However, compared with factor analysis it performs a very limited task. Although higher item reliability is better (alphas of .7 are considered satisfactory), really the reliability has to be assessed in terms of the purposes of the measurement. Scales for research purposes may have lower reliabilities than scales which are being used, say, by a clinician to make decisions about the future of an individual.

Inter-rater agreement is what it says, with one proviso – that is, it is agreement adjusted for the chance level of agreement. So it is possible that where raters agree in percentage terms very accurately, the value of kappa is low simply because both raters are using a particular category very frequently.

You can find out more about item reliability and inter-rater agreement in Chapter 36 of Howitt, D. and Cramer, D. (2011). *Introduction to Statistics in Psychology*, 5th edition. Harlow: Pearson.

31.6   The data to be analysed for item alpha reliability

The answers of 10 people to the four items of a questionnaire are shown in Table 31.1. These data will be used to illustrate two measures of item reliability known as alpha reliability and split-half reliability.

31.7   Entering the data

Step 1

In 'Variable View' of the 'Data Editor' name
the four variables 'Item1', 'Item2', 'Item3' and
'Item4'.
Remove the two decimal places by setting the
number to 0.

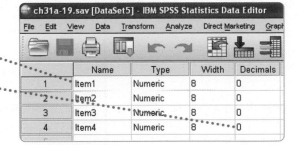

Step 2

In 'Data View' of the 'Data Editor'
enter the data in the first four
columns.

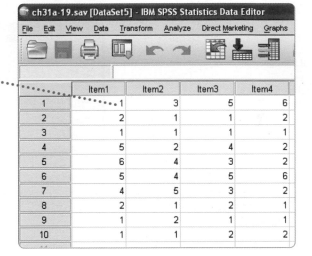

31.8   Alpha reliability

Step 1

Select 'Analyze',
'Scale'
and 'Reliability Analysis...'.

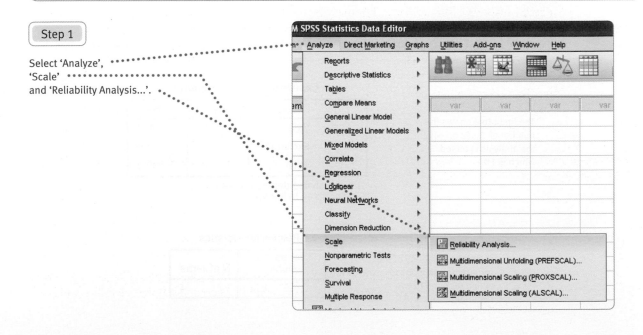

Step 2

Select the four items singly or together and the
➡ button to put them in the 'Items:' box.
Select 'Statistics…'.

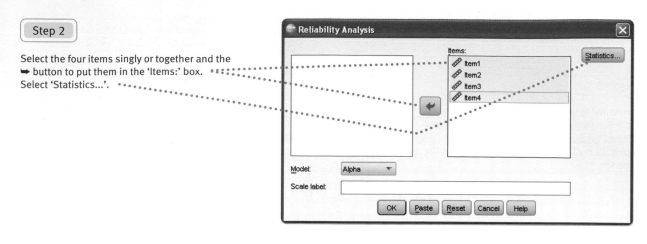

Step 3

Select 'Scale if item deleted'.
Select 'Continue'.
Select 'OK' in the previous box
which will reappear.

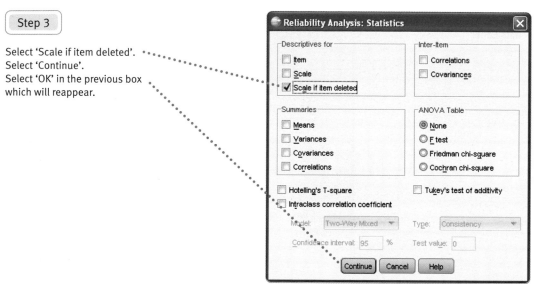

31.9   Interpreting the output

Case Processing Summary

The first table shows the
number of cases on which
the analysis is based
which is 10.

		N	%
Cases	Valid	10	100.0
	Excludedᵃ	0	.0
	Total	10	100.0

a. Listwise deletion based on all
variables in the procedure.

Reliability Statistics

The second table shows
the alpha reliability of the
four items and is .811,
which rounded to two
decimal places is .81.

Cronbach's Alpha	N of Items
.811	4

Item-Total Statistics

	Scale Mean if Item Deleted	Scale Variance if Item Deleted	Corrected Item-Total Correlation	Cronbach's Alpha if Item Deleted
Item1	7.60	18.933	.490	.840
Item2	8.00	19.556	.718	.731
Item3	7.70	17.789	.842	.671
Item4	7.90	18.767	.547	.806

This is the alpha reliability of the scale with one of the items dropped. From this column we can see that if we remove the first item ('Item1'), the alpha reliability of the remaining three items of the scale increases slightly to .840 from .811. Since this is a very small change, Item 1 is probably best retained.

REPORTING THE OUTPUT

One way of reporting the results of this analysis is as follows: 'The alpha reliability of the four item scale was .81, indicating that the scale had good reliability.' An alpha of .70 or above is considered satisfactory.

31.10 Split-half reliability

The previous data are reused for this analysis.

Select the ▼ button in the 'Model:' window and 'Split-half'. Select 'OK' (hidden behind the 'Model:' drop-down menu). (De-select 'Scale if item deleted' in 'Statistics...' if you wish.)

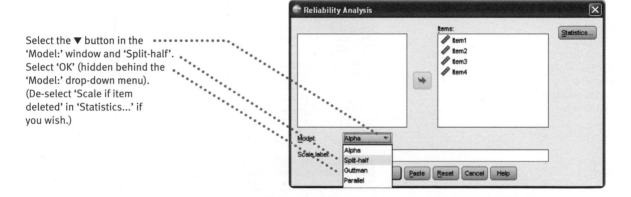

31.11 Interpreting the output

Reliability Statistics

Cronbach's Alpha	Part 1	Value	.777
		N of Items	2[a]
	Part 2	Value	.904
		N of Items	2[b]
	Total N of Items		4
Correlation Between Forms			.477
Spearman-Brown Coefficient	Equal Length		.646
	Unequal Length		.646
Guttman Split-Half Coefficient			.646

The last line of the second table shows the (Guttman) split-half reliability of the four items as .646, which rounded to two decimal places is .65.

a. The items are: Item1, Item2.
b. The items are: Item3, Item4.

REPORTING THE OUTPUT

One way of reporting the results of this analysis is as follows:

The split-half reliability of the four item scale was .65, indicating that the scale had only moderate reliability.

31.12 The data to be analysed for inter-rater agreement (kappa)

Kappa is used to measure the agreement between two raters, taking into account the amount of agreement that would be expected by chance. We will illustrate its computation for the data of Table 31.2, which shows the ratings by a forensic psychologist and a psychiatrist of 12 sex offenders in terms of the offenders being no risk (1), a moderate risk (2) or a high risk (3) to the public.

31.13 Entering the data

Step 1

In 'Variable View' of the 'Data Editor' name the first row 'Psychologist' and the second row 'Psychiatrist'.
Remove the two decimal places by setting the number to 0.

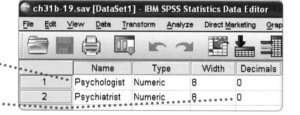

Step 2

In 'Data View' of the 'Data Editor' enter the data in the first two columns.

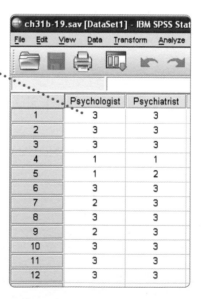

31.14 Kappa

Step 1

Select 'Analyze',
'Descriptive Statistics'
and 'Crosstabs...'.

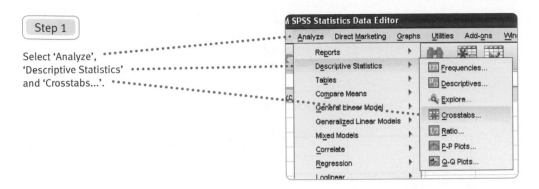

Step 2

Select 'Psychologist'
and the ➡ button beside
'Row(s):' to put it there.
Select 'Psychiatrist'
and the ➡ button beside
'Column(s):' to put it there.
Select 'Statistics...'.

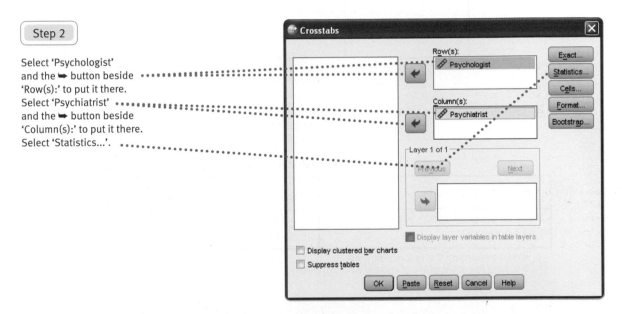

Step 3

Select 'Kappa'.
Select 'Continue'.
Select 'OK' in
previous box.

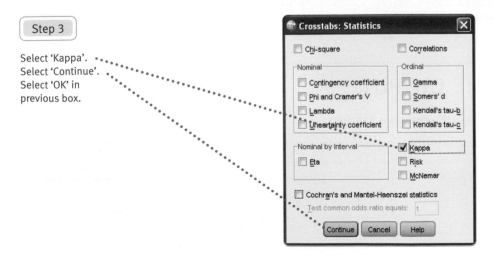

31.15   Interpreting the output

In the second table of the output the data have been arranged into a 2×2 contingency. The number of cases on which the forensic psychologist and the psychiatrist agree are shown in the diagonal cells of this table. They are 1 for the rating of 1, 0 for the rating of 2, and 8 for the rating of 3.

Psychologist * Psychiatrist Crosstabulation

Statistics:Count

		Psychiatrist			Total
		1	2	3	
Psychologist	1	1	1	0	2
	2	0	0	2	2
	3	0	0	8	8
Total		1	1	10	12

In the third table Kappa is shown as .400. Although Kappa is statistically significant with $p = .046$, it indicates only moderate agreement.

Symmetric Measures

		Value	Asymp. Std. Error[a]	Approx. T[b]	Approx. Sig.
Measure of Agreement	Kappa	.400	.219	2.000	.046
N of Valid Cases		12			

a. Not assuming the null hypothesis.
b. Using the asymptotic standard error assuming the null hypothesis.

Note that kappa allows for raters tending to use the same ratings most of the time. It is *not* a measure of percentage agreement.

REPORTING THE OUTPUT

One way of reporting the results of this analysis is as follows:

Kappa for the agreement between the ratings of the forensic psychologist and the psychiatrist was .40, which indicates only moderate agreement.

Summary of SPSS Statistics steps for reliability

Data

- Name the variables in 'Variable View' of the 'Data Editor'.
- Enter the data under the appropriate variable names in 'Data View' of the 'Data Editor'.

Analysis

- For alpha reliability, select 'Analyze', 'Scale' and 'Reliability Analysis ...'.
- Move appropriate items to the 'Items:' box.
- Select 'Statistics ...', 'Scale if item' deleted, 'Continue' and 'OK'.
- For kappa, select 'Analyze', 'Descriptive Statistics' and 'Crosstabs ...'.
- Move one rater to 'Row(s):' box and the other rater to the 'Column(s):' box.
- Select 'Statistics ...', 'Kappa', 'Continue' and 'OK'.

Output

- For alpha reliability, check number of cases and whether deleting an item substantially improves reliability.
- For kappa, check the distribution of ratings.

For further resources including data sets and questions, please refer to the website accompanying this book.

CHAPTER 32

Stepwise multiple regression

Overview

- Stepwise multiple regression is a way of choosing predictors of a particular dependent variable on the basis of statistical criteria.

- Essentially the statistical procedure decides which independent variable is the best predictor, the second best predictor, etc.

- The emphasis is on finding the best predictors at each stage. When predictors are highly correlated with each other and with the dependent variable, often one variable becomes listed as a predictor and the other variable is not listed. This does not mean that the latter variable is not a predictor, merely that it adds nothing to the prediction that the first predictor has not already done. Sometimes the best predictor is only marginally better than the second predictor and minor variations in the procedures may affect which of the two is chosen as the predictor.

- There are a number of multiple regression variants. Stepwise is usually a good choice though one can enter all variables simultaneously as an alternative (Chapter 33). Similarly, one can enter all of the variables simultaneously and gradually eliminate predictors one by one if eliminating does little to change the overall prediction.

- It is possible to enter variables as different groups for analysis. This is called hierarchical multiple regression and can, for example, be selected alongside stepwise procedures. The use of blocks is discussed in Chapter 35.

32.1 What is stepwise multiple regression?

Multiple regression is like simple or bivariate regression (Chapter 11) except that there is more than one predictor variable. Multiple regression is used when the variables are generally normally distributed. In other words, it is used when the criterion is a qualitative or score variable. In stepwise multiple regression the predictor variables are entered one variable at a time or step according to particular statistical criteria.

The first predictor to be considered for entry on the first step is the predictor that has the highest correlation with the criterion (Chapter 10). This predictor on its own will explain the most variance in the criterion. This correlation has to be statistically significant for it to be entered. If it is not significant, the analysis stops here with no predictors being entered.

The second predictor to be considered for entry on the second step is the one that explains the second highest proportion of the variance after its relation with the first predictor and the criterion is taken into account. In other words it is the predictor that has the highest part correlation with the criterion after the first predictor has been removed. Once again, this first-order part correlation has to be statistically significant for it to be entered. If it is not significant, the analysis stops after the first predictor has been entered.

The third predictor to be considered for entry on the third step is the predictor that has the highest part correlation after its association with the first two predictors and the criterion is taken into account. This second-order part correlation has to be statistically significant for it to be entered. If it is not significant, the analysis stops after the second step and does not proceed any further. If it is significant, then it is entered. At this stage, the second-order part correlations of the first two predictors with the criterion is examined. It is possible that one or both of these second-order part correlations are not significant. If this is the case, then the predictor with the non-significant second-order part correlation will be dropped from the analysis. The process continues until no other predictor explains a significant proportion of the variance in the criterion.

We will illustrate a stepwise multiple regression with the data in Table 32.1 which shows the scores of six children on the four variables of educational achievement, intellectual ability, school motivation and parental interest. We would not normally use such a small sample to carry out a stepwise multiple regression as it is unlikely that any of the correlations would be significant. So what we have done to make the correlations significant is to weight or multiply these six cases by 20 times so that we have a sample of 120, which is a reasonable number for a multiple regression.

The correlations among these four variables are shown in Table 32.2. (For a discussion of correlation see Chapter 10.) We can see that the predictor with the highest correlation is ability

Table 32.1	Data for stepwise multiple regression		
Educational achievement	**Intellectual ability**	**School motivation**	**Parental interest**
1	2	1	2
2	2	3	1
2	2	3	3
3	4	3	2
3	3	4	3
4	3	2	2

Table 32.2	A correlation matrix for the four variables			
	1	2	3	4
	Achievement	**Ability**	**Motivation**	**Interest**
Achievement				
Ability	.70			
Motivation	.37	.32		
Interest	.13	.11	.34	

which is .70 and which is statistically significant. So ability is the first predictor to be entered into the stepwise multiple regression.

The two first-order part correlations of achievement with motivation and interest controlling for ability have been calculated as .16 and .05 respectively. As the first-order part correlation for motivation is higher than that for interest and is statistically significant it is entered in the second step. The second-order part correlation for interest is .00 and is not significant. Consequently the analysis stops after the second step.

The proportion of variance explained by the first predictor is simply the square of its correlation with achievement which is $.70^2$ or about .49 or 49.0 per cent. The values can be found in the SPSS Statistics regression Model Summary table which is shown on p. 315. You will find Model 1 and Model 2 referred to. The first model includes just the first variable selected for inclusion in the regression, the second model includes the first and second variables selected for inclusion and so forth. However, it may be helpful to understand that the proportion of the variance explained by the first two predictors, for example, is their squared multiple correlation (R^2) which is calculated by multiplying the standardised partial beta regression coefficient by the correlation coefficient for each predictor and the criterion and summing these products. The standardised partial regression coefficient is normally more simply referred to as the standardised regression coefficient. It is .649 for ability and .164 for motivation. Consequently the squared multiple correlation is about .515. [$(.649 \times .70) + (.164 \times .37)$]. In other words, the two predictors taken together explain about 51.5 per cent of the variance in achievement. As ability explains about 49.0 per cent of the variance, motivation explains about a further 2.5 per cent.

You can use nominal (category) variables as predictors (independent variables) in regression using what is known as dummy coding. This is discussed in Chapters 37 and 38 on logistic regression. Basically it involves creating a variable for each category of the nominal (category) variable. If the categories are oranges, apples and bananas then one could create three dummy variables – oranges or not orranges, apples or not apples, and bananas or not bananas. These variables would only have two coding categories, obviously, and so can be used in correlations and hence in regression. You would have to create the dummy variables using Recode (Chapter 42). You also must understand that one of the dummy variables will *not* be used in the regression analysis because it contains the same information as the other two dummy variables. It does not matter which dummy variable you choose to omit. Figure 32.1 highlights the main steps in multiple regression.

32.2   When to use stepwise multiple regression

Stepwise multiple regression should be used when you want to find out what is the smallest number of predictors that you need which make a significant contribution in explaining the maximum amount of variance in the criterion variable, what these predictors are and how much of the variance of the criterion variable they explain. It may be helpful to use stepwise multiple regression when you have a large number of predictor variables and you want to get a quick impression of which of these explain most of the variance in the criterion variable.

32.3   When not to use stepwise multiple regression

Ideally it is better to use theoretical criteria rather than purely statistical ones in determining how you think your predictor variables are related to your criterion variable. This may be done by looking at particular variables or groups of variables at a time as is done in simple path analysis (Chapter 34) or by putting the variables in a multiple regression in a particular order. Hierarchical regression (which is dealt with in Chapter 35) is the procedure used for entering predictor variables in a particular sequence. Stepwise multiple regression would not generally be

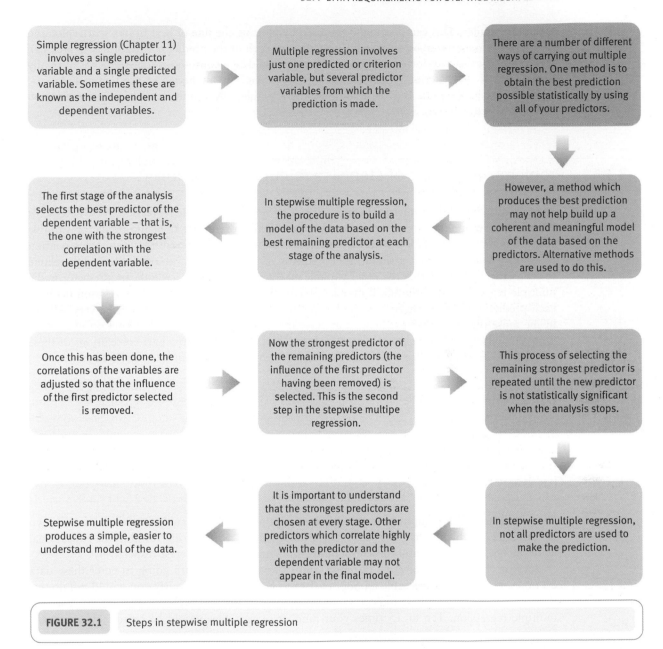

FIGURE 32.1   Steps in stepwise multiple regression

used when one or more of the predictor variables are nominal (category) variables involving three or more categories as these have to be converted into dummy variables. The dummy variables are best entered in a single step as a block to see what contribution this predictor variable makes to the criterion. This is done using hierarchical multiple regression.

32.4   Data requirements for stepwise multiple regression

Although multiple regression can handle categorical variables when they have been turned into dummy variables, a basic requirement for any form of simple or multiple regression is that the scatterplot of the relation between a predictor variable and the criterion should show homoscedasticity which means that the plot of the data points around the points of the line of best fit should be similar. This can be roughly assessed by plotting the line of best fit in a scatterplot and seeing if this assumption seems to have been met for each of the predictor variables (Chapter 11). It is important the criterion or dependent variable should be a continuous score which is normally distributed. If the criterion is a binomial variable, then it is far better to use binomial logistic regression (Chapter 38). If it is a multinomial variable, then multinomial logistic regression should be used (Chapter 37).

32.5   Problems in the use of stepwise multiple regression

Care needs to be taken in interpreting the results of a stepwise multiple regression. If two predictor variables show multicollinearity in that they are highly correlated, then the predictor variable which has the higher relation with the criterion variable will be entered first even if this relation is minutely bigger than that with the other predictor variable. As both variables are strongly related to each other, it is likely that the other variable may not be entered as a significant predictor in the multiple regression even though it may be an equally strong candidate. As the relation between each predictor and the criterion is similar, the results for this analysis may not be very reliable in the sense that the other predictor may have the higher relation with the criterion in another sample. Consequently it is important to look at the zero-order and the part correlations of these variables to see whether the difference in the size of their relation can be considered small.

You can find out more about stepwise multiple regression in Chapter 31 of Howitt, D. and Cramer, D. (2011). *Introduction to Statistics in Psychology*, 5th edition. Harlow: Pearson.

32.6   The data to be analysed

We will illustrate the computation of a stepwise multiple regression analysis with the data shown in Table 32.1, which consist of scores for six individuals on the four variables of educational achievement, intellectual ability, school motivation and parental interest respectively.

Because this is for illustrative purposes and to save space, we are going to enter these data 20 times to give us a respectable amount of data to work with. Obviously you would *not* do this if your data were real. It is important to use quite a lot of research participants or cases for multiple regression. Ten or 15 times your number of variables would be reasonably generous. Of course, you can use less for data exploration purposes.

32.7   Entering the data

Step 1

In 'Variable View' of the 'Data Editor', name the first five rows 'Achievement', 'Ability', 'Motivation', 'Interest' and 'Freq' respectively. Change 'Decimals' from '2' to '0'.

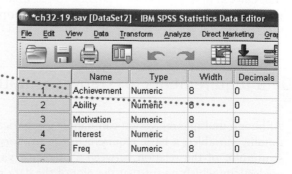

Step 2

Enter the data. The 'Freq' variable is for
weighting these six cases 20 times using the
'Weight Cases...' procedure (see Sections 3.7,
4.7, 8.8, 9.8, 16.7 or 17.7).
Save this data file to use for Chapters 33–35
and 52.

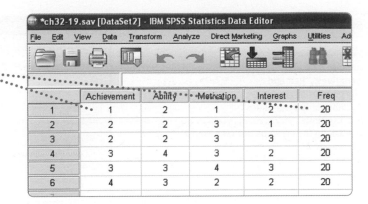

32.8 Stepwise multiple regression analysis

Step 1

Select 'Analyze', 'Regression' and 'Linear...'.

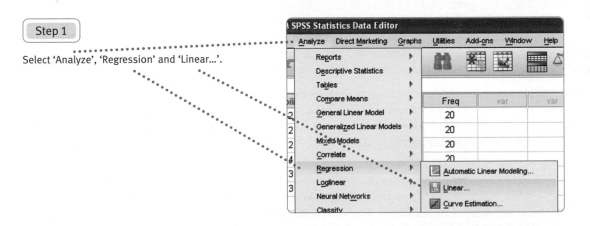

Step 2

Select 'Achievement' and the ➡ button beside
the 'Dependent:' box to put it there.
Select 'Ability', 'Motivation' and 'Interest'
and the ➡ button beside the
'Independent(s):' box to put them there.
Select the ▼ button beside
'Method:' and select 'Stepwise'.
Select 'Statistics...'.

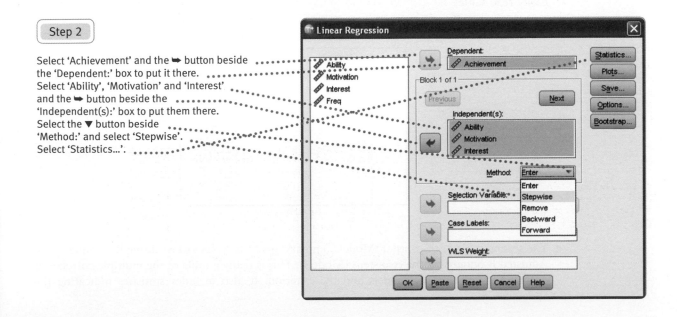

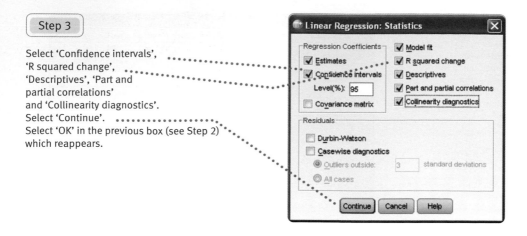

Step 3

Select 'Confidence intervals',
'R squared change',
'Descriptives', 'Part and
partial correlations'
and 'Collinearity diagnostics'.
Select 'Continue'.
Select 'OK' in the previous box (see Step 2)
which reappears.

32.9   Interpreting the output

- There is a great deal of information in the output. Multiple regression is a complex area and needs further study in order to understand all of its ramifications. In interpreting the results of this analysis we shall restrict ourselves to commenting on the following statistics: Multiple R, R Square, Adjusted R Square, B, Beta, R Square Change Part correlations and Collinearity Statistics. Most of these are dealt with in a simple fashion in *ISP* Chapter 31.

- In stepwise multiple regression, each new step is called a Model. In this example, two significant steps were involved. The first step (Model 1) uses the predictor Ability. The second step (Model 2) is built on this predictor with the addition of a second predictor Motivation. Generally, it is reasonable to concentrate on the highest numbered model.

- Notice how badly the first table in particular is laid out. If you double click on a table it will be enclosed in a rectangle. To move any but the first line, move the cursor to that line. When it changes to a double-arrow (↔), click the left button of the mouse and, holding the left button down, move the line to the position you want before releasing the button. By dragging the column dividers in this way you should be able to obtain a better and more easily read table.

Variables Entered/Removed[a]

Model	Variables Entered	Variables Removed	Method
1	Ability	.	Stepwise (Criteria: Probability-of-F-to-enter <= .050, Probability-of-F-to-remove >= .100).
2	Motivation	.	Stepwise (Criteria: Probability-of-F-to-enter <= .050, Probability-of-F-to-remove >= .100).

a. Dependent Variable: Achievement

- The second table of the output 'Model Summary' gives the values of Multiple R, R Square and Adjusted R Square for the two steps (Models). This is really a table of the multiple correlation coefficients between the models and the criterion. It also includes statistics indicating the improvement of fit of the models with the data. Each model in this example gives an improvement in fit. This can be seen from the final figures where the change in fit is significant for both Model 1 and Model 2. (The regression weights (B and Beta) are to be found in the fourth table of the output entitled 'Coefficients'.)

The most important thing is that the Sig. F Change indicates the improvement in fit for the two models is significant.

Model Summary

Model	R	R Square	Adjusted R Square	Std. Error of the Estimate	Change Statistics				
					R Square Change	F Change	df1	df2	Sig. F Change
1	.701ᵃ	.491	.487	.689	.491	113.786	1	118	.000
2	.718ᵇ	.515	.507	.675	.024	5.850	1	117	.01

a. Predictors: (Constant), Ability
b. Predictors: (Constant), Ability, Motivation

- The predictor that is entered on the first step of the stepwise analysis (Model 1) is the predictor which has the highest correlation with the criterion. In this example this predictor is 'Ability'. (Note 'a' immediately underneath the Model Summary table indicates this.)

- As there is only one predictor in the regression equation on the first step, Multiple R is the same as the correlation between Ability and Achievement (the dependent or criterion variable). In this case it is .701 or .70 to two decimal places.

- R Square is simply the multiple correlation coefficient squared, which in this instance is .491 or .49 to two decimal places. This indicates that 49 per cent of the variance in the criterion is shared with or 'explained by' the first predictor.

- Adjusted R Square is R Square which has been adjusted for the size of the sample and the number of predictors in the equation. The effect of this adjustment is to reduce the size of R Square, so Adjusted R Square is .487 or .49 to two decimal places.

- The variable which is entered second in the regression equation is the predictor which generally explains the second greatest significant proportion of the variance in the criterion. In this example, this variable is 'Motivation'.

- The Multiple R, R Square and Adjusted R Square for Model 2 are .718, .515 and .507 respectively which, rounded to two decimal places, are .72, .52 and .51. As might be expected, these values are bigger than for the corresponding figures for Model 1. This is to be expected because there is an additional predictor contributing to a better prediction.

- In Model 2, then, two variables ('Ability' and 'Motivation') explain or account for 51 per cent of the variance in the criterion.

- R Square Change presented under 'Change Statistics' in the second table shows the increase in the proportion of the variance in the criterion variable ('Achievement') by predictors that have been entered after the first predictor ('Ability'). In this case there is only one other predictor ('Motivation'). This predictor explains a further 2.4 per cent of the variance in the criterion.

- Examine the table headed 'Coefficients'. Find the column headed Beta in the table. The first entry is .701 for Model 1. This is exactly the same as the value of the multiple correlation above for Model 1. That is because Beta is the standardised regression coefficient which is the same as the correlation when there is only one predictor. It is as if all your scores had been transformed to z-scores before the analysis began.

Coefficients^a

Model		Unstandardized Coefficients		Standardized Coefficients	t	Sig.	95.0% Confidence Interval for B		Correlations			Collinearity Statistics	
		B	Std. Error	Beta			Lower Bound	Upper Bound	Zero-order	Partial	Part	Tolerance	VIF
1	(Constant)	.100	.234		.428	.669	-.363	.563					
	Ability	.900	.084	.701	10.667	.000	.733	1.067	.701	.701	.701	1.000	1.000
2	(Constant)	-.167	.254		-.656	.513	-.670	.337					
	Ability	.833	.087	.649	9.561	.000	.661	1.006	.701	.662	.615	.900	1.111
	Motivation	.167	.069	.164	2.419	.017	.030	.303	.369	.218	.156	.900	1.111

a. Dependent Variable: Achievement

The quickest way to access the output from multiple regression is to concentrate on the final model's output in the table labelled Coefficients. Ignore the row for constants, then concentrate on the remaining rows. These give the important predictors of the dependent variable. These are clearly Ability and Motivation, both of which contribute significantly. B weights are hard to interpret since they are dependent on the scale of measurement involved. Beta weights are analogous to correlation coefficients. Both Beta weights are positive so there is a positive relation between each predictor and the dependent variable. Ability has a correlation of .649 with Achievement and Motivation contributes an additional independent correlation of .164 with Achievement. The t-values plus their corresponding Sig. values indicates that the two independent variables contribute significantly to the prediction. The part correlation of Ability and Motivation are .615 and .156 respectively so they explain about 38% and 2% of the variance in Achievement when not taking account of the variance they share.

- For Model 2 Beta is .649 for the first predictor ('Ability') and .164 for the second predictor ('Motivation').

- The analysis stops at this point, as the third predictor ('Interest') does not explain a further significant proportion of the criterion variance. Notice that in the final table of the output entitled 'Excluded Variables', 'Interest' has a t-value of .000 and a significance level of 1.0000. This tells us that 'Interest' is a non-significant predictor of the criterion ('Achievement').

- The part correlations for Ability and Motivation are .615 and .156 respectively, which when squared means that they explain about 37.8 and 2.4 per cent of the variance in Achievement when not taking account of the variance they share together. As the total percentage of variance these two variables explain is about 51.5 per cent this means that the percentage of variance they share is about 11.3 per cent which is calculated by subtracting the sum of their separate variances (40.2) from the total variance (51.5).

- The tolerance level is shown in the 12th column of the table and is an index of the extent to which a predictor may be too highly correlated with other predictors to be entered into the multiple regression. This problem is known as multicollinearity or collinearity. A tolerance level of 0.1 or below indicates that the predictor may be too highly correlated with other predictors to be entered into the multiple regression. A tolerance level of above 0.1 means that the predictor may be entered into the multiple regression. As the tolerance level for both predictors is greater than 0.1, all three predictors may be entered into the multiple regression as there is no multicollinearity. The VIF (variance inflation factor) is simply 1/Tolerance so there is no need to report both statistics. A VIF of less than 10 indicates no multicollinearity.

REPORTING THE OUTPUT

- There are various ways of reporting the results of a stepwise multiple regression analysis. In such a report we should include the following kind of statement.

  In the stepwise multiple regression, intellectual ability was entered first and explained 49 per cent of the variance in educational achievement, $F(1, 118) = 113.76$, $p = .001$. School motivation was entered second and explained a further 2 per cent, $F(1, 117) = 5.85$, $p = .017$. Greater educational attainment was associated with greater intellectual ability and school motivation.

- A table is sometimes presented. There is no standard way of doing this but Table 32.3 is probably as clear as most.

| Table 32.3 | Stepwise multiple regression of predictors of educational achievement (only significant predictors are included) | | | | | |

Variable	Multiple *R*	*B*	Standard error b	Beta	*t*	Significance of *t*
Intellectual ability	0.70	0.83	0.09	0.65	9.56	.001
School motivation	0.72	0.17	0.07	0.16	2.42	.05

Summary of SPSS Statistics steps for stepwise multiple regression

Data

- Name the variables in 'Variable View' of the 'Data Editor'.
- Enter the data under the appropriate variable names in 'Data View' of the 'Data Editor'.

Analysis

- Select 'Analyze', 'Regression' and 'Linear …'.
- Move the criterion or dependent variable to the 'Dependent:' box and the predictors or independent variables to the 'Independent(s):' box.
- In the 'Method' box select 'Stepwise'.
- Select 'Statistics …' and then 'R squared change', 'Descriptives', 'Part and partial correlations', 'Collinearity diagnostics', 'Continue' and 'OK'.

Output

- Check sample size in 'Descriptive Statistics' table.
- In the 'Model Summary' table note the predictor and R Square Change in each step.
- In the 'Coefficients' table check that the Tolerance level is above 0.1.

For further resources including data sets and questions, please refer to the website accompanying this book.

CHAPTER 33

Simultaneous or standard multiple regression

Overview

- Simultaneous or standard multiple regression is a way of determining the relationship between each member of a set of predictor (or independent) variables and a criterion (dependent variable). This involves taking into account the fact that the predictor variables are related to each other as well as to the criterion. Basically what is calculated is the relationship between each predictor variable and the criterion variable after the relationship of the predictor variable to all other predictor variables has been removed (controlled for).

- This form of regression is not about building models but about choosing a set of predictors which independently predict the criterion significantly and collectively equate to an effective prediction of scores on the criterion variable.

- The relative size of the prediction of the criterion by each predictor in the set of predictors can be compared as well as the amount of variance in the criterion that each predictor explains independently of the other predictors.

- The relationship between a predictor and the criterion is given as a regression coefficient. These may be expressed in standardised form which makes it easier to compare one regression coefficient with another.

- Standardised regression coefficients vary from −1.00 to 1.00.

- Bigger standardised regression coefficients denote stronger associations between the criterion and a predictor.

- The proportion of variance in the criterion that is explained by or shared with the predictor is given by the square of the part correlation. A part correlation is like a partial correlation (Chapter 29) but only the association of one variable with the other predictor variables is removed. In other words, it is as if the correlation is between the original scores on one variable (the criterion) and the scores on the second variable (a predictor), with the other predictor variable's influences partialled out from the predictor variable in question.

- Predictors can be ordered in terms of the size of their standardised regression coefficients and the amount of variance of the criterion which they explain.

- The predictor having the highest standardised regression coefficient with the criterion, regardless of whether the sign of the coefficient is positive or negative, has the strongest relation with the criterion and explains the most variance in the criterion.

- The predictor with the second highest standardised regression coefficient with the criterion, regardless of the sign of the coefficient, has the second strongest relation with the criterion and explains the second greatest amount of variance in the criterion.

- The sign of the regression coefficient expresses the direction of the relation between the criterion and the predictor.

- A positive coefficient means that higher values on the predictor generally go with higher values on the criterion.

- A negative coefficient means that higher values on the predictor generally go with lower values on the criterion.

- Coefficients without a sign are assumed to be positive.

- Generally only predictors having a statistically significant coefficients are considered in the analysis.

33.1   What is simultaneous or standard multiple regression?

Multiple regression is generally used in psychology as a means of building models. So stepwise multiple regression allows the researcher to take predictors in turn of importance in order to find a set of variables which make for the best model to explain variation in scores on the criterion variable being studied. Stepwise multiple regression tends to result in a parsimonious set of predictor variables, which leads to a simple model to account for the criterion variable. This is because stepwise multiple regression chooses the best predictor first of all. It then chooses the second best predictor on the basis of the data having taken away statistically the influence of that best predictor. Simultaneous or standard multiple regression does something rather different. What it calculates is the independent predictive power of each of the predictor variables having taken into account the strength of the association of all of the other predictors with each predictor. This means that instead of a parsimonious number of predictors (such as emerge in stepwise multiple regression), the number of predictors at the start of the analysis remains the same at the end of the analysis. This does not mean that all of these predictors are significant predictors, however, and quite clearly the researcher is likely to concentrate on the significant predictors. While simultaneous or standard multiple regression has similarities with stepwise multiple regression (Chapter 32) and other forms of multiple regression, essentially all of the predictor variables are entered in a single block – that is, at the same time. Entering all predictor variables simultaneously in a single block enables the relative size of the association between each predictor and the criterion to be compared. In stepwise multiple regression, in contrast, the order of being chosen determines the relative size of the association between predictor and criterion variable.

This comparison is carried out in terms of the standardised (partial) regression or Beta coefficient which has a maximum value of ±1.00. The comparison can also be carried out in terms of the amount of variance in the criterion that is explained by or shared with each predictor. The proportion of variance accounted for is computed by squaring the part correlation. The proportion is often converted into a percentage by multiplying the proportion by 100.

The part correlation has great similarities with the partial correlation coefficient discussed in Chapter 29. The difference is that in partial correlation the influence of the third variable(s) on the two main variables is removed; in part correlation the influence of the third variable(s) is

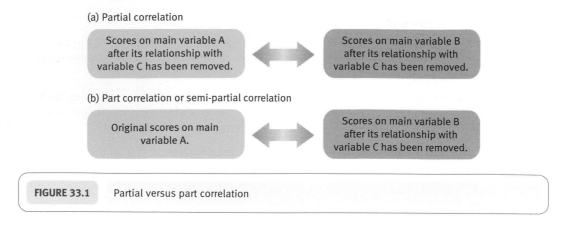

(a) Partial correlation

| Scores on main variable A after its relationship with variable C has been removed. | ⟷ | Scores on main variable B after its relationship with variable C has been removed. |

(b) Part correlation or semi-partial correlation

| Original scores on main variable A. | ⟷ | Scores on main variable B after its relationship with variable C has been removed. |

FIGURE 33.1   Partial versus part correlation

removed from just one of the two main variables. Since the part correlation involves partialling the third variable out from just one of the two main variables, part correlation is sometimes, not surprisingly, referred to as the semi-partial correlation. Conceptually the part correlation is the correlation of the actual scores on one of the two main variables with the adjusted scores on the other main variable. To stress the point, these adjusted scores have been adjusted for the correlations of the variable with a third variable(s). In simultaneous multiple regression, the scores on the criterion variable are the original scores whereas the scores on the predictor variable have been partialled. This is illustrated in Figure 33.1. In other words, standardised regression weights are similar to part correlation coefficients in that respect.

The reason that one might concentrate on the standardised regression coefficients is that these (like say the correlation coefficient) are directly comparable one with the other. It is quite different with unstandardised regression coefficients since these are used to predict actual values on the criterion variable. So the predictor that has the highest standardised regression coefficient with the criterion has the strongest relation with the criterion – that is, it is the most effective predictor. The predictor that has the next highest standardised regression coefficient has the next strongest association with the criterion and so is the second best predictor. So the predictors can easily be ordered in terms of the size of their standardised regression coefficients from largest to smallest. Generally speaking, only predictors that have a statistically significant regression coefficients are considered as variables that explain or predict the criterion.

Because the predictive power of each predictor variable is independent of (uncorrelated with) the predictive power of the other predictor variables, the predictive powers of each of the predictor variables add together. This makes the use of multiple regression extremely effective in ways that a series of simple regression analysis simply would not.

Multiple regression is used when the criterion variable is more or less normally distributed. In other words, it is used when the criterion is a quantitive or score variable. The predictor variables can be either binominal category (categorical) or score variables. If a categorical variable is used which has more than two categories, it needs to be converted into several dummy variables (see Chapter 38).

We will illustrate a simultaneous or standard multiple regression with the data in Table 33.1, which shows the scores of six children on the four score variables of educational achievement, intellectual ability, school motivation and parental interest. This table is the same as Table 32.1. We would not normally use such a small sample to carry out a multiple regression as it is unlikely that any of the correlations would be significant. So what we have done to make the correlations significant is to weight or multiply these six cases by 20 times so that we have a sample of 120, which is a reasonable number for a multiple regression. This does not affect the size of the correlation coefficients.

The correlations between these four variables are shown in Table 33.2 (see Chapter 10 for correlation). We need to look at each of these predictors in turn, paying attention to their correlations with the criterion and the other predictors. We will illustrate this procedure for the first

Table 33.1	Data for simultaneous or standard multiple regression		
Educational achievement	**Intellectual ability**	**School motivation**	**Parental interest**
1	2	1	2
2	2	3	1
2	2	3	3
3	4	3	2
3	3	4	3
4	3	2	2

Table 33.2	A correlation matrix for the four variables		
	1	**2**	**3**
1 Achievement			
2 Ability	.70		
3 Motivation	.37	.32	
4 Interest	.13	.24	.34

Table 33.3	Part correlations, squared part correlations, standardised regression coefficients (beta) and their statistical significance		
Predictors	**Part correlation**	**Squared part correlation**	**Beta**
Ability	.62	.38	.65***
Motivation	.15	.02	.15*
Interest	.00	.00	.00

*$p < .05$, ***$p < .001$ (two-tailed).

predictor of intellectual ability which has a correlation of .70 with educational achievement. However, intellectual ability is correlated with the other two predictors of school motivation and parental interest which are also correlated with the criterion of educational achievement. For example, intellectual ability has a correlation of .32 with school motivation and school motivation has a correlation of .37 with educational achievement. Consequently we do not know to what extent the correlation between the criterion of educational achievement and the predictor of intellectual ability is due to the correlation of the variables of educational achievement and intellectual ability with the predictors of school motivation and parental interest. This can be determined by calculating the part correlation between educational achievement and intellectual ability which controls for their associations with school motivation and parental interest.

Table 33.3 shows the part correlations between the criterion and each of the three predictors as well as the proportion of variance that is explained by each of the three predictors, the standardised (partial) regression or beta coefficients and their statistical significance. For example, the part correlation between the criterion of educational achievement and intellectual ability is .62 which squared is .38. This means that the amount of variance in educational achievement that is

Like other forms of multiple regression, simultaneous or standard multiple regression requires a criterion variable in the form of a score and predictor variables in the form of scores or binomial nominal categories.

The purpose of simultaneous/standard multiple regression is to provide regression coefficients for each predictor variable with the influence of all of the other predictor variables removed from that predictor.

By standardising the regression coefficients it becomes possible to compare their size numerically – that is, decide which explains the most variation in the criterion, etc.

This means that *all* predictor variables are used in the analysis and not just a few of the strongest predictors.

The regression coefficients are like part (or semi-partial) correlations between the predictor variable and the criterion variable removing the relationship between this predictor and all of the other predictors.

Regression coefficients should be checked for statistical significance as it is not usual to report or use further any non-significant ones.

FIGURE 33.2   Steps in simultaneous or standard multiple regression

explained just by intellectual ability is .38 having taken into account its relationships with the other variables. In other words, the proportion of variance in educational acheivement that is unique or specific to intellectual ability is .38. This proportion of variance is less than that which is shared between educational achievement and intellectual ability not taking into account its relationship with the other variables. This proportion is calculated by squaring the correlation of .70 between educational achievement and intellectual ability which is .49. If we subtract the proportion of variance of .38 from that of .49, then .11 (.49 − .38 = .11) of the variance between educational achievement and intellectual ability is shared with school motivation and parental interest. In other words, this proportion of variance is not unique or specific to intellectual ability. What proportion of this variance is shared with school motivation and with parental interest cannot be worked out from these figures. If we were interested in finding this out, which we are generally not, we would have to do two further simultaneous multiple regressions, one excluding parental interest and one excluding school motivation.

Figure 33.2 shows the essential steps in carrying out a simultaneous or standard multiple regression.

33.2   When to use simultaneous or standard multiple regression

Simultaneous or standard multiple regression should be used when you want to find out the association between the criterion and each predictor, taking into account the other predictors. If there was no correlation between the predictors we would not have to carry out a multiple regression as the relation between the criterion and each predictor would be represented by its zero-order correlation. However, it is very unsual for variables not to be correlated to some extent with each other. Simultaneous multiple regression takes these correlations into account, telling us what the association is between the criterion and the predictor which is unique or specific to that predictor. Simultaneous multiple regression is like a path analysis (Chapter 34) in which each predictor is considered to influence the criterion. Predictors which are not significantly associated with the criterion are thought not to affect the criterion.

33.3 When not to use simultaneous or standard multiple regression

Simultaneous or standard multiple regression should not be used in at least two situations. One is where we want to find out what is the minimum number of predictors which significantly explain the greatest amount of variance in the criterion. In this situation we would use stepwise multiple regression (Chapter 32). The other situation is where we want to enter the predictors in a particular sequence. Doing this effectively controls for the predictors entered earlier in the sequence. This form of multiple regression is called hierarchical multiple regression (Chapters 35 and 36). Each step in a hierarchical multiple regression is a simultaneous or standard multiple regression provided there is more than one predictor in that step and can be interpreted in that way.

33.4 Data requirements for simultaneous or standard multiple regression

Although multiple regression can handle categorical variables when they have been turned into dummy variables, a basic requirement for any form of simple or multiple regression is that the scatterplot of the relation between a predictor variable and the criterion should show homoscedasticity, which means that the plot of the data points around the points of the line of best fit should be similar. This can be roughly assessed by plotting the line of best fit in a scatterplot and seeing if this assumption seems to have been met for each of the predictor variables (Chapter 11). It is important the criterion or dependent variable should be a continuous score which is normally distributed. If the criterion is a binomial variable, then it is far better to use binomial logistic regression (Chapter 39). If it is a multinomial variable, then multinomial logistic regression should be used (Chapter 38).

33.5 Problems in the use of simultaneous or standard multiple regression

As with stepwise multiple regression, care needs to be taken in interpreting the results of a simultaneous or standard multiple regression. If two predictor variables show multicollinearity in that they are highly correlated, then the standardised regression coefficients may be greater than ±1.00, which they should not exceed, and/or the sign of the coefficients may be reversed. These factors make it difficult to interpret these coefficients. The extent to which multicollinearity exists in the data is given by two collinearity statistics provided by SPSS Statistics called tolerance and VIF (variance inflation factor). Tolerance and VIF are directly related in that VIF = 1/tolerance, so only one of these indices need be reported. A tolerance of 0.1 or less and a VIF of more than

10 are considered as indicating multicollinearity. Multicollinearity should be dealt with by removing one or more predictors. A good rationale should be provided for the predictors that are omitted.

> You can find out more about simultaneous or standard multiple regression in Chapter 31 of Howitt, D. and Cramer, D. (2011). *Introduction to Statistics in Psychology*, 5th edition. Harlow: Pearson.

33.6 The data to be analysed

We will illustrate the computation of a simultaneous or standard multiple regression analysis with the data shown in Table 33.1, which consist of scores for six individuals on the four variables of educational achievement, intellectual ability, school motivation and parental interest.

Because this is for illustrative purposes and to save space, we are going to enter the data 20 times to give us a respectable amount of data to work with. Obviously you would *not* do this if your data were real. It is important to use quite a lot of research participants or cases for multiple regression. Ten or 15 times your number of variables would be reasonably generous. Of course, you can use less for data exploration purposes.

33.7 Entering the data

If saved, use the data from Chapter 32. Otherwise enter the data as follows.

Step 1

In 'Variable View' of the 'Data Editor', name the first five rows 'Achievement', 'Ability', 'Motivation', 'Interest' and 'Freq' respectively. Change 'Decimals' from '2' to '0'.

	Name	Type	Width	Decimals
1	Achievement	Numeric	8	0
2	Ability	Numeric	8	0
3	Motivation	Numeric	8	0
4	Interest	Numeric	8	0
5	Freq	Numeric	8	0

Step 2

Enter the data.
The 'Freq' variable is for weighting these six cases 20 times using the 'Weight Cases...' procedure (see Sections 3.7, 4.7, 8.8, 9.8, 16.7 or 17.7).
Save this data file to use for Chapters 34 and 35.

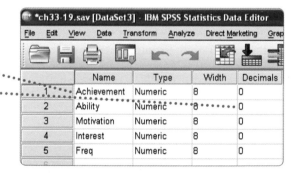

	Achievement	Ability	Motivation	Interest	Freq
1	1	2	1	2	20
2	2	2	3	1	20
3	2	2	3	3	20
4	3	4	3	2	20
5	3	3	4	3	20
6	4	3	2	2	20

33.8   Simultaneous or standard multiple regression analysis

Step 1

Select 'Analyze', 'Regression' and 'Linear...'.

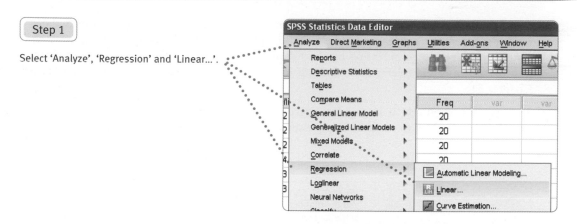

Step 2

Select 'Achievement' and the ➡ button beside the 'Dependent:' box to put it there.
Select 'Ability', 'Motivation' and 'Interest' and the ➡ button beside the 'Independent(s):' box to put them there.
Ensure 'Enter' 'Method:' is selected which is the default.
Select 'Statistics...'.

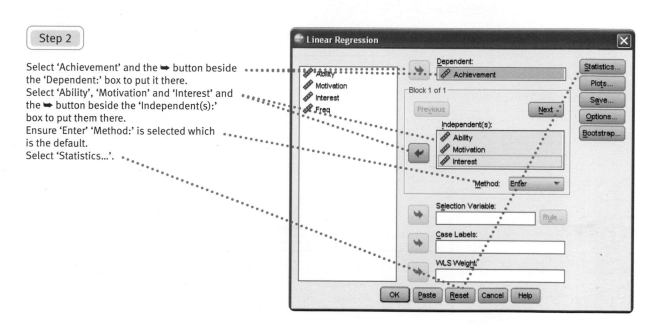

Step 3

Select 'Confidence intervals',
'R squared change', 'Descriptives',
'Part and partial correlations' and
'Collinearity diagnostics'.
Select 'Continue'.
Select 'OK' in the
previous box (see Step 2)
which reappears.

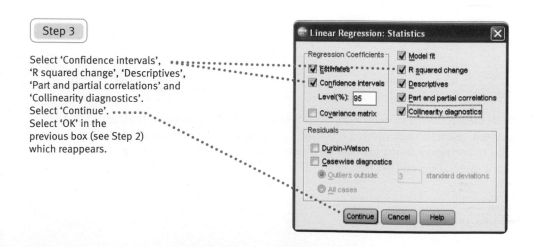

33.9 Interpreting the output

- There is a substantial amount of output. Multiple regression is a complex technique and needs further study in order to understand it fully. In interpreting the results of this analysis we shall restrict ourselves to commenting on the following statistics: Beta, its statistical significance, part correlations and tolerance level. Most of these are dealt with in a simple fashion in *ISP*, Chapter 31.

- The first table shows the means and standard deviations of the variables in the multiple regression. It is usual to report these descriptive statistics together with the correlations shown in the second table.

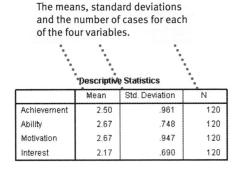

The means, standard deviations and the number of cases for each of the four variables.

Descriptive Statistics

	Mean	Std. Deviation	N
Achievement	2.50	.961	120
Ability	2.67	.748	120
Motivation	2.67	.947	120
Interest	2.17	.690	120

The correlations, their one-tailed significance level and the number of cases on which these are based is presented.

Correlations

		Achievement	Ability	Motivation	Interest
Pearson Correlation	Achievement	1.000	.701	.369	.127
	Ability	.701	1.000	.316	.108
	Motivation	.369	.316	1.000	.343
	Interest	.127	.108	.343	1.000
Sig. (1-tailed)	Achievement	.	.000	.000	.084
	Ability	.000	.	.000	.119
	Motivation	.000	.000	.	.000
	Interest	.084	.119	.000	.
N	Achievement	120	120	120	120
	Ability	120	120	120	120
	Motivation	120	120	120	120
	Interest	120	120	120	120

- The sixth table labelled 'Coefficients' is generally the most important table to look at. The standardised (partial) regression coefficients called Beta are presented in the fourth column. The biggest coefficient is for intellectual ability and is .649. The second biggest coefficient is for school motivation and is .164. The coefficient for parental interest is zero and so can be ignored.

● The statistical significance of the standardised coefficients is calculated by dividing the unstandardised or B coefficient by its standard error to give the t-value. So, for example, for intellectual ability the unstandardised or B coefficient of .833 which divided by its standard error of .088 gives a t-value of 9.47 which is similar to that of 9.52 shown in the table. The difference between the two values is due to rounding error. Our calculation is based on three decimal places while that of SPSS is based on more decimal places. The t-value is large and has a statistical significance of less than .001. In other it is statistically significant at less than .001.

● The part correlation is displayed in the 11th column of the table. The part correlation for intellectual ability is .615 which squared is about .38. So about .38 or 38 per cent of the variance in educational achievement is explained by or shared with intellectual ability taken into account its relations with the other variables.

● Note that the total amount of variance explained in this way by the three predictors of intellectual ability (.38), school motivation (.02) and parental interest (.00) is .40 which is less than that given by the R Square of .515 given in the fourth table. This is because the part correlation squared represents the variance that is unique to the criterion and the predictor and which is not shared by any other predictor. The R Square includes the variance that is shared between the criterion and all the predictors. So, for example, some of the variance between educational achievement and intellectual ability may be shared with the other predictors of school motivation and parental interest. The part correlation squared excludes this variance shared with the criterion and the other predictors.

● The tolerance level is shown in the 12th column of the table and is an index of the extent to which a predictor may be too highly correlated with other predictors to be entered into the multiple regression. This problem is known as multicollinearity or collinearity. A tolerance level of 0.1 or below indicates that the predictor may be too highly correlated with other predictors to be entered into the multiple regression. A tolerance level of above 0.1 means that the predictor may be entered into the multiple regression. As the tolerance level for all three predictors is greater than 0.1, all three predictors may be entered into the multiple regression as there is no multicollinearity. The VIF is simply 1/tolerance so there is no need to report both statistics. A VIF of less than 10 indicates no multicollinearity.

Coefficients[a]

Model		Unstandardized Coefficients		Standardized Coefficients	t	Sig.	95.0% Confidence Interval for B		Correlations			Collinearity Statistics	
		B	Std. Error	Beta			Lower Bound	Upper Bound	Zero-order	Partial	Part	Tolerance	VIF
1	(Constant)	-.167	.293		-.569	.571	-.747	.414					
	Ability	.833	.088	.649	9.520	.000	.660	1.007	.701	.662	.615	.900	1.111
	Motivation	.167	.073	.164	2.276	.025	.022	.312	.369	.207	.147	.804	1.244
	Interest	8.370E-18	.096	.000	.000	1.000	-.190	.190	.127	.000	.000	.882	1.133

a. Dependent Variable: Achievement

These are the standardised regression coefficients for the predictors.

These are the significance level of the standardised regression coefficients.

These are the part correlations for the predictors.

These are the tolerance level for the predictors.

REPORTING THE OUTPUT

● One way of concisely reporting the results of a simultaneous or standard multiple regression analysis is as follows:

> In a simultaneous multiple regression, intellectual ability had the strongest significant standardised regression coefficient with the criterion of educational achievement, Beta(116) = .65, $p < .001$, and explained about 38 per cent of the variance in educational achievement. School motivation had the second strongest significant standardised regression oefficient, Beta(116) = .15, $p < .05$, and explained about 2 per cent. Greater educational attainment was associated with greater intellectual ability and school motivation. The total variance explained by all three predictors is about 52 per cent. As the tolerance level for all predictors was greater than .840, multicollinearity was not considered a problem.

● It is usual to present a table which reports the means, standard deviations and correlations of the variables analyses as shown in Table 33.4. If alpha reliabilities are appropriate (Chapter 31), these are also often included in such a table. Means, standard deviations and correlations should be given to two decimal places. Two-tailed significance level can be obtained by conducting a correlation analysis (Chapter 10).

Table 33.4	Means, standard deviations and correlations for the four variables				
	M	**SD**	**1**	**2**	**3**
1 Achievement	2.50	0.96			
2 Ability	2.67	0.75	.70***		
3 Motivation	2.67	0.95	.37***	.32***	
4 Interest	2.17	0.69	.13	.11	.34***

***$p < .001$ (two-tailed).

● It may be useful to illustrate these results with the path diagram (see Chapter 34) shown in Figure 33.3, where arrows point from each of the predictors to the criterion. When this is done in journal articles, the correlation coefficients and the statistical symbols are not usually shown. However, when looking at or presenting your results it is useful to include them so that it is easier to see the effect of controlling for the other predictors. Asterisks indicate the coefficients are statistically significant. One asterisk indicates a significance level of .05 or less, two asterisks .01 or less and three asterisks .001 or less.

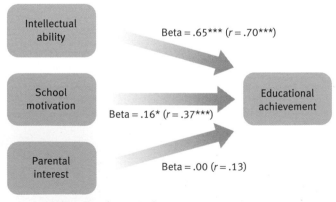

*$p < .05$, ***$p < .001$ (two-tailed).

| FIGURE 33.3 | Intellectual ability, school motivation, parental interest and educational achievement |

Summary of SPSS Statistics steps for simultaneous or standard multiple regression

Data

- Name the variables in 'Variable View' of the 'Data Editor'.
- Enter the data under the appropriate variable names in 'Data View' of the 'Data Editor'.

Analysis

- Select 'Analyze', 'Regression' and 'Linear ...'.
- Move the criterion or dependent variable to the 'Dependent:' box and the predictors or independent variables to the 'Independent(s):' box.
- Select 'Statistics ...' and then 'R squared change', 'Descriptives', 'Part and partial correlations', 'Collinearity diagnostics', 'Continue' and 'OK'.

Output

- Check sample size in 'Descriptive Statistics' table.
- In the 'Coefficients' table check that the Tolerance level is above 0.1 and note the standardised regression or Beta coefficient for each predictor and its significance level.

For further resources including data sets and questions, please refer to the website accompanying this book.

CHAPTER 34

Simple mediational analysis

Overview

- Mediational analysis examines how two variables precede or cause a third variable. Because this involves a sequence between the first two variable, the third variable may be called a consequent, effect, dependent, outcome or criterion variable. The other two variables also are given a variety of names including independent, predictor, antecedent or causal variables.

- One of these two predictor variables may be thought to intervene, mediate or explain the relation between the other predictor variable and the criterion variable. This variable may be called a mediating, mediator or intervening variable.

- The mediating variable may either partly explain the relation between the other predictor variable and the criterion variable or totally explain it. If it partly explains the relation, the other predictor will also partly explain the relation in the absence of additional mediating variables. If it totally explains the relation, the other predictor will only explain the relation through the mediating variable. In other words, the first variable will explain the relation indirectly rather than directly.

- For there to be a potential mediating effect, the correlation between the first predictor variable and the mediating variable and the correlation between the mediating and the criterion variable must be significant or substantial.

- A potential mediating effect is assessed with simultaneous or standard multiple regression in which the first predictor variable and the mediator variable are entered in a single step in a multiple regression.

- A total mediating effect may be considered to exist if the standardised partial regression coefficients (correlations) between the first predictor variable and the criterion variable and between the mediating and the criterion variable are non-significant and close to zero in the regression analysis.

- A partial mediating effect may be thought to occur if the following two conditions hold: (a) the standardised partial regression coefficient between the mediating and the criterion variable is significant or substantial and (b) the standardised partial regression coefficient between the first predictor variable and the criterion variable is significant or substantial but somewhat lower than the correlation between these two variables.

- More complex ways of conducting a mediational analysis are described in Chapters 52–54 (Chapters 53 and 54 are on the website).

34.1 What is simple mediational analysis?

A simple mediational analysis involves the statistics of correlations (Chapter 10) and multiple regression (Chapters 32 and 33). Both simple and multiple regression are used to explain one variable in terms of one or more than one other variable respectively. Various terms have been adopted to refer to the variable being explained. These include the consequent, criterion, dependent, effect or outcome variable. The terms employed to refer to the explanatory variables also vary, including terms such as causal, independent or predictor variables. We will generally use the term 'criterion variable' to refer to the variable being explained and the term 'predictor variable' to describe the variables measured to explain the criterion variable. The term 'explanation' usually refers to the size and the direction of the association between the predictor and the criterion variables and/or the percentage of the variance in the criterion variable that is explained or shared with the predictor variables. The greater the association, the greater the predictor variable explains the criterion variable. These two measures are related in that the stronger the association is between two variables, the greater the proportion or percentage of their variance they share. The terminology is summarised in Figure 34.1.

There are two important things to bear in mind:

- A mediator or intervening variable is one that is thought to wholly or partially explain the relation between a predictor and a criterion variable.

- As the mediator variable is also thought to explain the criterion variable, it is also a predictor variable but it is one whose effect on the criterion variable is assumed to come later on in the sequence of variables assumed to affect the criterion variable.

Suppose, for example, we think that more intelligent people are likely to do better at their schoolwork because of their intelligence. Now being intelligent does not necessarily mean that you will do better at school. You may be more interested in and spend more time doing activities other than schoolwork, in which case your schoolwork is likely to suffer. However, being more intelligent may mean that you are more interested in and spend more time doing schoolwork because you find it challenging, you think it relatively easy and you do well in it. So part or all of the relationship between intellectual ability and educational achievement may be due to how motivated you are to do well at school. The more motivated you are, the better you do at school.

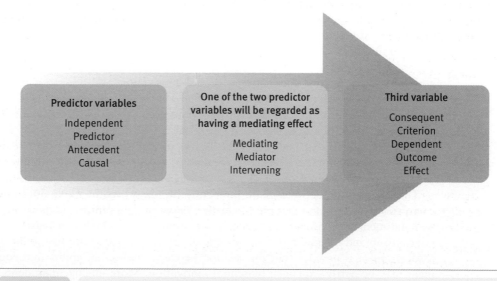

FIGURE 34.1   Equivalent terms for variables in mediational analysis

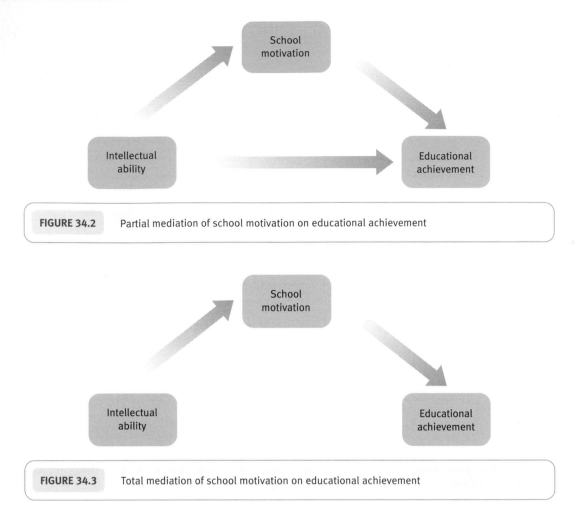

FIGURE 34.2   Partial mediation of school motivation on educational achievement

FIGURE 34.3   Total mediation of school motivation on educational achievement

It is important to draw out the relation between these three variables in terms of a simple path diagram as shown in Figure 34.2 where the predictor variable of intellectual ability is listed first and to the left on the page and the criterion variable of educational achievement is listed last and to the furthest right on the page. The mediator or intervening predictor variable of school motivation is placed between these other two variables and above them. The direction of the relation between the three variables is indicated by arrows pointing to the right. So there is a right pointing arrow between intellectual ability and educational achievement, between intellectual ability and school motivation and between school motivation and educational achievement. If we thought that the relation between intellectual ability and educational achievement was totally mediated by school motivation, then we would omit the arrow between intellectual ability and educational achievement as shown in Figure 34.3.

A simple mediational analysis will be illustrated with the same data that we used in Chapter 32 for the stepwise multiple regression except that we will ignore the fourth variable of parental interest. The data are shown in Table 34.1.

The correlations between these three variables is shown in Table 34.2 (see Chapter 10 for correlation). We can see that the correlation between the predictor variable of intellectual ability and the mediating variable of school motivation is .32, which is a reasonable size for such a relationship and is significant for a sample of this size. Because of this significant and substantial association, it is possible that the relation between intellectual ability and educational achievement may be partly explained by school motivation. More intelligent students may do better at school because they are more motivated to do better. Figure 34.4 shows the main steps in understanding simple mediational analysis.

Table 34.1	Data for a simple mediational analysis		
Educational achievement		**Intellectual ability**	**School motivation**
1		2	1
2		2	3
2		2	3
3		4	3
3		3	4
4		3	2

Table 34.2	A correlation matrix for the three variables	
	1	**2**
1 Achievement		
2 Ability	.70	
3 Motivation	.37	.32

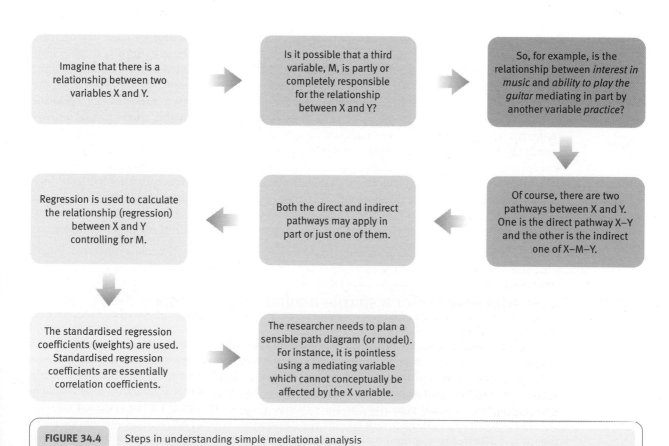

FIGURE 34.4   Steps in understanding simple mediational analysis

34.2   When to use simple mediational analysis

A simple mediational analysis should be used when the following three conditions are met:

- The mediator variable should provide a plausible explanation for the relation between the predictor and the criterion variable. So, in this case, age should not be used as a mediator variable since intellectual ability cannot affect age. For example, intellectual ability does not and cannot cause people to be older.

- There is a significant or substantial correlation between (a) the predictor and the mediator variable and (b) between the mediator and the criterion variable as this indicates that these variables are related. If there is not a substantial correlation between these variables, then this means that intellectual ability is not related to school motivation which is not related to educational achievement. In others words, school motivation cannot mediate the relation between intellectual ability and educational achievement because it is not related to either of these two variables.

- It is not necessary that there should also be a significant or substantial correlation between the predictor and the criterion variables as this relation may be reduced or suppressed by the mediator variable. This sometimes happens.

34.3   When not to use simple mediational analysis

A simple mediational analysis should not be carried out under the following four conditions. Firstly, no plausible explanation can be provided as to why the mediator variable may explain the relation between the predictor and the criterion variable. Secondly and thirdly, there is little or no relation between the predictor and the mediator variable and between the mediator and the criterion variable. Little or no relation between these variables suggests that these variables are unrelated and so cannot provide an explanation for the relation between the predictor and the criterion variable. Fourthly, some of the variables should not be very highly correlated as this may make interpretation difficult. Very high correlations between some of the variables may make the standardised regression coefficients greater than ±1.00 which they are not supposed to be. A perfect correlation of ±1.00 only exists when a variable is effectively being correlated with itself. This should not be the case when the two variables are expected to be different variables. Very high correlations may also reverse the sign of the partial standardised coefficient. This reversal may be due to the high correlation rather than to the actual relation between the three variables. In other words, it is due to the statistical formulae that are used to partial out or control for other variables.

34.4   Data requirements for a simple mediational analysis

A basic data requirement for any form of simple or multiple regression is that the scatterplot of the relation between the three variables should show homoscedasticity which means that the plot of the data points around the points of the line of best fit should be similar. This can be roughly assessed by plotting the line of best fit in a scatterplot and seeing if this assumption seems to have been met for each of the predictor variables (Chapter 11). It is important the criterion variable should be a continuous score which is normally distributed. If the criterion is a binomial variable, then it is far better to use binomial logistic regression (Chapter 39). If it is a multinomial variable, then multinomial logistic regression should be used (Chapter 38).

34.5   Problems in the use of simple mediational analysis

Care needs to be taken in interpreting the results of a simultaneous or standard multiple regression used in simple mediational analysis. If some of the variables are very highly correlated, then two potential problems may occur. One problem is the standardised regression coefficients may be greater than ±1.00, which indicates that the two variables are more than perfectly correlated. Two variables are only perfectly correlated if they are, in effect, the same variable, which should not be the case when the variables are different. Another problem is the sign of the partial regression coefficient may be reversed.

You can find out more about simple mediational analysis in Chapter 32 of Howitt, D. and Cramer, D. (2011). *Introduction to Statistics in Psychology*, 5th edition. Harlow: Pearson.

34.6   The data to be analysed

We will illustrate the computation of a simple mediational analysis using simultaneous multiple regression with the data shown in Table 34.1. The data consist of scores for six individuals on the three variables of educational achievement, intellectual ability and school motivation respectively.

Because this is for illustrative purposes and to conserve space, we are going to enter the data 20 times to give us a respectable amount of data to work with. Obviously you would *not* do this if your data were real. It is important to use quite a substantial number of research participants or cases for multiple regression. Ten or 15 times your number of variables would be reasonably generous. Of course, you can use less for data exploration purposes.

34.7   Entering the data

If saved, use the data from Chapter 32 or 33. Otherwise enter the data as follows.

Step 1

In 'Variable View' of the 'Data Editor', name the first five rows 'Achievement', 'Ability', 'Motivation', 'Interest' and 'Freq'. Change 'Decimals' from '2' to '0'.

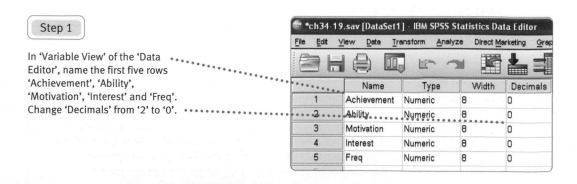

Step 2

Enter the data.
The 'Freq' variable is for weighting
these six cases 20 times using the
'Weight Cases...' procedure (see
Sections 3.7, 4.7, 8.8, 9.8, 16.7 or 17.7).
Save this data file to use for
Chapters 35 and 52.

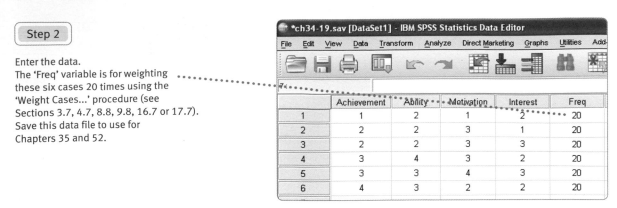

34.8 Simultaneous multiple regression analysis

Step 1

Select 'Analyze',
'Regression' and
'Linear...'.

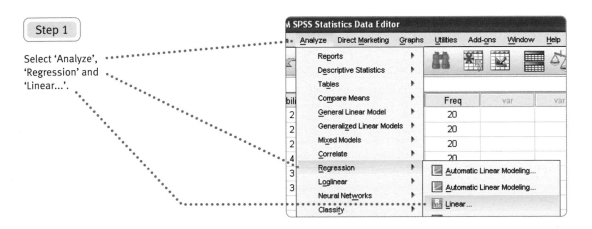

Step 2

Select 'Achievement' and the
➡ button beside the 'Dependent:'
box to put it there.
Select 'Ability' and 'Motivation'
and the ➡ button beside the
'Independent(s):' box to put
them there.
Select 'Statistics...'.

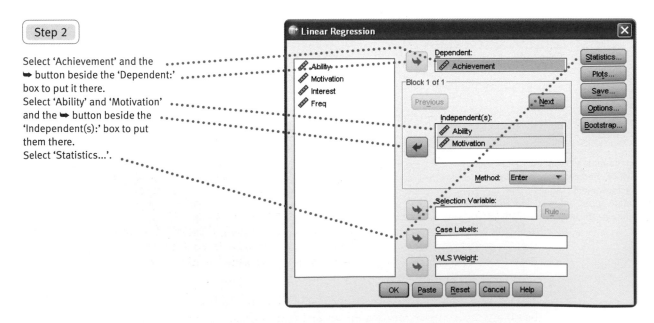

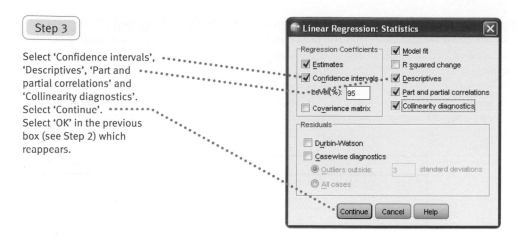

Step 3

Select 'Confidence intervals', 'Descriptives', 'Part and partial correlations' and 'Collinearity diagnostics'.
Select 'Continue'.
Select 'OK' in the previous box (see Step 2) which reappears.

34.9   Interpreting the output

- The main table in the output you have to look at is that headed 'Coefficients'. In this table you need to primarily look at the standardised regression coefficient Beta which is in the fourth column and the statistical significance of Beta which is in the sixth column. Both these regression coefficients are statistically significant. The standardised regression coefficient between 'Ability' and 'Achievement' is .649 while that between 'Motivation' and 'Achievement' is much smaller at .164.

- This table also shows the correlations between the effect variable of 'Achievement' and the two predictors of 'Ability' and 'Motivation'.

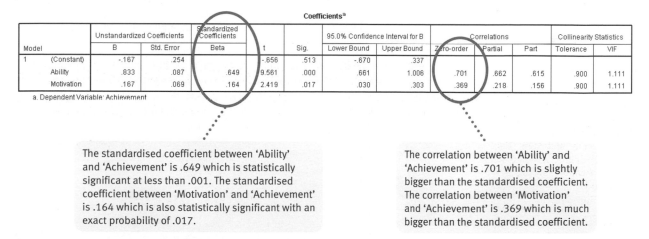

Coefficients[a]

Model	Unstandardized Coefficients B	Unstandardized Coefficients Std. Error	Standardized Coefficients Beta	t	Sig.	95.0% Confidence Interval for B Lower Bound	95.0% Confidence Interval for B Upper Bound	Correlations Zero-order	Correlations Partial	Correlations Part	Collinearity Statistics Tolerance	Collinearity Statistics VIF
1 (Constant)	-.167	.254		-.656	.513	-.670	.337					
Ability	.833	.087	.649	9.561	.000	.661	1.006	.701	.662	.615	.900	1.111
Motivation	.167	.069	.164	2.419	.017	.030	.303	.369	.218	.156	.900	1.111

a. Dependent Variable: Achievement

The standardised coefficient between 'Ability' and 'Achievement' is .649 which is statistically significant at less than .001. The standardised coefficient between 'Motivation' and 'Achievement' is .164 which is also statistically significant with an exact probability of .017.

The correlation between 'Ability' and 'Achievement' is .701 which is slightly bigger than the standardised coefficient. The correlation between 'Motivation' and 'Achievement' is .369 which is much bigger than the standardised coefficient.

- The 'Correlation' table shows the one-tailed statistical significance of these correlations. The two-tailed significance level may be obtained by doubling these significance levels or by using the 'Correlate' procedure to produce this significance level (Chapter 10). This makes no difference in this case as significance levels are reported to three decimal places and these significance levels remain below .001.

- The correlation coefficients, their significance level and the number of cases on which they are based are shown twice, once in the lower left triangle and again in the upper right triangle.

- The tolerance level is shown in the 12th column of the table and is an index of the extent to which a predictor may be too highly correlated with other predictors to be entered into the multiple regression. This problem is known as multicollinearity or collinearity. A tolerance level of 0.1 or below indicates that the predictor may be too highly correlated with other predictors to be entered into the multiple regression. A tolerance level of above 0.1 means that the predictor may be entered into the multiple regression. As the tolerance level for all three predictors is greater than 0.1, all three predictors may be entered into the multiple regression as there is no multicollinearity. The VIF (variance inflation factor) is simply 1/tolerance so there is no need to report both statistics. A VIF of less than 10 indicates no multicollinearity.

Correlations

		Achievement	Ability	Motivation
Pearson Correlation	Achievement	1.000	.701	.369
	Ability	.701	1.000	.316
	Motivation	.369	.316	1.000
Sig. (1-tailed)	Achievement	.	.000	.000
	Ability	.000	.	.000
	Motivation	.000	.000	.
N	Achievement	120	120	120
	Ability	120	120	120
	Motivation	120	120	120

The correlations between the three variables.

The one-tailed statistical significance of the correlations.

The number of cases on which the correlations and significance as based.

REPORTING THE OUTPUT

- One of the most succinct ways of reporting the results of this simple mediational analysis is as follows.

  As the relation between intellectual ability and educational achievement, $r(118) = .70$, two-tailed $p < .001$, was little affected when school motivation was controlled, $B = .65$, $t(117) = 9.56$, two-tailed $p < .001$, school motivation was not considered to mediate the relation between intellectual ability and educational achievement. Greater educational attainment was associated with greater intellectual ability.

- It may be useful to illustrate these results with the path diagram shown in Figure 34.5. When this is done in journal articles, the correlation coefficients and the statistical symbols are not usually shown. However, when examining or presenting your results you may find it useful to include them so that it is easier to see the effect of controlling for the mediator variable. Asterisks indicate the coefficients are statistically significant. One asterisk indicates a significance level of .05 or less, two asterisks .01 or less and three asterisks .001 or less.

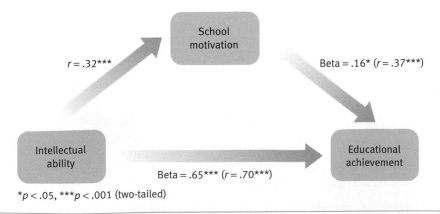

$*p < .05$, $***p < .001$ (two-tailed)

FIGURE 34.5 Intellectual ability, school motivation and educational achievement

Summary of SPSS Statistics steps for simple mediational analysis

Data

- Name the variables in 'Variable View' of the 'Data Editor'.
- Enter the data under the appropriate variable names in 'Data View' of the 'Data Editor'.

Analysis

- Select 'Analyze', 'Regression' and 'Linear ...'.
- Move the criterion or dependent variable to the 'Dependent:' box and the predictor and mediating variable to the 'Independent(s):' box.
- Select 'Statistics ...' and then 'R squared change', 'Descriptives', 'Part and partial correlations', 'Collinearity diagnostics', 'Continue' and 'OK'.

Output

- Check sample size in 'Descriptive Statistics' table.
- In the 'Coefficients' table note the standardised coefficients (Beta) and their statistical significance (Sig.).

For further resources including data sets and questions, please refer to the website accompanying this book.

CHAPTER 35

Hierarchical multiple regression

Overview

- Hierarchical multiple regression allows the researcher to decide which order to use for a list of predictors.

- This is achieved by putting the predictors or groups of predictors into blocks of variables. The computer will carry out the regression taking each block in the order that it was entered into SPSS Statistics, so it provides a way of forcing the variables to be considered in the sequence chosen by the researcher. Rather than let the computer decide on statistical criteria as in Chapter 32, the researcher decides which should be the first predictor, the second predictor, and so forth.

- A block may be a single predictor but it may be a group of predictors.

- This order of the blocks is likely to be chosen on theoretical grounds. One common procedure is to put variables which need to be statistically controlled in the first block. The consequence of this is that the control variables are partialled out before the rest of the blocks are analysed.

- Since the researcher is trying to produce models of the data, the multiple regression may be varied to examine the effects of, say, entering the blocks in a different order.

35.1 What is hierarchical multiple regression?

Hierarchical multiple regression is used to determine how much variance in the criterion, dependent or outcome variable is explained by predictors (independent variables) when they are entered in a particular sequence. The more variance that a predictor explains, the potentially more important that variable may be. The variables may be entered in a particular sequence on practical or theoretical grounds. An example of a practical situation is where we are interested in trying to predict how good someone might be in their job. We could collect a number of variables which we think might be related to how good they might be. This might include how well they did at school, how many outside interests they have, how many positions of responsibility they hold, how well they performed in interview and so on. Now some of these variables might be easy and cheap to obtain, such as finding out how well they did at school or how many outside interests they have, while other variables might be more costly to obtain, such as how well they performed at interview. We may be interested in whether using these more costly variables enable us to make much better predictions about how likely someone is to be good at their job than the cheaper ones. If they do not add much to how well we are able to predict job competence, then it might be better not to use them. So we could add the cheaper variables in the first step of our hierarchical multiple regression and see how much of the variance in job competence is generally explained. We could then add the more costly variables in the second step to see how much more of the variance in job competence these variables explain.

Turning to a more theoretical situation, we may be interested in trying to explain how much different variables or types of variables explain some criterion such as educational achievement. We may arrange the variables in a particular sequence. We may be interested first in how much basic demographic variables explain such as social class and age. Next we may wish to consider whether going to preschool made any difference to explaining educational achievement beyond social class and age. Third, we may wish to look at the effects of intellectual ability on educational achievement. Finally, we may want to examine the effects of other variables such as how motivated children are to do well at school and how interested their parents are in how well they do at school. Entering these variables in a sequence like this will enable us to see how much each group or block of variables adds to how well we can predict educational achievement.

Hierarchical multiple regression will be illustrated with the same data that we used in Chapter 32 except that we will add another variable, that of social class. The data are shown in Table 35.1. We will enter social class in the first step of the analysis, ability in the second step and motivation and interest in the third step.

To work out the proportion of variance in educational achievement that is explained at each step we calculate the squared multiple correlation (R^2) which is derived by multiplying the standardised (partial) beta regression coefficient by the correlation coefficient for each predictor and the criterion in that step and summing their products. The correlations between educational achievement and ability, motivation, interest and class are about .70, .37, .13 and .07 respectively. (All figures are calculated in Section 35.9.) The standardised regression coefficient for class in the first step is its correlation which is .07. Consequently class explains about 0.5 per cent (.07 × .07 × 100) of the variance in educational achievement which is effectively none. The standardised regression coefficients for class and ability in the second step are about −.14 and .74 respectively. So the variance explained by both class and ability is about 51 per cent {[(.07 × −.14) + (.70 × .74)] × 100}. If we subtract the per cent of variance in the first step from that in the second step we see that ability explains about a further 50.5 per cent. Finally the standardised regression coefficients for class, ability, motivation and interest is about −.439, .730, .185 and .314 respectively. Thus the variance explained by all four of these predictors in the third step is about 59 per cent {[(.07 × −.439) + (.70 × .730) + (.37 × .185) + (.13 × .314)] × 100}. If we

Table 35.1	Data for hierarchical multiple regression			
Educational achievement	Intellectual ability	School motivation	Parental interest	Social class
1	2	1	2	2
2	2	3	1	1
2	2	3	3	5
3	4	3	2	4
3	3	4	3	3
4	3	2	2	2

Hierarchical multiple regression builds on the basic ideas of multiple regression.

It is hierarchical in the sense that there is a structure to the order of the predictor variables entered into the analysis by the researcher.

Sometimes it is known as sequential multiple regression, which you may find a more apt name.

One sensible plan for the order of the blocks is to have the first block(s) consisting of variables that the researcher might wish to partial out or control for.

It is for the researcher to decide the order of the blocks but within a block there may be selection using the stepwise method, for example.

The structuring is achieved by using blocks of variables and each block has a stipulated order in the analysis. One could have one block per predictor but it is more usual to have several.

So the first block, for example, might consist of demographic variables such as age or occupational status.

There is no reason why the researcher should not experiment with different multiple regression models and not depend simply on the first one stipulated.

Note that the block as a whole has its own regression weights as well as the individual components of the block.

FIGURE 35.1   Steps in hierarchical multiple regression

subtract the per cent of variance in the second step from that in the third step we see that motivation and interest explain about a further 8.5 per cent. We can see that ability explains the most variance in educational achievement even when social class is taken into account. There is no reason to calculate these yourself as they can be found under Model Summary in the SPSS output discussed later in this chapter. Figure 35.1 outlines the main steps in hierarchical multiple regression.

35.2   When to use hierarchical multiple regression

Hierarchical multiple regression should be used when you want to enter the predictor variables in a particular sequence as described above. It is also used to determine whether there is a significant interaction between your predictor variables and the criterion. The interaction term is created by multiplying the predictor variables together and entering this term in a subsequent step in the multiple regression after the predictor variables have been entered. If the interaction term explains a significant proportion of the variance in the criterion, this implies that there is an interaction between these predictor variables. The nature of this interaction needs to be determined, as is described in the following chapter.

35.3   When not to use hierarchical multiple regression

Hierarchical multiple regression should not be used when there is no reason why the predictor variables should be entered in a particular order. If there are no good grounds for prioritising the variables, then it is better to enter them in a single block or step. This is known as a standard multiple regression.

35.4   Data requirements for hierarchical multiple regression

The data requirements for hierarchical multiple regression are the same as those for stepwise multiple regression which are outlined in Chapter 32. The criterion variable should be a quantitative variable which is normally distributed. The relation between each predictor and the criterion should show homoscedasticity in the sense that the plot of the data points around the best fitting regression line should be similar at each point along the regression line. A rough idea of whether this is the case can be obtained by producing a scatterplot with a line of best fit between each predictor and the criterion as shown in Chapter 11.

35.5   Problems in the use of hierarchical multiple regression

As in all regression and correlation analyses, it is important to look at the mean and standard deviation of the variables. If the mean is very low or high then this suggests that the variation in the variable may not be well spread. A variable which has a smaller standard deviation indicates that it has a smaller variance, which means that the relation between this variable and the criterion may be smaller than a predictor which has a larger variance when all other things are equal.

You can find out more about hierarchical multiple regression in Chapter 32 of Howitt, D. and Cramer, D. (2011). *Introduction to Statistics in Psychology*, 5th edition. Harlow: Pearson.

35.6   The data to be analysed

We will illustrate the computation of a hierarchical multiple regression analysis with the data shown in Table 35.1, which consist of scores for six individuals on the four variables of educational achievement, intellectual ability, school motivation and parental interest.

We have added a further variable, social class, which is on a scale of 1 to 5, with 5 being the lowest social class. Hierarchical analysis is used when variables are entered in an order predetermined by the researcher on a 'rational' basis rather than in terms of statistical criteria. This is done by ordering the independent variables in terms of blocks of the independent variables, called Block 1, Block 2, etc. A block may consist of just one independent variable or several. In this particular analysis, we will make Block 1 social class ('Class'), which is essentially a demographic variable which we would like to control for. Block 2 will be intellectual ability ('Ability'). Block 3 will be school motivation ('Motivation') and parental interest ('Interest'). The dependent variable or criterion to be explained is educational achievement ('Achievement').

In our example, the model essentially is that educational achievement is affected by intellectual ability and motivational factors such as school motivation and parental interest. Social class is being controlled for in this model since we are not regarding it as a psychological factor.

35.7 Entering the data

If saved, use the data from Chapter 32, 33 or 34 and add the sixth variable of 'Class'. Otherwise enter the data as follows.

Step 1

In 'Variable View' of the 'Data Editor', name the first six rows 'Achievement', 'Ability', 'Motivation', 'Interest', 'Freq' and 'Class'.
Change 'Decimals' from '2' to '0'.

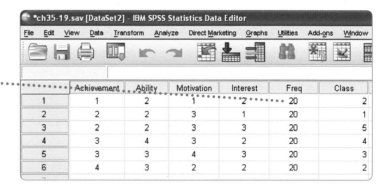

Step 2

Enter the data.
The 'Freq' variable is for weighting these six cases 20 times using the 'Weight Cases...' procedure (see Sections 3.7, 4.7, 8.8, 9.8, 16.7 or 17.7).

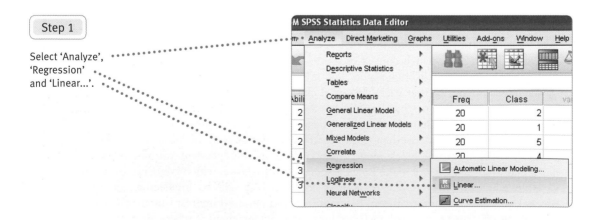

35.8 Hierarchical multiple regression analysis

Step 1

Select 'Analyze', 'Regression' and 'Linear...'.

Step 2

Select 'Achievement' and the ➡
button beside the 'Dependent:'
box to put it there.
Select 'Class' and the ➡ button
beside the 'Independent(s):' box
to put it there. Select 'Next'.

Step 3

Select 'Ability' and
the ➡ button beside the
'Independent(s):'
box to put it there.
Note this is 'Block 2 of 2'.
Select 'Next'.

Step 4

Select 'Motivation' and
'Interest' and the ➡ button
beside the 'Independent(s):'
box to put them there.
Select 'Statistics'.

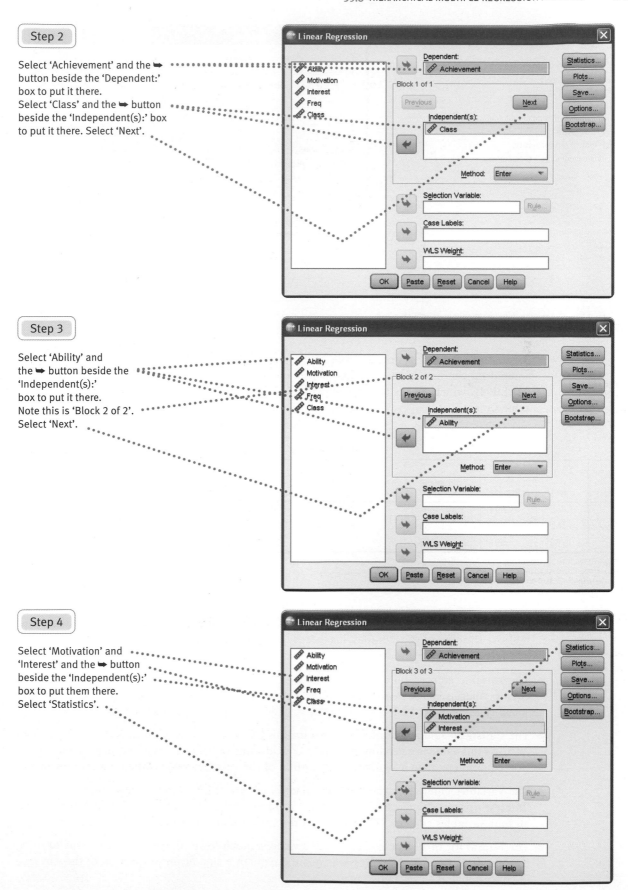

Step 5

Select 'Confidence intervals',
'R squared change',
'Descriptives', 'Part and
partial correlations' and
'Collinearity statistics'.
Select 'Continue'.
Select 'OK' in the previous
box (in Step 5) which reappears.

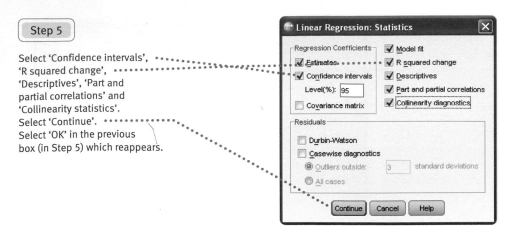

35.9 Interpreting the output

● As summarised in the fourth table of the output entitled 'Model Summary', the variable entered on the first block is 'Class' (social class). The R Square for this block is effectively 0.0 (.004), which means that social class explains 0 per cent of the variance of educational achievement.

Model Summary

Model	R	R Square	Adjusted R Square	Std. Error of the Estimate	Change Statistics				
					R Square Change	F Change	df1	df2	Sig. F Change
1	.065[a]	.004	-.004	.963	.004	.497	1	118	.482
2	.714[b]	.509	.501	.679	.505	120.333	1	117	.000
3	.769[c]	.591	.577	.626	.082	11.500	2	115	.000

a. Predictors: (Constant), Class
b. Predictors: (Constant), Class, Ability
c. Predictors: (Constant), Class, Ability, Motivation, Interest

● The statistical significance of the *F*-ratio of .497 for this block or model is .482. As this value is above the critical value of .05, this means that the regression equation at this first stage does not explain a significant proportion of the variance in educational achievement.

● The variable entered on the second block is 'Ability' (intellectual ability). The Adjusted R Square for this block or model is .501, which means that intellectual ability together with social class explains 50.1 per cent of the variance of educational achievement.

● The statistical significance of the *F*-ratio for this block is .000 which means that it is less than .001. As this value is much lower than the critical value of 0.05, the first two steps of the regression equation explain a significant proportion of the variance in educational achievement.

● The variables entered on the third and final block are 'Motivation' (school motivation) and 'Interest' (parental interest). The Adjusted R Square for this block is .577, which means that all four variables explain 57.7 per cent of the variance in educational achievement.

● The *F*-ratio for this block is .000. As this value is much lower than the critical value of 0.05, the first three steps in the regression equation explain a significant proportion of the variance in educational achievement.

- The simplest interpretation of the output comes from examining the fourth table entitled 'Coefficients' of the output. Especially useful are the Beta column and the Sig (of t) column. These tell us that the correlation (Beta) between 'Class' (social class) and 'Achievement' (educational achievement) is –.439 when the other predictors are taken into account. This correlation is significant at the .000 level which means that it is less than .001. This coefficient is now significant because the two variables of 'Ability' and 'Interest' suppress the zero-order coefficient between 'Class' and 'Achievement'. Having controlled for social class in Block 1, the correlation between 'Ability' (intellectual ability) and 'Achievement' (educational achievement) is .730. This is also significant at the .000 level. Finally, having controlled for 'Class' (social class) and 'Ability' (intellectual ability), the correlations for each of the variables in Block 3 (school motivation and parental interest) with educational achievement ('Achievement') are given separately.

- The part correlations for Class, Motivation, Ability and Interest are –.275, .667, .165 and .201 respectively, which when squared mean that they account for about 7.6, 44.5, 2.7 and 4.0 per cent of the variance in Achievement, not taking into account the variance that they share between them. As the total percentage of variance these four variables explain is about 59.1 per cent this means that the percentage of variance they share is 0.3 per cent (59.1 – 58.8).

- The tolerance level is shown in the 12th column of the table and is an index of the extent to which a predictor may be too highly correlated with other predictors to be entered into the multiple regression. This problem is known as multicollinearity or collinearity. A tolerance level of 0.1 or below indicates that the predictor may be too highly correlated with other predictors to be entered into the multiple regression. A tolerance level of above 0.1 means that the predictor may be entered into the multiple regression. As the tolerance level for all three predictors is greater than 0.1, all three predictors may be entered into the multiple regression as there is no multicollinearity. The VIF (variance inflation factor) is simply 1/tolerance so there is no need to report both statistics. A VIF of less than 10 indicates no multicollinearity.

Coefficients[a]

Model		Unstandardized Coefficients B	Std. Error	Standardized Coefficients Beta	t	Sig.	95.0% Confidence Interval for B Lower Bound	Upper Bound	Correlations Zero-order	Partial	Part	Collinearity Statistics Tolerance	VIF
1	(Constant)	2.369	.205		11.543	.000	1.963	2.776	·				
	Class	.046	.065	.065	.705	.482	-.083	.176	.065	.065	.065	1.000	1.000
2	(Constant)	.250	.241		1.036	.303	-.238	.728					
	Class	-.100	.048	-.140	-2.082	.040	-.195	-.005	.065	-.189	.135	.923	1.083
	Ability	.950	.087	.740	10.970	.000	.778	1.122	.701	.712	.711	.923	1.083
3	(Constant)	-.563	.284		-1.984	.050	-1.124	-.001					
	Class	-.313	.068	-.439	-4.615	.000	-.447	-.178	.065	-.395	-.275	.394	2.539
	Ability	.938	.084	.730	11.180	.000	.771	1.104	.701	.722	.667	.835	1.198
	Motivation	.188	.068	.185	2.769	.007	.053	.322	.369	.250	.165	.800	1.250
	Interest	.438	.130	.314	3.374	.001	.181	.694	.127	.300	.201	.411	2.426

a. Dependent variable: Achievement

The quickest way to access the output is to concentrate on the third model in the above output. The output indicates that when Class, Ability, Motivation and Interest are all entered that each has a significant association with the dependent variable Achievement. Class has a negative relationship whereas the others have a positive one.

REPORTING THE OUTPUT

● There are various ways of reporting the results of a hierarchical multiple regression analysis. In such a report we would normally describe the percentage of variance explained by each set or block of predictors (from the value of the R Square).
● One way of reporting these results is to state that:

> In a hierarchical multiple regression, potential predictors of Achievement were entered in blocks. Social class was entered first, then intellectual ability was added in the second block, and school motivation and parental interest were added in the final block. The final model indicated that social class was a negative predictor ($B = -.31$), intellectual ability was a positive predictor $B = .94$), and school motivation and parental interest were also positive predictors ($B = .19$ and $.44$). All predictors were significant at the 1 per cent level.

● One would also need to summarise the regression equation as in Table 35.2.

Table 35.2	Hierarchical multiple regression of predictors of educational achievement		
Blocks	**B**	**Standard error B**	**Beta**
Block 1: Social class	−.31	.07	−.44*
Block 2: Intellectual ability	.94	.08	.73*
Block 3: School motivation	.19	.07	.19*
Parental interest	.44	.13	.31*

*Significant at .01.

Summary of SPSS Statistics steps for hierarchical multiple regression

Data

● Name the variables in 'Variable View' of the 'Data Editor'.
● Enter the data under the appropriate variable names in 'Data View' of the 'Data Editor'.

Analysis

● Select 'Analyze', 'Regression' and 'Linear . . .'.
● Move the criterion or dependent variable to the 'Dependent:' box and the other predictors or independent variables to the 'Independent(s):' box in the appropriate blocks.
● In the 'Method' box check 'Enter' has been selected.
● Select 'Statistics . . .' and then 'Descriptives', 'Collinearity diagnostics', 'Continue' and then 'OK'.

Output

● In the 'Coefficients' table note the size and direction of the standardised or beta coefficient for the predictors together with their tolerance value.

For further resources including data sets and questions, please refer to the website accompanying this book.

CHAPTER 36

Moderator analysis with continuous predictor variables

Overview

- A moderator variable is one where there is a different relationship between the main variables in a study at different levels of the moderator variable.

- A simple example of a moderator variable could be gender. Imagine that a study finds a strong relationship between educational achievement and intelligence. Subsequently it is found that the relationship is much stronger in girls than in boys. This indicates that gender is a moderator variable since the main findings differ for the different levels (categories) of the third variable gender.

- A clear distinction should be made between moderator variables and mediating variables. The later is a variable which brings about the relationship between the two main variables in an analysis.

- The method of detecting moderator variable effects differs according to the statistical nature of the two main variables and the third variable. If all three variables are score variables then hierarchical multiple regression using interactions is the appropriate analysis method. Since score variables are the most likely data in psychological research, this chapter concentrates on the hierarchical multiple regression approach.

- If all of the variables are nominal category variables then log-linear analysis is the appropriate method of analysis. If there is a mixture of score and nominal category variables it is usually possible to adopt the ANOVA approach though some manipulation of the variables may be required.

- Calculating the hierarchical multiple regression involves the preliminary step of converting all of the variables into z-scores otherwise problems can occur in the analysis.

- Furthermore, an interaction term has to be generated by multiplying the z-scores of the moderator variable by the scores on one of the other variables.

- The analysis proceeds hierarchically since one needs to remove the independent effects of the moderator variable and the other variable. Once this has been done, the analysis can turn to the question of whether the interaction term has any relationship with the variable which has been used as the dependent variable. If it has, then a moderator effect has been found.

36.1   What is a moderator variable analysis?

A moderator variable can be seen when the relationship between two main variables in a study is substantially different for different levels of a third variable – the moderator variable. For example, there may be a relationship between delinquency and the parenting skills of a child's mother. This is of interest in itself but on exploring the data further, we may find that things differ according to a third variable such as the family's financial circumstances. It may be that the relationship between delinquency and parenting skills is much stronger where the family has a low income than where it has a higher income. In these circumstances, one might come to the view that the third variable income is moderating the relationship between delinquency and parenting skills. This is illustrated in the example in Figure 36.1.

Moderator variables are not the same thing as mediator variables. Mediator variables, in this case, would be third variables which mediate between the two main variables. For example, the relationship between parenting skills and delinquency may be the result of parenting skills affecting how accepted/rejected the child feels by its parents. The feeling of rejection then encourages delinquent behaviour in the child. This mediating effect is illustrated in Figure 36.2. What should be clear is that the moderator variable and the mediator variable are very different.

The process of detecting moderator variable effects varies according to the nature of the variables in the study. There are basically two types of variable: scores and nominal category variables. Scores dominate the data collected by psychologists so the typical situation will be one in which the two main variables and the potential moderator variables are all score variables. The recommended way of identifying moderator variables in this case is to use hierarchical multiple regression (Chapter 35). If all three variables were nominal category variables then the appropriate way of testing for moderator variables would be to use logistic regression (Chapters 38 and 39). If you have a mixture of score variables and nominal category variables then it might be

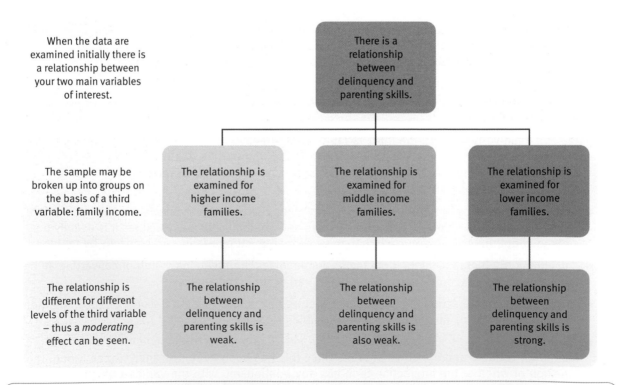

FIGURE 36.1   Illustrating one circumstance in which there is a moderator effect of a third variable

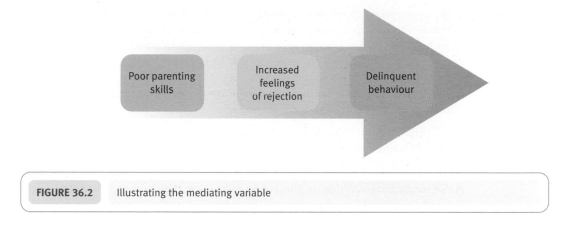

FIGURE 36.2	Illustrating the mediating variable

possible to use ANOVA to test for moderator effects. Indeed, some researchers use ANOVA to test for moderator effects simply by forcing some of the score variables into a small number of categories. Of course, if you are intending to use ANOVA in this way, you need to make sure that one of the variables is a score variable which will be used as the dependent variable in ANOVA.

In all three approaches, basically we are using the presence of interactions as the indication of moderator effects being present. For ANOVA and for log-linear we have discussed interactions

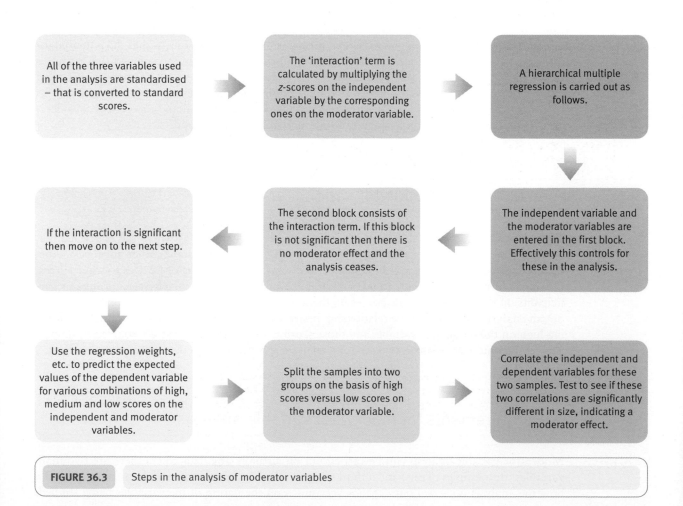

FIGURE 36.3	Steps in the analysis of moderator variables

in the appropriate chapters (Chapters 23 and 37). However, we have not discussed interactions in relation to multiple regression, so we will concentrate on this.

So long as you are familiar with hierarchical multiple regression (Chapter 35), the procedure described in this chapter to identify moderator effects when all of your variables are measured as scores should be fairly straightforward.

When using hierarchical multiple regression there are two important stages before the regression analysis can be carried out (Figure 36.3):

- All three variables should be converted into standardised or z-scores. This is easily done using SPSS (Chapter 7). If standard scores are not used then there may be problems in the analysis because of multicollinearity between the variables.

- An interaction variable or interaction term is created by multiplying the z-scores for the independent variable by the corresponding ones on the moderator variable. Again this is easily done using SPSS.

A fuller explanation of these steps can be found in Chapter 38 of Howitt, D. and Cramer, D. (2011). *Introduction to Statistics in Psychology*, 5th edition. Harlow: Pearson.

36.2   When to use moderator variable analysis

The use of moderator variable analysis needs, of course, to be selective. In modern psychological research, the number of variables measured is often substantial. This means that the potential for carrying out moderator variable analyses is enormous. But we would recommend that its use is rather more selective than this would imply. The researcher needs to ask themselves just what moderator effects would be of interest to understanding the main relationship in question. This would translate into a small number of analyses based on conceptual and theoretical considerations than an empiricist data-bashing overkill approach would demand.

36.3   When not to use moderator variable analysis

The use of moderator variable analysis is appropriate in all situations where the above requirements are met. Moderator variable analysis does call for the use of some rather advanced statistical procedures which may tax the statistical skills and understanding of novices to some extent, of course. It may be preferable to use simpler approaches which effectively select groups of participants on the basis of their responses on the moderator variable. For example, one could carry out an analysis separately of the relationship between the two main variables for individuals scoring high on the moderator variable and those scoring low on the moderator variable. A comparison of the outcomes of these two separate analyses would help you detect a moderator effect.

36.4   Data requirements for moderator variable analysis

Moderator variable analysis can be carried out on virtually any data where there are two main variables (or multiples of this) and a third variable which is a potential moderator variable. Quite how the analysis is carried out depends very much on the nature of the variables being studied:

If all three variables are score variables	If all three variables are nominal category variables	If there is a mixture of score variables and nominal category variables
● The recommended approach is hierarchical multiple regression using interactions as described in the analysis section of this chapter.	● The recommended approach here is to use log-linear analysis with three independent variables. The moderator effects will show as interactions.	● The recommended approach is to use two-way ANOVA. Since the independent variables have to be nominal category variables then it may be necessary to turn a score variable into a variable having just a few categories by recoding a score variable. A score variable is needed for the dependent variable. Once again moderator effects will show as an interaction.

FIGURE 36.4   The different types of moderator variable analysis for different sorts of data

are they score variables, are they nominal category variables or are they a mixture? In this chapter we explain how to deal with the typical case in psychology where all of the three variables are score variables. See Figure 36.4 for the range of alternatives.

36.5   Problems in the use of moderator variable analysis

The data for most forms of test of significance may be used when trying to identify moderator variable effects if the various approaches mentioned in this chapter are used – and, of course, the potential moderator variable is part of the data that has been collected. Some imagination is required, at times, in order to make the data fit the analytic approach taken.

You can find out more about moderator analysis with continuous predictor variables in Chapter 38 of Howitt, D. and Cramer, D. (2011). *Introduction to Statistics in Psychology*, 5th edition. Harlow: Pearson.

36.6   The data to be analysed

We will illustrate the computation of a moderator analysis with continuous predictor variables with the data shown in Table 36.1, which consist of scores for nine individuals on the three variables of depression, stress and social support. Higher scores on these three variables indicate higher levels of these three characteristics.

As with the data for the other multiple regressions because this is for illustrative purposes and to save space, we are going to enter these data 20 times to give us a respectable amount of data

Table 36.1	Data for moderator variable analysis	
Depression	**Stress**	**Support**
4	9	37
6	17	30
6	24	15
7	18	22
6	23	33
5	28	24
6	36	32
7	36	26
17	37	21

to work with. Obviously you would *not* do this if your data were real. It is important to use quite a number of research participants or cases for multiple regression. Ten or 15 times your number of variables would be reasonably generous. Of course, you can use less for data exploration purposes.

36.7   Entering the data

Step 1

In 'Variable View' of the 'Data Editor', name the first four rows 'Depression', 'Stress', 'Support' and 'Freq'.
Change 'Decimals' from '2' to '0'.

Step 2

Enter the data.
The 'Freq' variable is for weighting these nine cases 20 times using the 'Weight Cases...' procedure (see Sections 3.7, 4.7, 8.8, 9.8, 16.7 or 17.7).

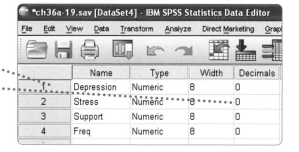

36.8 Standardising the variables

Step 1

Select 'Analyze', 'Descriptive Statistics' and 'Descriptives...'.

Step 2

Select 'Depression', 'Stress' and Support' and the ➡ button to put these variables in the 'Variable(s):' box.
Select 'Save standardized values as variables'.
Select 'OK'.

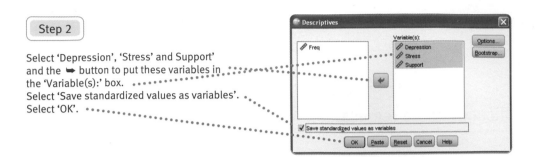

Step 3

The Z or standardized scores are in the fifth, sixth and seventh columns of 'Data View' in the 'Data Editor' and are called 'ZDepression', 'ZStress' and 'ZSupport'.

36.9 Computing the interaction term

Step 1

Select 'Transform' and 'Compute Variable... '.

Step 2

Type a name for the new variable in the box under 'Target Variable:' (e.g. 'Zinteraction').
Either type in or select the terms of the expression in the 'Numeric Expression:' box.
Select 'OK'.
Select 'Paste' to save this procedure as a syntax command.

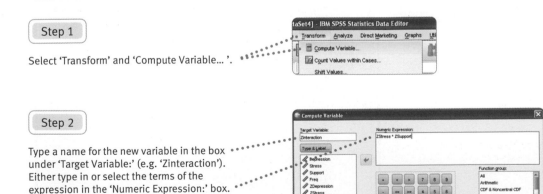

Step 3

The new variable and its values is entered into
the 'Data Editor'. Check that the values are
what they should be for a few cases.
Save the file if you want to use it again.

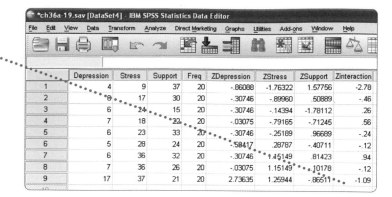

36.10   Hierarchical multiple regression analysis

Step 1

Select 'Analyze', 'Regression' and 'Linear...'.

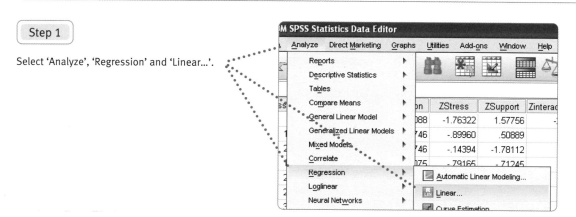

Step 2

Select 'ZDepression' and the ➡ button beside
the 'Dependent:' box to put it there.
Select 'ZStress' and 'Zsupport'.
Select the ➡ button beside the
'Independent(s):' box to put them there.
Select 'Next'.

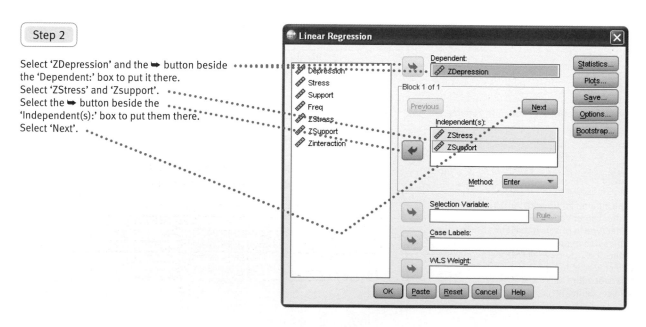

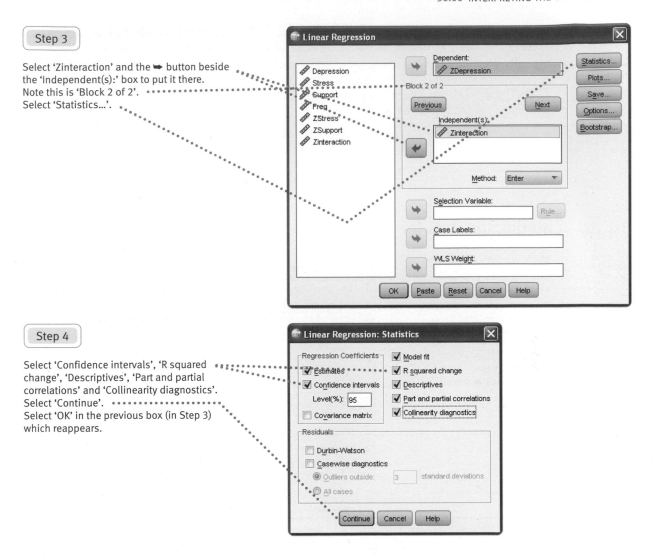

Step 3

Select 'Zinteraction' and the ➡ button beside the 'Independent(s):' box to put it there. Note this is 'Block 2 of 2'. Select 'Statistics...'.

Step 4

Select 'Confidence intervals', 'R squared change', 'Descriptives', 'Part and partial correlations' and 'Collinearity diagnostics'. Select 'Continue'. Select 'OK' in the previous box (in Step 3) which reappears.

36.11 Interpreting the output

- The second line in the fourth table of the output entitled 'Model Summary' shows whether there is a significant moderator or interaction effect or not. If the moderator or interaction effect is significant as it is here, there is a moderator or interaction effect. If the moderator or interaction effect is not significant, there is no moderator or interaction effect.

Model Summary

Model	R	R Square	Adjusted R Square	Std. Error of the Estimate	R Square Change	F Change	df1	df2	Sig. F Change
					Change Statistics				
1	.588[a]	.346	.339	.81315584	.346	46.855	2	177	.000
2	.761[b]	.579	.572	.65407023	.233	97.572	1	176	.000

a. Predictors: (Constant), Zscore(Support), Zscore(Stress)
b. Predictors: (Constant), Zscore(Support), Zscore(Stress), Zinteraction

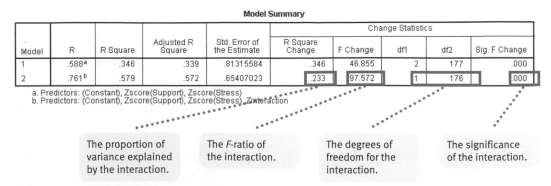

The proportion of variance explained by the interaction.

The *F*-ratio of the interaction.

The degrees of freedom for the interaction.

The significance of the interaction.

- The statistical significance of the F-ratio ('Sig. F Change') for the moderator or interaction effect ('Model 2') is less than .000. Consequently, the moderator or interaction effect is statistically significant at less than .001. The F-ratio ('F Change') of this effect is 97.572. The degrees of freedom are 1 for the moderator or interaction effect and 176 for the error term. The proportion of variance explained by the moderator or interaction effect ('R Square Change') is .223 or about 22 per cent.

- If the moderator or interaction is not significant, the analysis stops here. If it is significant it proceeds.

36.12 Entering the data for predicting the criterion values

From the output for Model 2 in the 'Coefficients' table shown below, enter the constant (a), the unstandardised regression coefficients for the predictor (b1), the moderator (b2) and the interaction (b3) and the three values of one standard deviation above (1), below (−1) and at the mean (0) for the predictor X and the moderator M.

*ch36b-19.sav [DataSet6] - IBM SPSS Statistics Data Editor

File Edit View Data Transform Analyze Direct Marketing Graphs Utilities Add-ons Window

	a	b1	b2	b3	Xstress	Msupport
1	-.192	.676	-.424	-.576	1	-1
2	-.192	.676	-.424	-.576	1	0
3	-.192	.676	-.424	-.576	1	1
4	-.192	.676	-.424	-.576	0	-1
5	-.192	.676	-.424	-.576	0	0
6	-.192	.676	-.424	-.576	0	1
7	-.192	.676	-.424	-.576	-1	-1
8	-.192	.676	-.424	-.576	-1	0
9	-.192	.676	-.424	-.576	-1	1

Coefficients[a]

Model		Unstandardized Coefficients		Standardized Coefficients	t	Sig.	95.0% Confidence Interval for B		Correlations			Collinearity Statistics	
		B	Std. Error	Beta			Lower Bound	Upper Bound	Zero-order	Partial	Part	Tolerance	VIF
1	(Constant)	6.467E-17	.061		.000	1.000	-.120	.120					
	Zscore(Stress)	.455	.065	.455	7.032	.000	.327	.582	.540	.467	.427	.884	1.131
	Zscore(Support)	-.249	.065	-.249	-3.859	.000	-.377	-.122	-.404	-.279	-.235	.884	1.131
2	(Constant)	-.192	.052		-3.657	.000	-.295	-.088					
	Zscore(Stress)	.676	.057	.676	11.936	.000	.564	.787	.540	.669	.584	.746	1.341
	Zscore(Support)	-.424	.055	-.424	-7.723	.000	-.533	-.316	-.404	-.503	-.378	.792	1.262
	Zinteraction	-.567	.057	-.582	-9.878	.000	-.680	-.453	-.074	-.597	-.483	.688	1.453

a. Dependent Variable: Zscore(Depression)

These are the values for the constant and the three predictors to be entered in the data file above.

36.13   Computing the predicted criterion values

Step 1

Select 'File', 'New' and 'Syntax'.

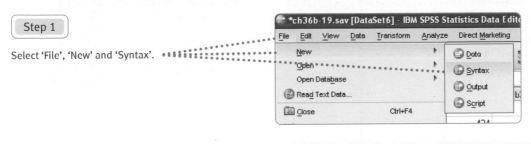

Step 2

Type in the syntax commands in the Syntax editor.

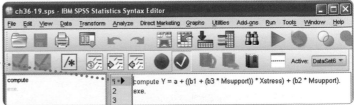

```
compute Y = a + ((b1 + (b3 * Msupport)) * Xstress) + (b2 * Msupport).
exe.
```

Step 3

Select 'Run' and 'All' to run this syntax command. Alternatively, highlight the command and select the ▶ button.

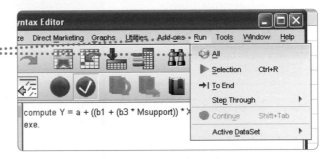

```
compute Y = a + ((b1 + (b3 * Msupport)) * X
exe.
```

Step 4

The predicted depression scores are in the seventh column of Data View.

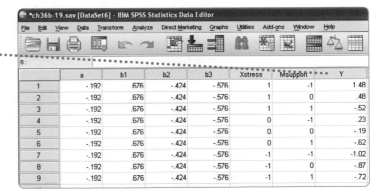

	a	b1	b2	b3	Xstress	Msupport	Y
1	-.192	.676	-.424	-.576	1	-1	1.48
2	-.192	.676	-.424	-.576	1	0	.48
3	-.192	.676	-.424	-.576	1	1	-.52
4	-.192	.676	-.424	-.576	0	-1	.23
5	-.192	.676	-.424	-.576	0	0	-.19
6	-.192	.676	-.424	-.576	0	1	-.62
7	-.192	.676	-.424	-.576	-1	-1	-1.02
8	-.192	.676	-.424	-.576	-1	0	-.87
9	-.192	.676	-.424	-.576	-1	1	-.72

36.14   Plotting the predicted criterion values

Step 1

Select 'Analyze', 'General Linear Model' and 'Univariate...'.

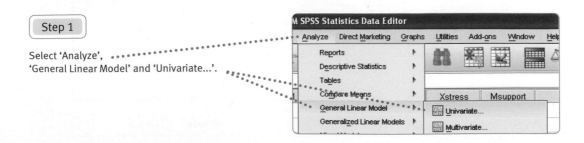

Step 2

Select 'Y' and the ➡ button beside the
'Dependent Variable:' box to put it there.
Select 'Xstress' and 'Msupport' either singly or
together and the ➡ button beside
'Fixed Factor(s):' to put them there.
Select 'Plots...'.

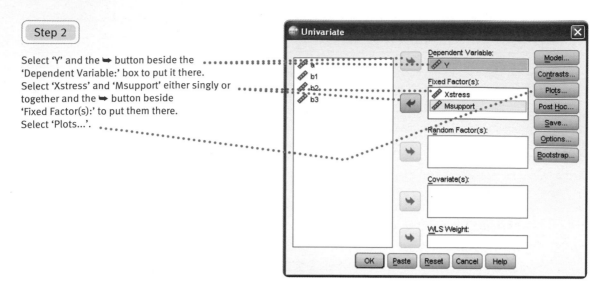

Step 3

Select 'Xstress' and the ➡ button beside the
'Horizontal Axis:' box to put it there.
Select 'Msupport' and the ➡ button beside the
'Separate Lines:' box to put it there.
Select 'Add'.
Select 'Continue'.
Select 'OK' in the previous box which reappears.

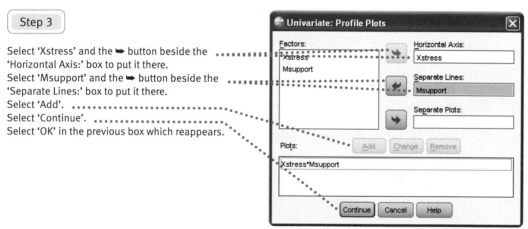

Step 4

Edit the graph if desired as shown in
Section 23.10.

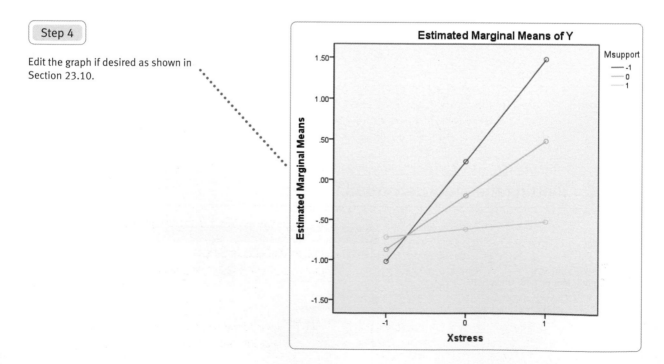

REPORTING THE OUTPUT

- One way of reporting the multiple regression results is as follows:

  Baron and Kenny (1986) have suggested a moderator effect is most appropriately tested with multiple regression. Such an effect is indicated if the interaction of the two predictor variables explains a significant increment in the variance of the criterion variable while the two predictor variables are controlled. Aitken and West (1991) recommended that the criterion and the two predictor variables be standardised. Following these recommendations, a significant proportion of the variance in depression was accounted for by the interaction of stress and social support after the individual variables comprising the interaction were controlled, R^2 change = .04, $p < .01$. To interpret the significant interaction three separate unstandardised regression lines were plotted between standardised stress, standardised social support, and the standardised level of depression at the mean and at one standard deviation above and below the mean of standardised social support. The relation between stress and depression was strongest at low levels of social support.

  Aitken, L.S. & West, S.G. (1991). *Multiple Regression: Testing and Interpreting Interactions*. Newbury Park, CA: Sage.

  Baron, R.M. & Kenny, D.A. (1986). The moderator-mediator variable distinction in social psychological research: Conceptual, strategic, and statistical considerations. *Journal of Personality and Social Psychology, 51,* 1173–1182.

- The plot of the predicted values of depression may be included.

Summary of SPSS Statistics steps for moderator analysis with continuous predictor variables

Data

- Name the variables in 'Variable View' of the 'Data Editor'.
- Enter the data under the appropriate variable names in 'Data View' of the 'Data Editor'.

Standardising variables and computing the interaction

- Select 'Analyze', 'Descriptive Statistics' and 'Descriptives ...'.
- Move the variables to be standardised to the 'Variable(s):' box and select 'Save standardized values as variables'.
- To compute the interaction term, select 'Analyze', 'Descriptive Statistics' and 'Descriptives'.
- Select 'Transform' and 'Compute Variable ...'.
- Type interaction name in 'Target Variable:' box and multiply the two standardised predictors in the 'Numeric Expression:' box, and then select 'OK'.

Hierarchical multiple regression analysis

- Select 'Analyze', 'Regression' and 'Linear ...'.
- In the 'Method' box check 'Enter' has been selected.
- Move the criterion or dependent variable to the 'Dependent:' box and the two predictors or independent variables to the 'Independent(s):' box.
- Select 'Next' and move the interaction term to the 'Independent(s):' box.
- Select 'Statistics ...' and then 'Descriptives', 'Collinearity diagnostics', 'Continue' and then 'OK'.

Output

- In the 'Coefficients' table note whether the *F*-ratio for the moderator or interaction effect is significant.
- If the *F*-ratio is not significant, the analysis stops. If it is significant, proceed.

Computing predicted values for use in the graph

- To draw the graph showing predicted criterion values for the three levels of one standard deviation above, below and at the mean for the predictor X and the moderator M, enter these values in 'Data View' of the 'Data Editor' together with the constant (a) and the unstandardised regression coefficients for the predictor ($b1$), the moderator ($b2$) and the interaction ($b3$).
- In the Syntax editor, type in and run the following commands:

```
compute Y = a + ((b1 + (b3 * M)) * X) + (b2 * M).
exe.
```

Plotting the values

- Select 'Analyze', 'General Linear Model' and 'Univariate ...'.
- Move the dependent variable to the 'Dependent Variable:' box.
- Move the independent variables to the 'Fixed Factor(s):' box.
- Select 'Plots ...'.
- Move the independent variable (Factor) to the 'Horizontal Axis:' box and the moderator to the 'Separate Lines:' box.
- Select 'Add', 'Continue', then 'OK'.

Output

- Edit the graph as desired.

For further resources including data sets and questions, please refer to the website accompanying this book.

PART 6

Advanced qualitative or nominal techniques

CHAPTER 37

Log-linear analysis

Overview

- Log-linear analysis is used to analyse frequency tables (contingency or cross-tabulation) consisting of three or more variables. It can therefore be regarded as an extension of the chi-square test discussed in Chapter 16.

- Its purpose is to determine which of the variables and their interactions best explain (or reproduce) the observed frequencies in the table.

- Variables and their interactions on their own and in combination are known as models in log-linear analysis.

- Goodness-of-fit test statistics are used to assess the degree of correspondence between the model and the data. Statistical significance when comparing the model with the data indicates that the model fails to account totally for the observed frequencies. Statistical non-significance means that the model being analysed fits the observed frequencies. If more than one model fits the data well, the model having the fewer or fewest variables and interactions is the simplest one and may be the preferred model.

- Hierarchical log-linear analysis described in this chapter begins with the saturated model (which includes all possible main effects and interactions) and examines the effect of removing these in steps. If a main effect or interaction is removed then the question is whether this affects the statistical fit of the model to the data. If its removal produces a significant change (reduction) in fit then the main effect or interaction should be reinstated in the model as it is having an important influence.

- Generally, main effects are unimportant in interpreting the model since they merely reflect different proportions of cases in each category.

- Likelihood ratio chi-square is employed as the test statistic in log-linear analysis.

37.1 What is log-linear analysis?

Log-linear analysis can be regarded as being simply an extension of chi-square (see Chapter 16). However, rather than being limited to two nominal (category) independent variables as in the case of chi-square, log-linear analysis allows the researcher to study the associations between several variables. Chi-square can only deal with a two-way contingency (cross-tabulation) table whereas log-linear can deal with three-way, four-way and so forth tables. Contingency (or cross-tabulation) tables record the frequencies in each of the categories produced by the various independent variables. An example is shown in Table 37.1. The table shows the relationship between having been sexual abused, physically abused and gender in a sample of psychiatric hospital patients. Of course, this table is more difficult to interpret than the two-way examples we saw in Chapter 16. This is not simply because of the extra variable. It is also because we would not expect equal numbers in the cells if there was no relationship between the three variables since the categories of the independent variables do not have equal sample sizes, for one reason. But it becomes an even more complex situation in reality because, for example, two of the three independent variables may have an association which is different from the association between all three of the independent variables. There is no intuitively obvious way of working out the expected frequencies in these circumstances. Actually, we are faced with a mathematically extremely complex situation compared with the two-way chi-square we studied earlier. Without computers, the analysis is not possible.

Log-linear analysis does use chi-square but it is a slightly different form from the Pearson chi-square used in Chapter 16. This is known as the likelihood ratio chi-square. Log-linear analysis also uses natural logarithms (Napierian logarithms) which give the log-linear method its name. Neither of these are particularly helpful in understanding log-linear output from SPSS Statistics.

Log-linear analysis involves things called interactions which are not dissimilar from interactions in ANOVA. They are simply aspects of the data which cannot be accounted for on the basis of the independent variables acting independently (or the influence of lower level interactions). In a sense, particular combinations of categories from two or more independent variables tend to result in relatively high or low cell frequencies in the data table. The number of interactions depends on the number of independent variables. If the analysis has three independent variables then there will be three main effects – one for each of the independent variables acting alone. But, in addition, there are interactions between two or more of the independent variables. If we call the independent variables (and so also the main effects) A, B and C, the interactions are A*B, B*C, A*C and A*B*C. That is, there are three two-way interactions and one three-way interaction. The three-way interaction is what cannot be accounted for by the main effects and the three two-way interactions. Step up a notch to four independent variables and there are 11 interactions.

Table 37.1	A three-way contingency table showing the relationship between gender, sexual abuse and physical abuse in a sample of psychiatric hospital patients			
Sexual abuse	**Physical abuse**	**Gender**		**Margin totals**
		Female	**Male**	
Sexually abused	Physical abuse	20	30	50
	No physical abuse	40	25	65
Not sexually abused	Physical abuse	35	55	90
	No physical abuse	45	50	95
Margin totals		140	160	300

Actually the chi-square itself (Chapter 16) works on interactions but nobody refers to them as such. When we discussed the two-way chi-square, we investigated the frequencies on the basis of the independent variables acting alone. What was left over after taking away the influences of these was used to calculate chi-square. This is another way of saying that chi-square assesses the interaction of the two independent variables.

In log-linear analysis there is a great deal of emphasis on models. A model is simply an attempt to explain the empirical data that we obtain on the basis of the influence of the independent variables. In log-linear analysis we try to explain the frequencies obtained in the research on the basis of the independent variables and their interactions. Goodness-of-fit is another commonly used term in log-linear analysis. It simply refers to the extent to which the observed frequencies are modelled (or predicted) by the variables in the model.

It is common in log-linear analysis to start with what is known as the saturated model and work backwards. This is the method that we demonstrate in this chapter. The saturated model is the one which perfectly fits the data so it includes all of the independent variables plus all of their interactions. Of course, some of these may have absolutely no influence on the data we obtain. So what happens next is that aspects of the model are removed one at a time to see whether removing them actually makes a significant difference to the model's fit to the data. If taking away a main effect or an interaction does not change the fit of the model to the actual data, then this main effect or interaction is doing nothing in the model – that is, it is making no contribution since when it is removed it makes no difference to the model.

It is only the interactions which are of interest since the main effects simply indicate whether the proportions of the sample in each category of the main effect differ. Does it matter, for example, whether the genders differ from each other in terms of numbers in the sample? The answer is almost certainly not except in the unusual circumstances where there has been proper random sampling from the population. It is more important to know if there is an interaction between gender and being physically abused.

So, in the analysis we carry out on Table 37.1, basically we begin with all possible main effects and interactions and then discard any which make no difference to the fit of the model to the data. The way in which such main effects and interactions are identified is to examine the size of the likelihood-ratio chi-square when that component is removed. If the likelihood-ratio chi-square is statistically significant then this means that removing that component (main effect or interaction) makes a significant difference to the fit of the model to the data. Hence, that component should be retained in the data. Where removing a component results in a non-significant change in the fit of the model to the data then that component can be discarded from the analysis. This is done a step at a time. The final model consists of the original (saturated or full) model minus any component which makes no significant difference if it is removed.

In a sense, log-linear analysis is very much like three-way ANOVAs, etc. (see Chapter 23). The big difference is that it deals with frequencies in cells of a contingency table rather than the means of the scores in the cells. Figure 37.1 depicts the main steps in a log-linear analysis.

37.2   When to use log-linear analysis

Log-linear analysis is used in exactly the same circumstances as a two-way chi-square to deal with the situation where the researcher has three or more nominal (category) independent variables which may be associated in some way. Many psychologists prefer to collect data primarily in the form of scores so may never need log-linear analysis. Score data cannot be used in log-linear analysis.

Of course, it is tempting to think that nominal (category) data are simpler data than score data. This may encourage the less numerate novice researcher to employ nominal (category) data and thus snare themselves into the trap of needing log-linear analysis, which is among the more difficult statistical techniques to understand.

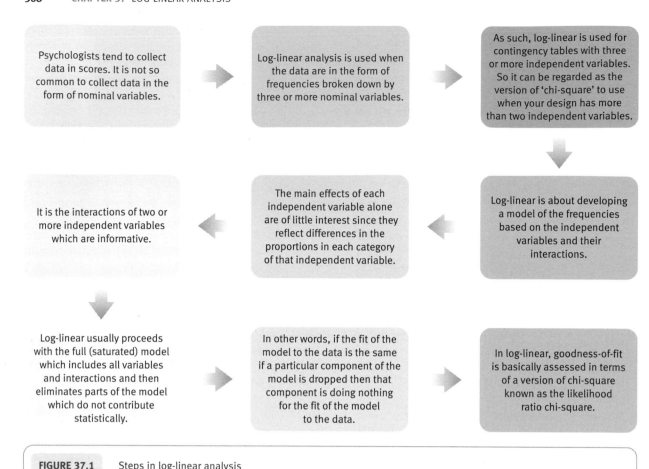

FIGURE 37.1   Steps in log-linear analysis

37.3   When not to use log-linear analysis

If any of your variables are score variables then do not use log-linear analysis.

Be careful to consider quite what you need out of a log-linear analysis which could not be obtained by some other form of statistical analysis. For example, in the case of the study described in this chapter does the researcher really want to know whether there is a difference between males and females in terms of their histories of physical and sexual abuse? If this were the case, then binomial logistic regression or multinomial logistic regression (Chapters 38 and 39) might be capable of supplying the analysis required. One could even mix score and nominal variables in this instance. It is possible to use interactions with some forms of logistic regression.

37.4   Data requirements for log-linear analysis

Three or more nominal (category) variables are required – score variables may not be used. Be parsimonious in terms of the number of variables you include as each additional one adds significantly more complexity to the output.

37.5 Problems in the use of log-linear analysis

Log-linear analysis is not the simplest statistic to understand because of the variety of new and difficult concepts it introduces. It is probably simply unsuitable for total novices. So we would suggest that you do not contemplate using it until you have a more advanced level of statistical skill and experience in deciphering SPSS output. Only choose it when there is no alternative analysis which would do an equally good job.

You can find out more about log-linear analysis in Chapter 40 of Howitt, D. and Cramer, D. (2011). *Introduction to Statistics in Psychology*, 5th edition. Harlow: Pearson.

37.6 The data to be analysed

The computation of a log-linear analysis is illustrated with the data in Table 37.1. This table shows the frequency of sexual and physical abuse in 140 female and 160 male psychiatric patients. To analyse a table of data like this one with SPSS we first have to input the data into the Data Editor and weight the cells by the frequencies of cases in them.

37.7 Entering the data

Step 1

In 'Variable View' of the 'Data Editor', name the first four rows 'Sexual', 'Physical', 'Gender' and 'Freq'. Name the values of 'Sexual' and 'Physical' as '1' for abuse and '2' for non-abuse. Change 'Decimals' from '2' to '0'.

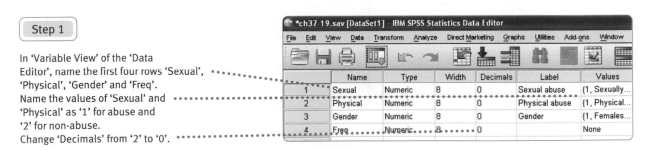

Step 2

Enter the data.
The 'Freq' variable is for weighting these eight cases 20 times using the 'Weight Cases...' procedure (see Sections 3.7, 4.7, 8.8, 9.8, 16.7 or 17.7).

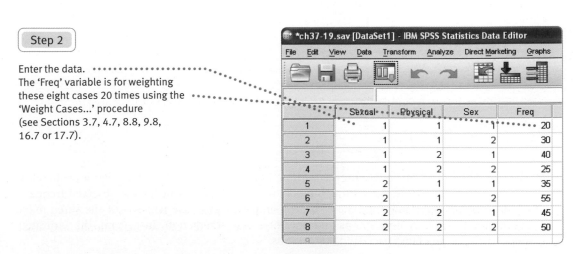

37.8   Log-linear analysis

Step 1

Select 'Analyze',
'Loglinear' and
'Model Selection...'.

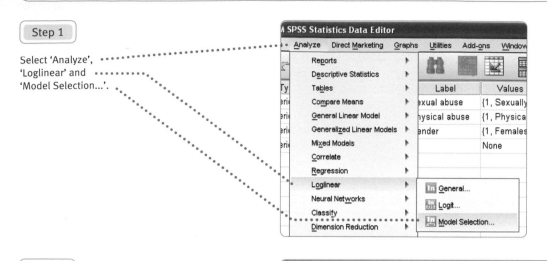

Step 2

Select singly or together 'Sexual',
'Physical' and 'Gender' and the ➡
button beside the 'Factor(s):'
box to put them there.
Select 'Define Range'.
As all three variables have the same
range, they can be defined at the
same time.

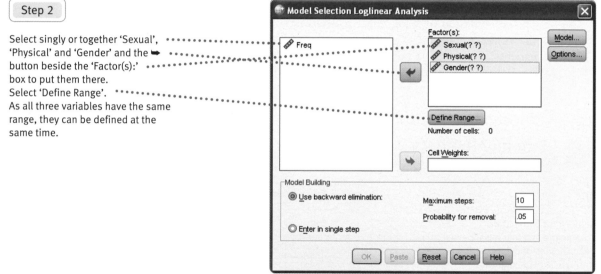

Step 3

Type '1' in the 'Minimum:' box.
Type '2' in the 'Maximum' box.
Select 'Continue'.
Select 'OK' in the previous box
which will reappear.

37.9   Interpreting the output

- The likelihood ratio chi-square for the saturated or full model is .000 which has a probability of 1.000. In other words, the saturated model provides a perfect fit for the observed frequencies and so is non-significant. The saturated model in this case consists of the three main effects, three two-way interactions and one three-way interaction. In general, the saturated model includes all main effects and interactions.

- However, the saturated model includes *all* components whether or not they individually contribute to explaining the variation in the observed data. So it is necessary to eliminate components in turn to see whether this makes the model's fit worse. If it does, this component of the model is kept for the final model.

- SPSS begins with the full model and eliminates each effect in turn to determine which effects make the least significant change in the likelihood ratio chi-square.

> The likelihood ratio chi-square for the full or saturated model is zero and provides a perfect fit to the data.

Step Summary

Step[a]		Effects	Chi-Square[c]	df	Sig.	Number of Iterations
0	Generating Class[b]	Sexual*Physical*Gender	.000	0	.	
	Deleted Effect 1	Sexual*Physical*Gender	1.185	1	.276	3
1	Generating Class[b]	Sexual*Physical, Sexual*Gender, Physical*Gender	1.185	1	.276	
	Deleted Effect 1	Sexual*Physical	.454	1	.501	2
	2	Sexual*Gender	1.963	1	.161	2
	3	Physical*Gender	5.461	1	.019	2
2	Generating Class[b]	Sexual*Gender, Physical*Gender	1.638	2	.441	
	Deleted Effect 1	Sexual*Gender	2.272	1	.132	2
	2	Physical*Gender	5.770	1	.016	2
3	Generating Class[b]	Physical*Gender, Sexual	3.910	3	.271	
	Deleted Effect 1	Physical*Gender	5.770	1	.016	2
	2	Sexual	16.485	1	.000	2
4	Generating Class[b]	Physical*Gender, Sexual	3.910	3	.271	

a. At each step, the effect with the largest significance level for the Likelihood Ratio Change is deleted, provided the significance level is larger than .050.
b. Statistics are displayed for the best model at each step after step 0.
c. For 'Deleted Effect', this is the change in the Chi-Square after the effect is deleted from the model.

> This is the best-fitting model. It has a likelihood ratio chi-square of 3.91 which is not significant.

- The best-fitting model is presented last. In our example, this includes the interaction of physical abuse and gender and the main effect of sexual abuse. This model has a likelihood ratio chi-square of 3.91 (rounded to two decimal places), three degrees of freedom and a probability level of .271. In other words, it is not significant which means that the observed data can be reproduced with these two effects.

These are the counts or frequencies for the data plus the expected frequencies under this model. The residuals are just the differences between the data and the data predicted by the model.

Cell Counts and Residuals

Sexual abuse	Physical abuse	Gender	Observed Count	Observed %	Expected Count	Expected %	Residuals	Std. Residuals
Sexually abused	Physically abused	Females	20.000	6.7%	21.083	7.0%	-1.083	-.236
		Males	30.000	10.0%	32.583	10.9%	-2.583	-.453
	Not physically abused	Females	40.000	13.3%	32.583	10.9%	7.417	1.299
		Males	25.000	8.3%	28.750	9.6%	-3.750	-.699
Not sexually abused	Physically abused	Females	35.000	11.7%	33.917	11.3%	1.083	.186
		Males	55.000	18.3%	52.417	17.5%	2.583	.357
	Not physically abused	Females	45.000	15.0%	52.417	17.5%	-7.417	-1.024
		Males	50.000	16.7%	46.250	15.4%	3.750	.551

There are two statistics used to test the goodness-of-fit of the final model. These are the likelihood ratio chi-square and Pearson chi-square. The likelihood ratio chi-square is the test more commonly used because it has the advantage of being linear so chi-square values may be added or subtracted.

Goodness-of-Fit Tests

	Chi-Square	df	Sig.
Likelihood Ratio	3.910	3	.271
Pearson	3.953	3	.267

To interpret these two effects, we need to present the data in terms of a one-way table for sexual abuse and a two-way table for physical abuse and gender. We can do this using 'Chi-square ...' for the one-way table (see Section 16.13) and 'Crosstabs ...' for the two-way table (see Section 16.9). These two tables are shown in Table 37.2. The one-way table shows that more

Table 37.2 Contingency tables for sexual abuse and the interaction of physical abuse and gender

Sexual abuse

	Observed N	Expected N	Residual
Sexually abused	115	150.0	−35.0
Not sexually abused	185	150.0	35.0
Total	300		

Physical abuse * Gender Crosstabulation

			Females	Males	Total
			Gender		
Physical Abuse	Physically Abused	Count	55	85	140
		Expected Count	65.3	74.7	140.0
		Residual	−10.3	10.3	
	Not physically Abused	Count	85	75	160
		Expected Count	74.7	85.3	160.0
		Residual	10.3	−10.3	
Total		Count	140	160	300
		Expected Count	140.0	160.0	300.0

psychiatric patients have not been sexually abused than have been sexually abused. The two-way table indicates that males are more likely to be physically abused than females.

It is possible to see the contribution of each component to the final mode in the step just before the final step. The final step in this example is Step 4 in the step summary table. These entries essentially indicate the change (reduction) in the goodness-of-fit chi-square if each component is taken away. Thus in Step 3, it can be seen that Physical*Gender has a likelihood ratio Chi-Square Change of 5.770 which is significant (.016). Sexual has a value of 16.485 which is very significant (.000). Obviously these two effects cannot be eliminated from the model because of their significant contribution.

In a hierarchical model, components of an interaction may be significant. Since Physical*Gender has a significant contribution to the model, Physical and Gender may themselves be significant main effects. Select 'Model' in the Model Selection Loglinear Analysis dialog box (Step 2 in Section 37.8). The window that appears will allow you to test these main effects by stipulating models containing only these particular main effects.

REPORTING THE OUTPUT

One way of describing the results found here is as follows:

A three-way frequency analysis was performed to develop a hierarchical linear model of physical and sexual abuse in female and male psychiatric patients. Backward elimination produced a model that included the main effect of sexual abuse and the interaction effect of physical abuse and gender. The model had a likelihood ratio, $\chi^2(3) = 3.91$, $p = .27$, indicating a good fit between the observed frequencies and the expected frequencies generated by the model. About 38 per cent of the psychiatric patients had been sexually abused. About 53 per cent of the males had been physically abused compared with about 39 per cent of the females.

Summary of SPSS Statistics steps for log-linear analysis

Data

- Name the variables in 'Variable View' of the 'Data Editor'.
- Enter the data under the appropriate variable name in 'Data View' of the 'Data Editor'.

Analysis

- Select 'Analyze', 'Loglinear' and 'Model Selection ...'.
- Move variables to the 'Factor(s):' box.
- Select 'Define Range' and define range for each variable.
- Select 'Continue' and then 'OK'.

Output

- In the 'Step Summary' table note the best-fitting model which is presented last and which should not be significant.

For further resources including data sets and questions, please refer to the website accompanying this book.

CHAPTER 38

Multinomial logistic regression

Overview

- Logistic regression is a form of multiple regression (see Chapter 32). It identifies the variables which collectively distinguish cases that belong to different categories of a nominal (or category) variable. For example, it could be used to identify the differentiating characteristics of psychology, sociology and physics students.

- Put another way, logistic regression identifies groups of variables that accurately classify people according to their membership of the different categories of a nominal variable.

- Binomial logistic regression is used if there are just two categories of the variable to be predicted. This is dealt with in Chapter 39. Multinomial logistic regression is used if there are three or more categories.

- The predictors (i.e. the independent variables) may be score variables, nominal (category or categorical) variables or a mixture of both.

- The best predictors of which category a case belongs to have significant b weights (or regression weights). This is much the same as for the more familiar multiple regression procedures described in Chapter 32.

- The b weights in logistic regression are actually applied to the natural logarithm of something termed the *odds ratio*, which is the ratio of the frequencies for two alternative outcomes. This logarithm is also known as the *logit*. Hence the term logistic regression. The odds ratio is simply the likelihood of being in one category rather than any of the other categories. There is little need for most researchers to calculate these values themselves so the logit is mainly of conceptual rather than practical importance.

- It is more important to understand the concept of *dummy variable*. This is a device by which a nominal variable may be dealt with numerically. If the nominal variable has just *two* categories then these may be coded numerically as 0 and 1.

- However, if the nominal category has *three* or more categories then the process is slightly more complex. Essentially, the data are coded for the presence or absence of each of the *three* (or more) categories. In effect, three (or more) new variables are created. So if the three categories are called A, B and C, three new variables are created:

(i)   The individual is in Category A (or not).
(ii) The individual is in Category B (or not).
(iii) The individual is in Category C (or not).

Each of these three variables is a different dummy variable.

- However, one dummy variable from the set is always excluded from the analysis. It does not matter which one. The reason is that this dummy variable contains no new information which is not contained in the other dummy variables.

- SPSS Statistics will generate dummy variables automatically for the dependent variable but needs to be informed which of the predictor variables are nominal (category) variables.

- Classification tables generated by SPSS indicate the prediction of category membership based on the predictor variables. This is a good indication of how good the prediction is because the number of correct classifications is given in the table.

- Logistic regression analyses contain numerous goodness-of-fit statistics based on chi-square. These serve a number of functions, but most importantly they indicate the improvement in fit of the predicted category membership to the actual category membership. A useful predictor should improve the fit of the predicted membership to the actual categories cases belong to.

- Multinomial logistic regression must be used when there are three or more categories for the dependent (predicted or criterion) variable. If there are just *two* categories for the dependent variable, then binomial logistic regression is normally used (see Chapter 39).

38.1   What is multinomial logistic regression?

Multinomial regression is a form of regression analysis. However, unlike simple regression (Chapter 11) and multiple regression (Chapter 32) it does not use predictor (independent) variables in order to predict scores on a dependent variable. Instead it uses predictor variables to predict which groups individuals belong to. That is, it predicts to which nominal category an individual belongs. Of course, the main function of regression is mostly to simply describe the relationship between a group of independent variables and a dependent variable – that is, to say what the significant predictors of the dependent variable are out of a set of predictors.

Multinomial logistic regression can use score variables as predictors but also nominal (category) variables too. Normally one would use a mixture of score variables and category variables but the technique is very flexible in this regard. One cannot easily do this with multiple regression but multinomial logistic regression makes this simple.

Multinomial logistic regression is so called firstly because the dependent variable consists of three or more nominal categories. There is also binomial logistic regression (Chapter 39) which only handles nominal dependent variables which have just two categories. There is a greater range of options with binomial logistic regression which actually make it more complex to use on SPSS. So it is better, as a learning experience, to start with multinomial logistic regression. As might be anticipated, logistic regression employs logarithms in the calculation. Actually, what one is predicting is something called the *logit*. The logit is the natural logarithm of something called the *odds* or *odds ratio*. This is closely related to probability but is expressed differently. If a sample consists of 30 men and 40 women then the odds of selecting a man by chance would be

30/40 = 0.75. In other words, there is 0.75 men for every woman in the sample. In truth, to carry out logistic regression on SPSS calls for absolutely no knowledge of logarithms or odds ratios. So if they are meaningless to you then do not worry too much.

Much more crucial in logistic regression is to understand the concept of *dummy variables*. These are needed when a nominal (category) variable has three or more categories. In order to analyse such data one has to create a new binary (yes–no) nominal variable for each category of the nominal (category) variable. (Actually one fewer but see later.) If you had three categories of the nominal variable colour consisting of red, yellow and blue then you would create three binary nominal variables from this: (1) red and not red, (2) yellow and not yellow, and (3) blue and not blue. These three categories contain as much information as was contained in the original three category nominal variable. Actually, one of the categories is redundant since it contains no additional information. So if someone tells you that they are thinking of a colour which is red, yellow or blue then you need to ask only two questions to find out definitely the colour. These questions might be 'Is it red?' and 'Is it blue?' If the person says no to both of these questions then you know that the colour they are thinking of is yellow. Actually, of course, it does not matter which two colours you ask about. For this reason, in multinomial logistic regression you will find that SPSS gives you one fewer dummy variables than score variables.

Table 38.1 contains data for which a multinomial logistic regression would be appropriate since the purpose of the research is to predict which type of sex offender a man is from six

Table 38.1	Data for multinomial logistic regression						
Age	DAS*	Mother hostile	Father hostile	Children's home	Physical abuse	Sexual abuse	Type of offence
1 younger	low	high	low	no	yes	no	rapist
2 younger	low	high	low	no	yes	yes	rapist
3 older	low	high	low	no	yes	yes	rapist
4 older	high	high	high	yes	no	no	incest
5 older	high	high	high	yes	yes	yes	rapist
6 younger	low	high	low	no	no	no	rapist
7 older	high	low	high	no	yes	yes	rapist
8 older	high	low	high	yes	no	no	incest
9 younger	low	low	high	yes	no	yes	incest
10 older	high	high	low	no	yes	yes	incest
11 older	high	low	low	yes	no	yes	incest
12 younger	high	low	high	no	yes	no	rapist
13 older	high	low	high	yes	no	yes	incest
14 older	high	high	low	yes	yes	yes	incest
15 older	low	high	high	no	yes	yes	incest
16 younger	high	high	low	yes	no	no	paedophile
17 older	high	low	high	yes	no	yes	paedophile
18 older	low	high	high	no	no	yes	paedophile
19 younger	high	low	high	yes	yes	yes	paedophile
20 older	low	low	high	yes	no	no	paedophile
etc.							

*Depression Anxiety Stress scale.

predictor variables. He can be a rapist, incest offender or a paedophile. The variable Type of offence is clearly a nominal (category) variable with three categories. This would be turned into two dummy variables by SPSS. These would be rapist versus non-rapist and incest offender versus non-incest offender possibly, but which two of the three dummy variables are used depends on how the data are coded and the options chosen in SPSS. It makes no difference which of the three is omitted. The predictor variables in this case are all binary nominal (category) variables in this example but score variables could have been used as predictors just as nominal (category) variables with three or more categories could be used. The dependent variable used must not be a score.

The multinomial logistic regression analysis basically gives regression weights for the predictor variables which are much the same as the weights found in multiple regression. The difference is that in multinomial logistic regression the weights are used to predict group (category) membership on the dependent variable. There will be more than one dependent variable for the simple reason that the categories of the dependent variable will be converted to dummy variables by SPSS. So the output looks more complex than for a multiple regression. The Wald test is used to assess the statistical significance of the regression weight.

All predictors which consist of three or more nominal categories are turned into dummy variable predictors by SPSS – except for the one which is left out because it is redundant. At first, unless you anticipate the production of these dummy variables it is easy to get confused, so take care. There are two more important features of the multinomial logistic regression output which help the interpretation of the output:

● Logistic regression produces pseudo R-square statistics which are analogous to the R-square statistics in multiple regression. They give an indication of how well the model (the predictor variables) fit the data.

● Logistic regression produces a classification table which indicates the number of each group of the dependent variable which are accurately (and not) accurately predicted to be in their actual group (as well as those wrongly predicted to be in other groups). These tables give a concrete indication of what is happening in the analysis.

Figure 38.1 highlights the main steps in multinomial logistic regression.

38.2 When to use multinomial logistic regression

Multinomial logistic regression is a very flexible regression procedure. The most important thing is that the dependent variable is a nominal (category) variable with three or more categories. It will work with a binomial dependent variable though it is probably better to use the binomial logistic regression which has more advanced features in SPSS for this. The predictor variables can be any mixture of score variables and nominal (category) variables. Multinomial logistic regression is useful when trying to differentiate a nominal (category) dependent variable's categories in terms of the predictor variables. In the present example, the question is what differentiates rapists, incest offenders and paedophiles.

38.3 When not to use multinomial logistic regression

There are few reasons not to use multinomial logistic regression where the dependent variable is a nominal (category) variable with three or more categories. The predictors used can be scores and nominal (category) variables so it is very flexible in this respect.

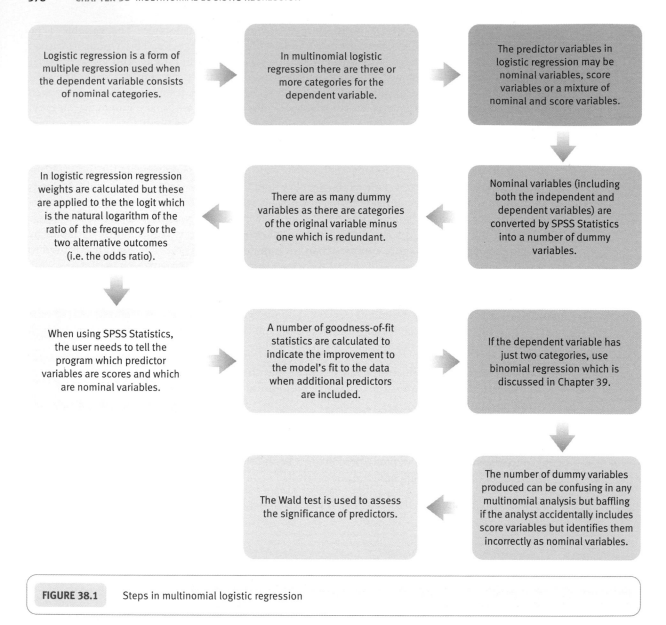

FIGURE 38.1   Steps in multinomial logistic regression

In circumstances where binomial logistic regression is appropriate then this should be used in preference simply because SPSS has more powerful procedures for the two-category case.

38.4   Data requirements for multinomial logistic regression

The dependent variable should consist of three or more nominal categories. We would not advise the use of dependent variables with more than say five or six categories simply because of the complexity of the output. SPSS will cope with more categories – though the researcher may not cope with the output.

The independent variables ideally should not correlate highly (as in any regression) and it is wise to examine the interrelationships between the independent variables when planning which of them to put into the logistic regression analysis.

38.5   Problems in the use of multinomial logistic regression

Multinomial logistic regression can produce a great deal of SPSS output. However, much of it is not necessary and a good interpretation can be based on a limited number of tables. We find it particularly helpful to keep an eye on the classification table as this is a simple and graphic summary of the analysis.

SPSS procedures for the multinomial logistic regression and the binomial logistic regression are rather different in terms of how one goes about the analysis. This can cause confusion. Even the terminology used is different.

The major problem in our experience centres around stipulating which are score variables and which are nominal (category) variables – or failing to differentiate between the two. The worst situation is where SPSS believes that a score variable is a nominal (category) variable. The problem is that each value of the score variable is regarded as a different nominal category so numerous dummy variables are created, to the consternation of the researcher.

You can find out more about multinomial logistic regression in Chapter 41 of Howitt, D. and Cramer, D. (2011). *Introduction to Statistics in Psychology*, 5th edition. Harlow: Pearson.

38.6   The data to be analysed

The use of multinomial logistic regression can be illustrated using the data described in Table 38.1 (*ISP*, Table 41.2). These data are from a fictitious study of the differences between rapists, incestuous sex offenders and paedophiles. This means that the categories of offender equate to a nominal or category variable with three different values. In this example all of the predictor variables – age, DAS (Depression, Anxiety and Stress scale), mother hostile, father hostile, children's home, physical abuse and sexual abuse – are nominal (category) variables with just two different values in each case. It must be stressed that any type of variable may be used as a predictor in multinomial logistic regression. However, the researcher needs to indicate which are score variables in the analysis.

38.7   Entering the data

These data are entered into SPSS in the usual way, with each variable being represented by a column. For learning purposes, the data have been repeated 10 times in order to have a realistic data set for the analysis, but to limit the labour of those who wish to reproduce the analysis exactly.

Step 1

In 'Variable View' of the 'Data Editor', name the first nine rows 'age', 'das', 'etc. Name the values of variables as in Table 38.1 (e.g. '1' = 'low' and '2' = 'high'). Change 'Decimals' from '2' to '0'.

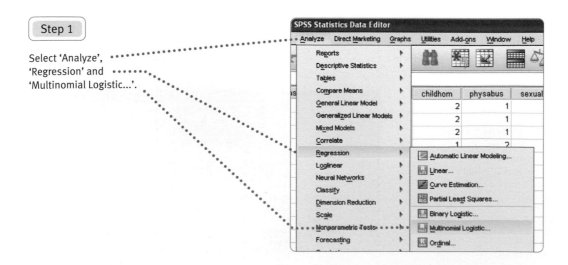

Step 2

Enter the data.
The 'Freq' variable is for weighting these 20 cases 10 times using the 'Weight Cases...' procedure (see Sections 3.7, 4.7, 8.8, 9.8, 16.7 or 17.7).

In *all* multinomial logistic regression analyses, dummy variables are created by SPSS. This is always the case for the predicted variable (offence type in this case), but there may be also nominal (category) predictor variables with more than two categories. If this is the case, SPSS generates new variables (dummy variables) for inclusion in the analysis. So do not be surprised to find variables reported in the output which were not part of the data input. SPSS does not show the dummy variables that it creates in the data spreadsheet – they are referred to in the output, however. SPSS creates appropriate dummy variables for the dependent (criterion) variable automatically based on the number of different values (categories) of that variable. The predictor or independent variables are also dummy coded if they are defined by the researcher as being nominal (category or categorical) variables. Dummy variables are discussed in the overview and first section at the start of this chapter and also in the accompanying statistics text (*ISP*, Chapter 41).

Until Release 12 of SPSS, multinomial logistic regression was available only in a rather unsatisfactory number of options. However, from Release 12 onwards, SPSS includes a stepwise version in which variables are selected as predictors in order of their independent predictive powers. In this guide we will analyse the data using stepwise multinomial logistic regression. The findings are slightly different in detail from those of the accompanying statistics text (*ISP*), but not substantially so.

38.8   Stepwise multinomial logistic regression

Step 1

Select 'Analyze', 'Regression' and 'Multinomial Logistic...'.

Step 2

Select 'typoff' and the ➡ button
beside the 'Dependent:' box to
put it there.
Select the other seven variables
either singly or together
(excluding 'freq') and the ➡
button beside the 'Factor(s):'
box to put them there.
Select 'Model:'.

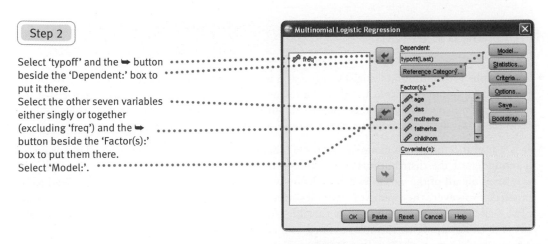

Step 3

Select 'Custom/Stepwise'.
Select ▼ next to 'Interaction'.
Select each of the seven variables in turn or
altogether and the ➡ button (hidden behind the
drop-down menu) beside 'Stepwise Terms:' to
put them there.
Select 'Continue'.
Select 'Statistics...' in previous box which will
reappear.

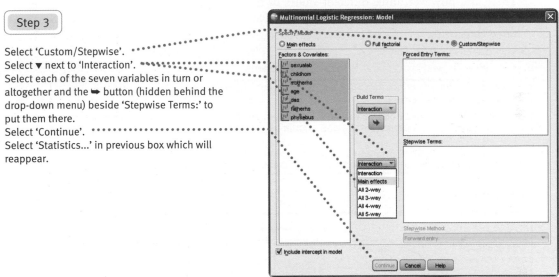

Step 4

Select 'Cell probabilities',
'Classification table' and
'Goodness-of-fit'.
Select 'Continue'.
Select 'OK' in the main box.

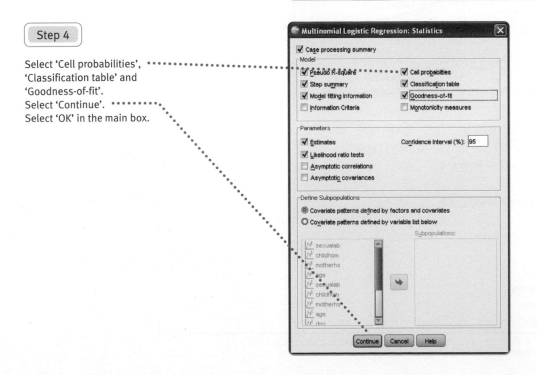

38.9   Interpreting the output

- The output for multinomial logistic regression is quite substantial. Of course, it is possible to reduce the amount of output but then the researcher needs to be clear just what aspects of the output are not necessary for their purposes. Since it is easier to ignore surplus tables than to redo the analysis then it is better to err on the side of too much output.

- It is a useful reminder to examine the table labelled Case Processing Summary. This provides a reminder of the distributions of the categories of each of the variables in the analysis. In our example all of the variables simply have two categories so no dummy variables need to be created. However, the predictor variables may have three or more categories in which case SPSS will create appropriate dummy variables for all but one of the categories of that variable. (This does not apply to variables defined as a covariate which are treated as score variables.) There are, however, three different values of the criterion variable of offence type – rapists, incestuous offenders and paedophiles – and SPSS will create two dummy variables from these three categories.

Warnings

> There are 37 (64.9%) cells (i.e., dependent variable levels by subpopulations) with zero frequencies.
>
> Unexpected singularities in the Hessian matrix are encountered. This indicates that either some predictor variables should be excluded or some categories should be merged.
>
> The NOMREG procedure continues despite the above warning(s). Subsequent results shown are based on the last iteration. Validity of the model fit is uncertain.

Case Processing Summary

		N	Marginal Percentage
type of offence	rapist	70	35.0%
	Incestuous child abuser	80	40.0%
	paedophile	50	25.0%
age up to 29 or 30 and above	younger	70	35.0%
	older	130	65.0%
depression and anxiety scale	low score	80	40.0%
	high score	120	60.0%
mother's hostility to offender as child	low hostility	90	45.0%
	high hostility	110	55.0%
father's hostility to offender as child	low hostility	80	40.0%
	high hostility	120	60.0%
spent time in children's home	yes	110	55.0%
	no	90	45.0%
physically abused as a child	yes	100	50.0%
	No	100	50.0%
sexually abused as a child	yes	130	65.0%
	no	70	35.0%
Valid		200	100.0%
Missing		0	
Total		200	
Subpopulation		19[a]	

This table shows the distribution for each of the variables in the analysis. The offence distribution is circled.

a. The dependent variable has only one value observed in 18 (94.7%) subpopulations.

- The Step Summary table essentially gives the sequence of Variables entered into the stepwise multiple regression.

- Remember that in stepwise analyses the predictors are selected in terms of their (distinct) predictive power. So the best predictor is selected first, adjustments made, and the remaining best predictor selected second, and so forth. There is a Step 0 which contains only the intercept of the regression line. Steps 1 to 4 in this example add in the strongest predictors in turn. The variable 'childhom' (children's home) is added in Step 1, age in Step 2. 'physabus' (physical abuse) in Step 3 and 'sexualab' (sexual abuse) in Step 4. Each of these produces a significantly better fit of the predicted (modelled) data to the actual data. This can be seen from the lowering values of the –2 log likelihood values (which are chi-square values) given in the table. Each of these changes is significant in this example, meaning that none of the predictors may be dropped without worsening the accuracy of the classification.

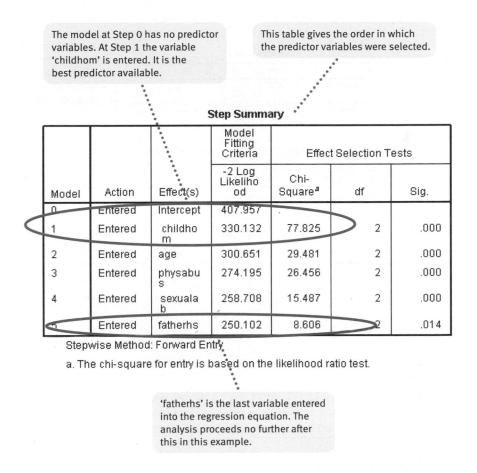

The model at Step 0 has no predictor variables. At Step 1 the variable 'childhom' is entered. It is the best predictor available.

This table gives the order in which the predictor variables were selected.

Step Summary

Model	Action	Effect(s)	Model Fitting Criteria -2 Log Likelihood	Effect Selection Tests Chi-Square[a]	df	Sig.
0	Entered	Intercept	407.957	.		
1	Entered	childhom	330.132	77.825	2	.000
2	Entered	age	300.651	29.481	2	.000
3	Entered	physabus	274.195	26.456	2	.000
4	Entered	sexualab	258.708	15.487	2	.000
5	Entered	fatherhs	250.102	8.606	2	.014

Stepwise Method: Forward Entry

a. The chi-square for entry is based on the likelihood ratio test.

'fatherhs' is the last variable entered into the regression equation. The analysis proceeds no further after this in this example.

- The Model Fitting table gives the value of the –2 log-likelihood chi-square for the fit of the model (i.e. the significant predictors plus the intercept). This value is significant in this case. It merely is an indication that the model (predictor variables) does not completely predict the actual data. In other words, the prediction is less than complete or partial. Clearly, there are other factors which need to be taken into account to achieve a perfect fit of the model to the data. This will normally be the case.

The Intercept Only Model includes *none* of the predictor variables. The Final model is the one using the best set of predictors. Notice that there is a significant improvement in fit to the data by using the Final model.

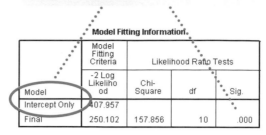

Model Fitting Information

Model	Model Fitting Criteria	Likelihood Ratio Tests		
	-2 Log Likelihood	Chi-Square	df	Sig.
Intercept Only	407.957			
Final	250.102	157.856	10	.000

Goodness-of-Fit

	Chi-Square	df	Sig.
Pearson	230.126	26	.000
Deviance	246.629	26	.000

This table indicates how well the predicted data fits the actual data. There is a significant difference between the two – that is, the prediction is less than perfect.

- The table of the Pseudo R-square statistics merely confirms this. The R-square statistic gives the combined correlation of a set of predictors with the predicted variable for score data. The Pseudo R-square is analogous to this in interpretation but is used when it is not possible to accurately compute the R-square statistic itself as in the case of logistic regression. As can be seen, there are three different methods of calculation used. They are all indicators of the combined relationship of the predictors to the category variable. A value of 0 means no multiple correlation, a value of 1.00 means a perfect multiple correlation. Values of around 0.5 are fairly satisfactory as they indicate an overall combined correlation of the predictor variables with the predicted variable of around 0.7. (This is obtained by taking the square root of 0.5).

Pseudo R-Square

Cox and Snell	.546
Nagelkerke	.617
McFadden	.365

This table gives three estimates of the 'multiple correlation' between the predictor variables and sex offender category membership. It is interpreted like a squared correlation coefficient. The values in this case are moderate (1.00 would indicate a perfect classification) confirming that the prediction is less than perfect.

- The table of the Likelihood Ratio Tests tells us what happens to the model if we remove each of the predictor variables in turn. The model is merely the set of predictors which emerge in the analysis. In this case we have four predictors as already described. In each case there is a significant decrement in the fit of the predicted data to the actual data following the dropping of any of the predictors. In other words, each of the predictors is having a significant effect and normally should be retained. Of course, should the researcher have good reason then any predictor can be dropped though it is recommended that inexperienced researchers do not do this.

Likelihood Ratio Tests

Effect	Model Fitting Criteria	Likelihood Ratio Tests		
	-2 Log Likelihood of Reduced Model	Chi-Square	df	Sig.
Intercept	250.102	.000	0	.
sexualab	264.656	14.555	2	.001
childhurn	273.076	22.975	2	.000
age	271.748	21.647	2	.000
fatherhs	258.708	8.606	2	.014
physabus	291.381	41.280	2	.000

The chi-square statistic is the difference in -2 log-likelihoods between the final model and a reduced model. The reduced model is formed by omitting an effect from the final model. The null hypothesis is that all parameters of that effect are 0.

This is a very important table which indicates whether removing each of the predictor variables from the prediction reduces significantly the fit of the prediction to the actual data. As can be seen, in this case removing any of the four predictors adversely effects the fit of the predicted data to the actual data.

- The table of the Parameter Estimates basically gives the Intercept and the regression weights for this multinomial regression analysis. The intercept value is clear at .088 for Rapists. But notice a number of things. The dependent variable (offence type) which has three categories has been turned into two dummy variables Rapists (versus the other two groups) and Incestuous Child Molester (versus the other two groups).

This table gives the B weights that the computer will use in making its predictions. The B weights are useful to report but the researcher would not need to actually do any calculations involving them.

Parameter Estimates

typoff type of offence[a]		B	Std. Error	Wald	df	Sig.	Exp(B)	95% Confidence Interval for Exp (B) Lower Bound	Upper Bound
1 rapist	Intercept	.088	.787	.012	1	.911			
	[physabus=1]	20.173	.654	951.829	1	.000	5.765E8	1.601E8	2.077E9
	[physabus=2]	0[b]	.		0	.			
	[fatherhs=1]	1.024	.672	2.323	1	.127	2.783	.746	10.380
	[fatherhs=2]	0[b]	.		0	.			
	[age=1]	-.160	.621	.066	1	.797	.852	.252	2.878
	[age=2]	0[b]	.		0	.			
	[childhorn=1]	-1.967	.606	10.535	1	.001	.140	.043	.459
	[childhorn=2]	0[b]	.		0	.			
	[sexualab=1]	-18.398	.000	.	1	.	.023E-8	1.023E-8	1.023E-8
	[sexualab=2]	0[b]	.		0	.			
2 Incestuous child abuser	Intercept	-.206	.596	.119	1	.730			
	[physabus=1]	.613	.536	1.307	1	.253	1.846	.645	5.281
	[physabus=2]	0[b]	.		0	.			
	[fatherhs=1]	1.481	.556	7.090	1	.008	4.399	1.478	13.087
	[fatherhs=2]	0[b]	.		0	.			
	[age=1]	-1.955	.507	14.845	1	.000	.142	.052	.383
	[age=2]	0[b]	.		0	.			
	[childhom=1]	.380	.503	.572	1	.450	1.462	.546	3.917
	[childhom=2]	0[b]	.		0	.			
	[sexualab=1]	.488	.476	1.053	1	.305	1.629	.641	4.138
	[sexualab=2]	0[b]	.		0	.			

a. The reference category is: 3 paedophile.
b. This parameter is set to zero because it is redundant.

The standard error is 0.00 so the Wald value would be infinitely large. No significance level can actually be computed although this variable is highly statistically significant.

When interpreting the output the values of the variable should be carefully checked. The B weight is −1.955 for the category 1.00 of the variable age – the other category is given a weight of 0.00. The value label 1.00 in this case means that the offender was younger. However, there is a negative value of B. This means that Incestuous child abusers are partly distinguished from the other two groups by being older. This is a little confusing at first but is simply the way SPSS operates.

● Remember that the number of dummy variables is given by the number of categories minus one. There are three offender categories so two dummy variables. The dummy variables are created by taking one of the three offender categories and contrasting this with the remaining offender categories. In our example, we have Rapists versus Incestuous Child Molesters *and* Paedophiles, Incestuous Child Molesters versus Rapists *and* Paedophiles, and Paedophiles versus Rapists and Incestuous Child Molesters. The choice of which of the possible dummy variables to select is arbitrary and can be varied by selecting 'Custom/Stepwise' in Step 3 of Section 38.8. Also notice that the variables have been given two regression weights – a different one for each value. However, one of the pair is always 0 which essentially means that no contribution to the calculation is made by these values. The significance of each of these regression weights is given in one of the columns of the table. Significance is based on the Wald value which is given in another column. Do not worry too much if you do not understand this too clearly as you will not need to do any actual calculations.

> This is explained in more detail in Chapter 41 of Howitt, D. and Cramer, D. (2011). *An Introduction to Statistics in Psychology*, 5th edition. Harlow: Pearson.

● The classification table is very important and gives the accuracy of the predictions based on the Parameter Estimates. This cross-tabulation table indicates what predictions would be made on the basis of the significant predictor variables and how accurate these predictions are. As can be seen, the predictions are very accurate for Rapists and to a lesser degree for the Incestuous Child Molesters. However, the classification is poor for Paedophiles.

The classification table indicates in a simple form how good the prediction is. As can be seen, the predictors are very good at predicting rapists correctly (85.7% correct). The predictors do a poor job at predicting paedophiles (only 40% of paedophiles correctly identified).

Classification

Observed	Predicted			Percent Correct
	1 rapist	2 Incestuous child abuser	3 paedophile	
1 rapist	60	10	0	85.7%
2 Incestuous child abuser	20	50	10	62.5%
3 paedophile	0	30	20	40.0%
Overall Percentage	40.0%	45.0%	15.0%	65.0%

● The Observed and Predicted Frequencies Table is probably most useful to those who have a practical situation in which they wish to make the best prediction of the category based on the predictor variables. The table gives every possible pattern of the predictor variables (SPSS calls them covariates) and the actual classifications in the data for each pattern plus the most likely outcome as predicted from that pattern. In other words, the table could be used to make predictions for individual cases with known individual patterns.

This table gives all the possible combinations of the values of the predictor variables. As such, it can be used to trace the most likely predicted outcome for any combination of the predictor variables. That is, an individual offender's predictor pattern could be used to find the most likely prediction for that pattern.

Observed and Predicted Frequencies

Sexually abused as a child	physically abused as a child	spent time in children's home	father's hostility to offender as child	mother's hostility to offender as child	depression and anxiety scale	age up to 29 or 30 and above	Type of offence	Frequency Observed	Frequency Predicted	Pearson Residual	Percentage Observed	Percentage Predicted
yes	yes	yes	low hostility	high hostility	high score	older	rapist	0	1.387	-1.269	.0%	13.9%
							Incestuous child abuser	10	7.078	2.032	100.0%	70.8%
							paedophile	0	1.535	-1.347	.0%	15.3%
			high hostility	low hostility	high score	younger	rapist	0	3.601	-2.372	.0%	36.0%
							Incestuous child abuser	0	3.192	-2.165	.0%	31.9%
							paedophile	10	3.207	4.603	100.0%	32.1%
				high hostility	high score	older	rapist	10	1.387	7.882	100.0%	13.9%
							Incestuous child abuser	0	7.078	-4.922	.0%	70.8%
							paedophile	0	1.535	-1.347	.0%	15.3%
		no	low hostility	high hostility	low score	younger	rapist	10	8.501	1.328	100.0%	85.0%
							Incestuous child abuser	0	.702	-.869	.0%	7.0%
							paedophile	0	.797	-.930	.0%	8.0%
						older	rapist	10	6.281	2.433	100.0%	62.8%
							Incestuous child abuser	0	2.987	-2.064	.0%	29.9%
							paedophile	0	.732	-.889	.0%	7.3%
					high score	older	rapist	0	6.281	-4.110	.0%	62.8%
							Incestuous child abuser	10	2.987	4.845	100.0%	29.9%
							paedophile	0	.732	-.889	.0%	7.3%
			high hostility	low hostility	high score	older	rapist	10	6.281	2.433	100.0%	62.8%
							Incestuous child abuser	0	2.987	-2.064	.0%	29.9%
							paedophile	0	.732	-.889	.0%	7.3%
				high hostility	low score	older	rapist	0	6.281	-4.110	.0%	62.8%
							Incestuous child abuser	10	2.987	4.845	100.0%	29.9%
							paedophile	0	.732	-.889	.0%	7.3%
	No	yes	low hostility	low hostility	high score	older	rapist	0	.000	.000	.0%	.0%
							Incestuous child abuser	10	6.785	2.177	100.0%	67.9%
							paedophile	0	3.215	-2.177	.0%	32.1%
			high hostility	low hostility	low score	younger	rapist	0	.000	.000	.0%	.0%
							Incestuous child abuser	10	3.130	4.685	100.0%	31.3%
							paedophile	0	6.870	-4.685	.0%	68.7%
					high score	older	rapist	0	.000	.000	.0%	.0%
							Incestuous child abuser	10	13.571	-1.710	50.0%	67.9%
							paedophile	10	6.429	1.710	50.0%	32.1%
		no	high hostility	high hostility	low score	older	rapist	0	.000	.000	.0%	.0%
							Incestuous child abuser	0	6.514	-4.323	.0%	65.1%
							paedophile	10	3.486	4.323	100.0%	34.9%
no	yes	no	low hostility	high hostility	low score	younger	rapist	10	10.000	.000	100.0%	100.0%
							Incestuous child abuser	0	.000	.000	.0%	.0%
							paedophile	0	.000	.000	.0%	.0%
			high hostility	low hostility	high score	younger	rapist	10	10.000	.000	100.0%	100.0%
							Incestuous child abuser	0	.000	.000	.0%	.0%
							paedophile	0	.000	.000	.0%	.0%
	No	yes	low hostility	high hostility	high score	younger	rapist	0	1.523	-1.340	.0%	15.2%
							Incestuous child abuser	0	2.141	-1.651	.0%	21.4%
							paedophile	10	6.336	2.405	100.0%	63.4%
			high hostility	low hostility	low score	older	rapist	0	.701	-.868	.0%	7.0%
							Incestuous child abuser	0	5.675	-3.622	.0%	56.7%
							paedophile	10	3.624	4.194	100.0%	36.2%
					high score	older	rapist	0	.701	-.868	.0%	7.0%
							Incestuous child abuser	10	5.675	2.761	100.0%	56.7%
							paedophile	0	3.624	-2.384	.0%	36.2%
				high hostility	high score	older	rapist	0	.701	-.868	.0%	7.0%
							Incestuous child abuser	10	5.675	2.761	100.0%	56.7%
							paedophile	0	3.624	-2.384	.0%	36.2%
		no	low hostility	high hostility	low score	younger	rapist	10	6.374	2.385	100.0%	63.7%
							Incestuous child abuser	0	.835	-.954	.0%	8.3%
							paedophile	0	2.791	-1.967	.0%	27.9%

The percentages are based on total observed frequencies in each subpopulation.

REPORTING THE OUTPUT

There is no conventional method of reporting the findings succinctly for a multinomial logistic regression. If the technique has been used for previous studies in the chosen area of research then the reporting methods used in those studies might be adopted.

For the data analysed in this chapter, it is clear that there is a set of predictors which work fairly effectively in part. One way of reporting these findings would be as follows:

> Multinomial logistic regression was performed to establish what characteristics distinguish the three different types of offender. Out of the seven predictor variables included in the analysis, four were shown to distinguish the different types of offender to a degree. Rapists were correctly identified as such in 85.7 per cent of cases, Incestuous Child Abusers were correctly identified in 62.5 per cent of instances, but paedophiles were correctly identified in only 40.0 per cent of instances. Paedophiles tended to be misclassified as Incestuous Child Abusers very often. The predictors which distinguished rapists from others the best were time spent in a children's home ($b = -1.97$), physical abuse ($b = 20.17$) and sexual abuse ($b = -18.40$). The last is not reported as significant as such in the table. No significance is given. However, paedophiles were best distinguished from the other two groups by age ($b = -1.96$) and father's hostility ($b = -1.48$).

Summary of SPSS Statistics steps for multinomial logistic regression

Data

- Name the variables in 'Variable View' of the 'Data Editor'.
- Enter the data under the appropriate variables names in 'Data View' of the 'Data Editor'.

Analysis

- Select 'Analyze', 'Regression', and 'Multinomial Logistic …'.
- Move the dependent variable to the 'Dependent:' box and the independent variables to the 'Factor(s):' box.
- Select 'Model', 'Custom/Stepwise', move each of the predictors to the 'Stepwise Terms:' box and then select 'Continue'.
- Select 'Statistics …', 'Cell probabilities', 'Classification table', 'Goodness-of-fit', 'Continue' and 'OK'.

Output

- The 'Step Summary' table presents the significant predictors in order with the best first.
- The 'Parameter Estimates' table gives the regression weights for the predictors together with their statistical significance.

For further resources including data sets and questions, please refer to the website accompanying this book.

CHAPTER 39

Binomial logistic regression

Overview

- Binomial logistic regression may be regarded as a special case of multinomial logistic regression, which was described in Chapter 38. The major difference is that it differentiates the characteristics of people in just *two* different groups.

- SPSS Statistics has a much more extensive repertoire of regression techniques for binomial logistic regression although Releases 12 and onwards of SPSS have introduced a little more flexibility to its multinomial logistic regression procedures.

- Binomial logistic regression is a form of multiple regression. It identifies patterns of variables which can effectively differentiate between the members of two different categories. That is, binomial logistic regression predicts category membership as opposed to a score as in the case of multiple regression (Chapter 32). For example, one could use it to examine the pattern of variables which best differentiates male from female participants in a study of reasons for seeking medical advice. There may be a different pattern of reasons why men go to a doctor from the pattern of reasons why women go to a doctor.

- Another way of putting this is that binomial logistic regression uses predictor variables to predict the most likely category of the dependent variable to which different individuals belong.

- The binomial logistic regression procedure calculates *b* weights (or regression weights) much as in multiple regression (Chapter 32). The big difference is that in binomial logistic regression these *b* weights are *not* applied to predict scores. Instead they are applied to something called the logit which is the natural logarithm of the odds ratio. The odds ratio is rather like a probability. It is simply the ratio of the numbers in one category to the number of cases in the other category. The logit (and natural logarithms for that matter) generally is not crucial to a researcher's use of binomial logistic regression so does not require detailed understanding.

- Although the categories are always a binomial (binary or two-value) category variable, the other variables in the analysis may be score variables, nominal (category or categorical) variables or a mixture of both.

- Any category variable which has more than two categories is automatically turned into a set of dummy variables by SPSS. Each dummy variable consists of one of the categories of the variable versus all of the rest of the categories. There are as many dummy variables as categories. For technical reasons, one of the possible dummy variables is arbitrarily omitted from the binomial logistic regression. This is because it contains no new information which is not already contained in the other dummy variables.

- Binomial regression, for example, can proceed using stepwise procedures, forward entry of poor predictors, backward elimination of poor predictors and so forth. The choice of the type of model to use is often a rather subtle matter. Stepwise procedures gradually build up the model by selecting on a step-by-step basis variables which are good at differentiating between the members of the two categories of the dependent variable. Forward entry means that at each step the computer finds the best remaining variable for differentiating between the two categories. If it does not meet certain requirements in terms of predictive (or classificatory) power then it is not entered into the model. For many purposes, stepwise forward entry is a good choice in psychology.

- Much of the binomial logistic regression output from SPSS consists of indicators of how well the modelled data (the category membership predicted by the predictor variables) fits the actual data (the actual category which the individual belongs to). These are based on chi-square for the most part. A useful predictor should improve the fit of the predicted membership to the actual categories cases belong to.

- More directly understood are the classification tables generated by SPSS which indicate how well the predictions match the actual classifications.

39.1   What is binomial logistic regression?

Binomial logistic regression is more or less the same as multinomial logistic regression with the sole difference that the dependent variable consists of two nominal categories. Apart from that most things are identical – except for, unfortunately, the SPSS procedure which is frustratingly different for the two. Any type of variable – scores or nominal (category) – may be used as the predictor variables in binomial logistic regression. However, in our example in Table 39.1, we have used very simple variables such as age, previous imprisonment, treatment, contrition, marital status and whether or not the offender is a sex offender. We have also used a binomial variable as the dependent variable, of course. In this case it is where the offender goes on to be a repeat offender (i.e. recidivist). All of these are binomial nominal variables though they do not have to be, except for the dependent variable. We can use score variables, which are defined as covariates in SPSS, as some or all of our predictor variables. We can also use nominal (category) variables with three or more different values (categories). In this case, SPSS recodes the variable into a number of dummy variables automatically as was the case with multinomial logistic regression in the previous chapter. This means that the set of predictor variables is actually greater than the initial number of variables in the data. If we had a variable such as type of crime with the categories sex crime, violent crime and theft then, in theory, we could create three dummy variables. These would be:

Dummy variable 1 = sex crime or not

Dummy variable 2 = violent crime or not

Dummy variable 3 = theft or not

	Recidivism	Age	Previous prison	Treatment	Contrite	Married	Sex offender
			Table 39.1 Data for the study of recidivism – the data from 19 cases is reproduced five times to give realistic sample sizes but only to facilitate explanation				
1	yes	younger	yes	no	no	no	yes
2	yes	older	yes	no	no	no	yes
3	yes	older	yes	yes	no	no	yes
4	yes	older	yes	yes	no	yes	no
5	yes	younger	yes	no	no	no	no
6	yes	younger	no	yes	yes	no	no
7	yes	older	no	yes	yes	yes	yes
8	yes	younger	yes	no	no	no	yes
9	yes	younger	no	no	no	yes	yes
10	yes	older	no	no	no	no	no
11	no	younger	no	yes	yes	no	no
12	no	older	no	yes	yes	no	no
13	no	older	yes	yes	yes	yes	yes
14	no	younger	no	yes	yes	yes	yes
15	no	younger	no	yes	yes	no	yes
16	no	younger	no	no	yes	yes	no
17	no	older	no	no	no	yes	no
18	no	older	yes	yes	yes	no	no
19	no	older	yes	yes	yes	no	no

It really is as simple as that to create dummy variables. There is just one complication to note which confirms what was discussed in the previous chapter. One of the dummy variables contains no new information if we know a person's values for the other two dummy variables. So, if we know that someone has not committed a sex crime and we know that they have not committed a violent crime then they must have committed a theft. Since they are *all* offenders, this is a simple logical assessment. Because one of the dummy variables is always repetitive and redundant, in logistic regression there is always one fewer than the maximum number of dummy variables. SPSS chooses this arbitrarily unless the user stipulates otherwise.

Table 39.2 demonstrates one way of coding the data ready for the computer analysis. It is best to enter data numerically into SPSS so each value of each variable is given a numerical code. 1s and 0s have been used in every case to indicate the presence or absence of a characteristic. The values would be given value labels in SPSS so as to make the output as clear as possible.

Binomial logistic regression in SPSS has a good variety of analysis options, most of which are unavailable on multinomial logistic regression. Beginners will find the range a little bewildering and many of the options will be of no interest to many researchers. Choosing the appropriate options (what some would refer to as the appropriate model) depends very much on the purpose of one's analysis and only the researcher knows this. Generally speaking, if the sole purpose of the analysis is to obtain the best set of predictors to categorise people then it is sufficient to enter all of the predictors at the same time – SPSS merely takes the variables in the order they are listed. However, more typically the researcher is trying to build up a theoretical/conceptual explanation as in this example of why some offenders reoffend. In these circumstances, the researcher may

Table 39.2		Data from Table 39.1 coded in binary fashion as 0 and 1 for each variable					
	Recidivism	Age	Previous prison	Treatment	Contrite	Married	Sex offender
1	1	0	1	0	0	0	1
2	1	1	1	0	0	0	1
3	1	1	1	1	0	0	1
4	1	1	1	1	0	1	0
5	1	0	1	0	0	0	0
6	1	0	0	1	1	0	0
7	1	1	0	1	1	1	1
8	1	0	1	0	0	0	1
9	1	0	0	0	0	1	1
10	1	1	0	0	0	0	0
11	0	0	0	1	1	0	0
12	0	1	0	1	1	0	0
13	0	1	1	1	1	1	1
14	0	0	0	1	1	1	1
15	0	0	0	1	1	0	1
16	0	0	0	0	1	1	0
17	0	1	0	0	0	1	0
18	0	1	1	1	1	0	0
19	0	1	1	1	1	0	0

wish to give some variables priority in the analysis. This involves entering the predictors hierarchically in blocks. Furthermore, one needs to identify the best predictors. The variables may be entered in step-by-step form, in which the best predictor is taken first, then the next remaining best predictor entered second (allowing for correlations between predictors), and so forth until no significant improvement in prediction is achieved.

An alternative is to enter all of the predictors initially, then drop predictors one by one (backwards elimination). If dropping the weakest predictor significantly reduces the accuracy of the prediction then it should be retained. However, it can be eliminated *if* dropping it makes no change to the accuracy of the prediction. The process goes on step by step by examining the effect of removing the weakest predictor in the remaining set. There is only one complication of note – that is eliminated predictors can return to the set of predictors if the elimination of another variable results in the previously eliminated variable increasing its independent predictive power.

Backwards elimination is *not* better than other approaches. It largely does the job in a different way. It does not necessarily give exactly the same findings as other methods either – it is simply one reasonable and appropriate approach to the data.

Binary logistic regression produces weights for a set of predictor variables. The Wald test is the test of significance of these. The computer can also classify participants in terms of the accuracy of the predictions made as to which of the two categories of the dependent variable they belong to. You may find these categorisation tables very helpful in understanding what is happening in the analysis. Figure 39.1 highlights the steps in binomial logistic regression.

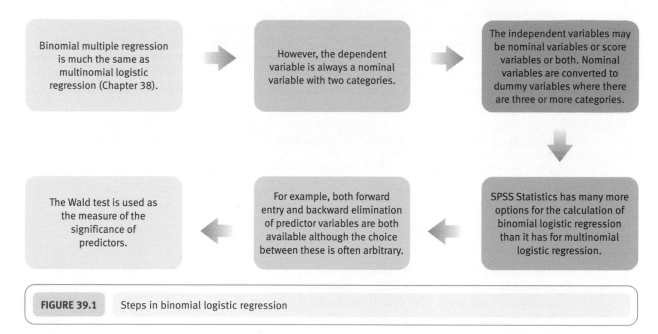

FIGURE 39.1   Steps in binomial logistic regression

39.2   When to use binomial logistic regression

Binomial logistic regression helps the researcher decide which of a set of predictor variables best discriminates two groups of participants. For example, which of our variables best discriminates male and female participants. More generally, it can be regarded as an example of a regression procedure but designed for situations where the dependent variable simply has two categories.

39.3   When not to use binomial logistic regression

There are no major circumstances which would argue against the use of binomial logistic regression except where there is just a single predictor variable. Also your predictor variables should not be highly correlated with each other as in any regression. Avoid throwing every independent variable into the analysis without careful thought in advance. Usually it is better to be selective in the choice of predictors in any form of regression.

39.4   Data requirements for binomial logistic regression

The dependent variable should have just two nominal categories, whereas the predictor variables can be any sort of variable including score variables. A participant may be only one category of the independent and dependent variable or contribute just a single score to score variables.

39.5   Problems in the use of binomial logistic regression

As with multinomial logistic regression, binomial logistic regression produces a great deal of SPSS output for most analyses, just a part of which may be useful to a researcher. Learning to identify the crucial aspects of the output is as essential here as with any other sort of SPSS analysis.

There is a confusing dissimilarity between how multinomial logistic regression and the binomial logistic regression are computed on SPSS and they do not use consistent terminology.

Score variables can cause problems if the researcher fails to tell SPSS that a particular variable is not a nominal (category) variable but it is a score. This is because SPSS will consider each different score mistakenly as a separate nominal category. Thus huge amounts of output may occur when the proper identification is not made.

> You can find out more about binomial logistic regression in Chapter 42 of Howitt, D. and Cramer, D. (2011). *Introduction to Statistics in Psychology*, 5th edition. Harlow: Pearson.

39.6   The data to be analysed

The example we will be working with is a study of the variables which might be used to help assess whether a prisoner is likely to reoffend or not reoffend after leaving prison. Reoffending is known as recidivism. The data from the study are shown in Table 39.2 (*ISP*, Table 42.5). For pedagogical purposes and the convenience of those who wish to follow our steps on a computer, the 19 cases are reproduced five times. As can be seen, recidivism is a binomial category variable – in a given period of time, a prisoner either reoffends or does not. Since the purpose of our analysis is to find the pattern of variables which predict which of these two categories a prisoner will fall then this is an obvious set of data for the binomial logistic regression.

39.7   Entering the data

Step 1

In 'Variable View' of the 'Data Editor', name the first eight rows 'recidivist', 'age', etc. Name the values of variables as in Table 39.1 (e.g. '1' = 'no' and '2' = 'yes'). Change 'Decimals' from '2' to '0'.

Step 2

Enter the data. The 'Freq' variable is for weighting these 19 cases five times using the 'Weight Cases...' procedure (see Sections 3.7, 4.7, 8.8, 9.8, 16.7 or 17.7).

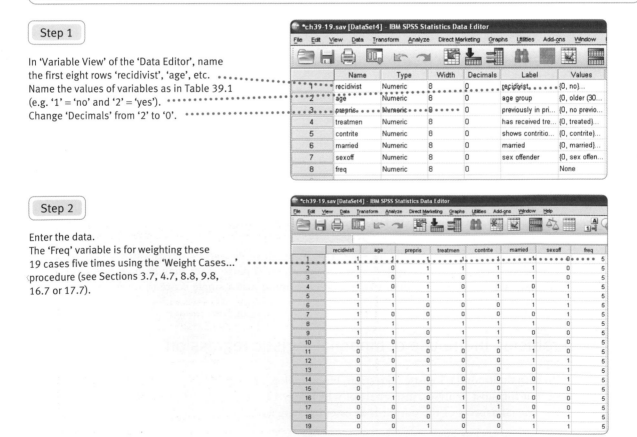

39.8 Binomial logistic regression

Step 1

Select 'Analyze',
'Regression' and
'Binary Logistic...'.

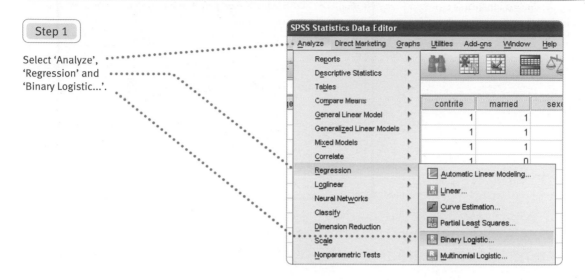

Step 2

Select 'recidivist' and the ➡ button beside the
'Dependent:' box to put it there.
Select the other six variables (excluding 'freq')
and the ➡ button beside the 'Covariates:'
box to put them there.
Select 'Categorical...'.

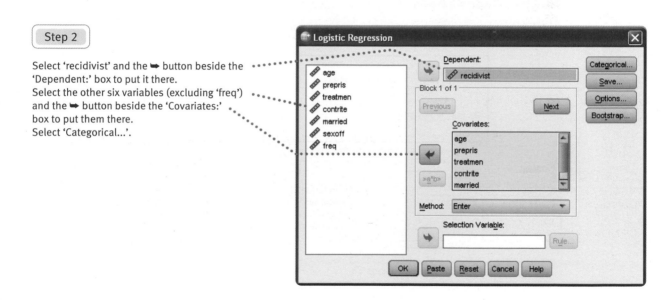

Step 3

Select all six variables (as they are nominal
variables in this example) and the ➡ button
beside the 'Categorical Covariates:'
box to put them there.
Select 'Continue'.

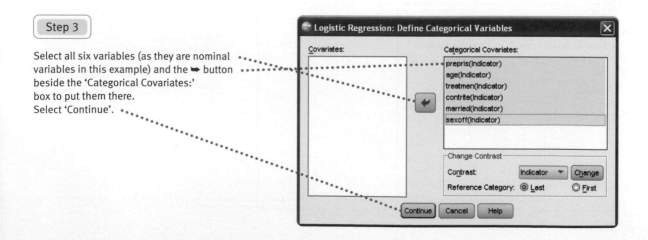

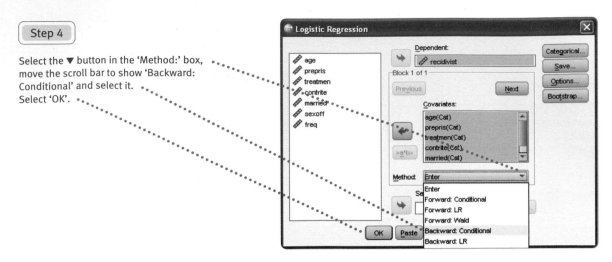

Step 4

Select the ▼ button in the 'Method:' box, move the scroll bar to show 'Backward: Conditional' and select it. Select 'OK'.

39.9   Interpreting the output

Logistic Regression

Case Processing Summary

Unweighted Cases[a]		N	Percent
Selected Cases	Included in Analysis	19	100.0
	Missing Cases	0	.0
	Total	19	100.0
Unselected Cases		0	.0
Total		19	100.0

a. If weight is in effect, see classification table for the total number of cases.

Dependent Variable Encoding

Original Value	Internal Value
no	0
yes	1

Check this table – it makes clear just how the dependent variable has been coded by SPSS (the Internal Value). In this case they are the same as the values that we entered but it is well worth checking otherwise serious errors can be made.

Categorical Variables Codings

		Frequency	Parameter coding (1)
sex offender	sex offender	10	1.000
	not sex offender	9	.000
previously in prison	no previous	11	1.000
	previous prison	8	.000
has received treatment in prison	treated	11	1.000
	not treated	8	.000
shows contrition for offence	contrite	10	1.000
	not contrite	9	.000
married	married	7	1.000
	not married	12	.000
age group	older (30 plus)	10	1.000
	younger	9	.000

This table breaks down each variable and gives numbers in each category. Thus there are nine in the unweighted sample who are not sex offenders.

Block 0 is before any predictors have been taken into account. Generally you can skip this section.

Block 0: Beginning Block

Classification Table[a],[b]

			Predicted		
			recidivist		
Observed			no	yes	Percentage Correct
Step 0	recidivist	no	50	0	100.0
		yes	45	0	.0
	Overall Percentage				52.6

a. Constant is included in the model.
b. The cut value is .500

This table gives the actual numbers of recidivists and non-recidivists. There are 45 recidivists and 50 non-recidivists.

The constant is more or less the same as the intercept in multivariate regression.

Variables in the Equation

		B	S.E.	Wald	df	Sig.	Exp(B)
Step 0	Constant	-.105	.205	.263	1	.608	.900

No predictors have been included at this stage so there cannot be any in the regression equation.

Variables not in the Equation

			Score	df	Sig.
Step 0	Variables	age(1)	2.299	1	.129
		prepris(1)	21.159	1	.000
		treatmen(1)	6.345	1	.012
		contrite(1)	31.714	1	.000
		married(1)	.452	1	.501
		sexoff(1)	.293	1	.588
	Overall Statistics		50.804	6	.000

Block 1: Method = Backward Stepwise (Conditional)

Block 1 contains the key variables.

Omnibus Tests of Model Coefficients

		Chi-square	df	Sig.
Step 1	Step	70.953	6	.000
	Block	70.953	6	.000
	Model	70.953	6	.000
Step 2[a]	Step	-.210	1	.647
	Block	70.743	5	.000
	Model	70.743	5	.000

a. A negative Chi-squares value indicates that the Chi-squares value has decreased from the previous step.

Step 1 includes all of the predictors.

Step 2 and any later steps each involve the elimination of a variable – that is they are the backward elimination steps.

The variables in the equation table (see later) indicate which predictors are in the model at each step.

Model Summary

Step	-2 Log likelihood	Cox & Snell R Square	Nagelkerke R Square
1	60.482[a]	.526	.702
2	60.692[a]	.525	.701

a. Estimation terminated at iteration number 20 because maximum iterations has been reached. Final solution cannot be found.

These are indexes of the fit of the model to the actual data. The nearer the two (pseudo) R-Square measures are to 1.0 the better the fit of the model at that step to the data. The smaller the −2 Log likelihood value the better the fit of the model to the data.

Notice that the final model (Step 2) shows an improved fit – a variable has been allowed to re-enter the regression equation.

Classification Table[a]

			Predicted		
			recidivist		
	Observed		no	yes	Percentage Correct
Step 1	recidivist	no	45	5	90.0
		yes	5	40	88.9
	Overall Percentage				89.5
Step 2	recidivist	no	45	5	90.0
		yes	5	40	88.9
	Overall Percentage				89.5

a. The cut value is .500

This table indicates the accuracy of the classification at each step.

The least significant predictor in Step 1 is chosen for elimination in Step 2. 'treatmen' and 'sexoff' both have significances of .998 but 'treatmen' occurs first and is selected.

Variables in the Equation

		B	S.E.	Wald	df	Sig.	Exp(B)
Step 1[a]	age(1)	-2.726	.736	13.702	1	.000	.065
	prepris(1)	-1.086	.730	2.215	1	.137	.337
	treatmen(1)	19.362	8901.292	.000	1	.998	2.564E8
	contrite(1)	-41.459	11325.913	.000	1	.997	.000
	married(1)	-.307	.674	.208	1	.648	.735
	sexoff(1)	-20.641	7003.092	.000	1	.998	.000
	Constant	23.802	7003.092	.000	1	.997	2.174E10
Step 2[a]	age(1)	-2.699	.731	13.625	1	.000	.067
	prepris(1)	-1.153	.708	2.648	1	.104	.316
	treatmen(1)	19.428	8895.914	.000	1	.998	2.739E8
	contrite(1)	-41.375	11337.365	.000	1	.997	.000
	sexoff(1)	-20.475	7028.411	.000	1	.998	.000
	Constant	23.542	7028.411	.000	1	.997	1.675E10

a. Variable(s) entered on step 1: age, prepris, treatmen, contrite, married, sexoff.

It is important to note that the regression weights are only applied to the value of the category variable coded as 1. So the negative regression weight for 'sexoff(1)' of −20.475 actually indicates that sex offenders are *less* likely to reoffend on release. Great care is needed to know what codings are given to the different values of variables. This is made worse because nominal variables not coded as 0 and 1 may be recoded by SPSS into those values.

Variables not in the Equation

			Score	df	Sig.
Step 2[a]	Variables	married(1)	.209	1	.648
	Overall Statistics		.209	1	.648

a. Variable(s) removed on step 2: married.

This table gives the significance of the variables not included in the model.

REPORTING THE OUTPUT

One way of reporting the findings of this analysis is as follows:

Using the conditional backward elimination model, characteristics differentiating prisoners who reoffend after release from those who do not reoffend were investigated. The final regression model indicated that younger offenders, those with a previous history of prison, those who were not contrite about their offences, and those who were not sex offenders were more likely to reoffend. Age was a significant predictor at the .05 level. The Cox and Snell pseudo r-square was .53, indicating that the fit of the model to the data was only moderate. This model was almost equally accurate for reoffending (88.9 per cent correct) as for non re-offending (90.0 per cent).

Summary of SPSS Statistics steps for binomial logistic regression

Data

- Name the variables in 'Variable View' of the 'Data Editor'.
- Enter the data under the appropriate variables names in 'Data View' of the 'Data Editor'.

Analysis

- Select 'Analyze', 'Regression' and 'Binary Logistic ...'.
- Move the dependent variable to the 'Dependent:' box, the predictors to the 'Covariate(s):' box and select 'Categorical ...'.
- Move the categorical predictors to the 'Categorical Covariate(s):' box and select 'Continue'.
- Select the 'Method:' of entry from the dropdown menu next to this option (e.g. 'Backward conditional') and 'OK'.

Output

- Check the way the dependent variable has been coded in the 'Dependent Variable Encoding' table.
- Note the predictors in the 'Variables in the Equation' table for the final model and the direction of the association for the regression weights.
- Note the percentage of cases correctly identified in the 'Classification Table' for this model.

For further resources including data sets and questions, please refer to the website accompanying this book.

PART 7

Data handling procedures

CHAPTER 40

Reading ASCII or text files into the 'Data Editor'

Overview

- Sometimes you have a computer file of data which you wish to use on SPSS Statistics. This chapter tells you how to use data not specifically entered into the SPSS 'Data Editor' spreadsheet.

- Although student work will rarely require the use of ASCII files, there are a number of databases of archived data which researchers may wish to analyse.

- It is worth noting that sometimes data from other spreadsheets can be entered into SPSS. For example, data from Excel spreadsheets can simply be copied and pasted into SPSS though not the spreadsheet in its entirety. SPSS data can be saved as an Excel file.

40.1 What is an ASCII or text data file?

SPSS Statistics is obviously one of many different computer programs for analysing data. There are circumstances in which researchers might wish to take data sets which have been prepared for another computer and run those data through SPSS Statistics. It can be expensive in time and/or money to re-enter data, say, from a big survey into the 'Data Editor' spreadsheet. Sometimes, the only form in which the data are available is as an archived electronic data file. The original questionnaires may have been thrown away. No matter the reason for using an imported data file, SPSS Statistics can accept files in other forms. In particular, data files are sometimes written as simple text or ASCII files, as these can be readily transferred from one type of computer to another. ASCII stands for American Standard Code for Information Interchange. To analyse an ASCII data file you first need to read it into the 'Data Editor'.

Suppose, for example, that you had an ASCII data file called 'data.txt' which consisted of the following numbers:

```
1118
2119
3218
```

Obviously you cannot sensibly use an ASCII file until you know exactly where the information for each variable is. However, we do know where and what the information is for this small file.

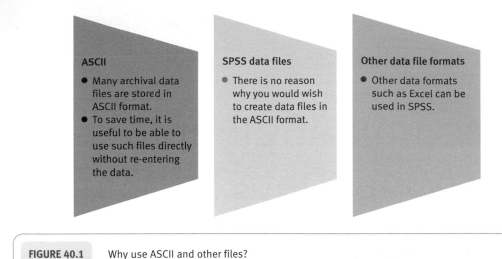

FIGURE 40.1   Why use ASCII and other files?

The figures in the first column simply number the three different participants for whom we have data. The values in the second column contain the code for gender, with 1 representing females and 2 males. While the values in the third and fourth column indicate the age of the three people. Figure 40.1 highlights issues in using ASCII and other files. We would carry out the following procedure to enter this ASCII data file into the 'Data Editor'.

40.2 Entering data into an ASCII or text data file

Step 1

Type data into, say, a Word document.

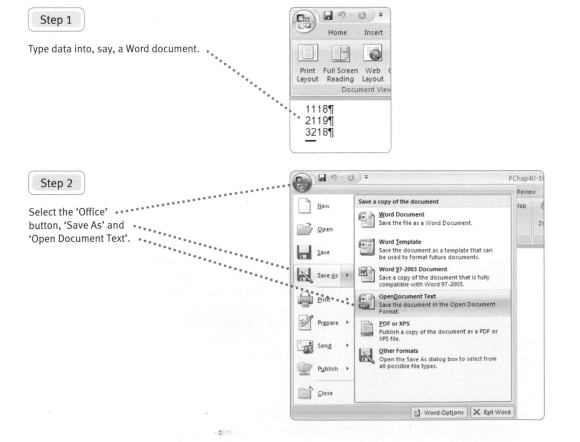

Step 2

Select the 'Office' button, 'Save As' and 'Open Document Text'.

Step 3

Select directory to save text
in ('Save in: DataFiles').
Type 'File name:' (ch40-19) and
select file type ('Plain Text' in
'Save as file type').

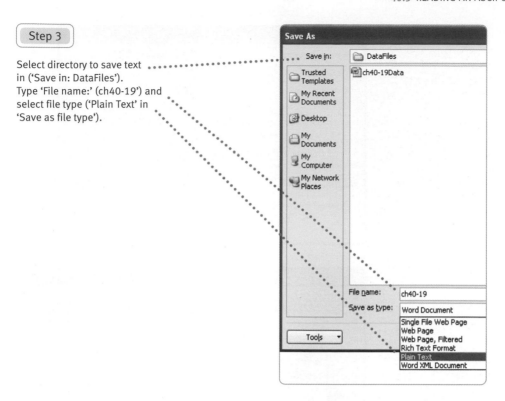

Step 4

Select 'OK'.

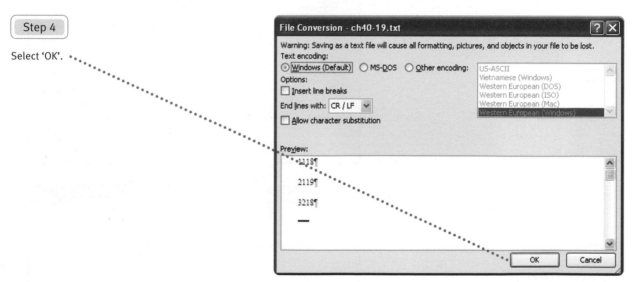

40.3 Reading an ASCII or text data file

Step 1

Select 'File' and
'Read Text Data...'.

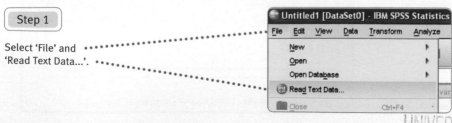

Step 2

Select the ▼ button of
the 'Look in:' box to find the data
file (eg ch40-19.txt).
Select the file.
Select 'Open'.

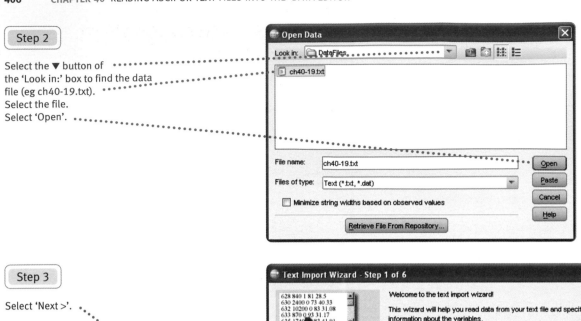

Step 3

Select 'Next >'.

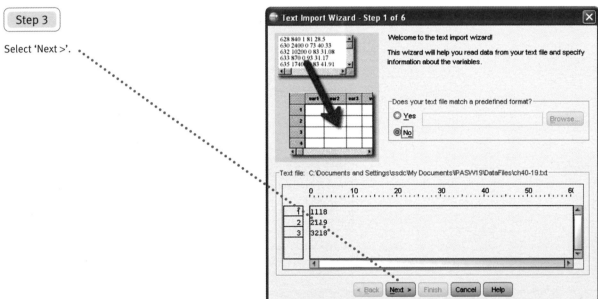

Step 4

Select 'Next >'.

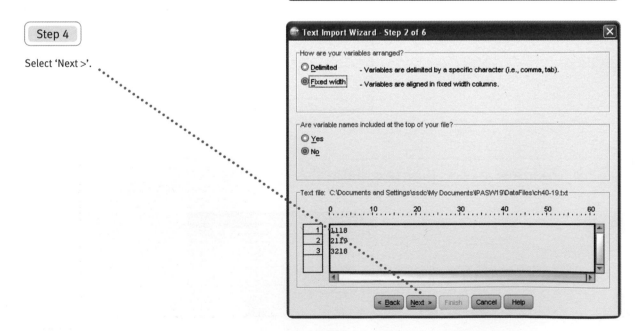

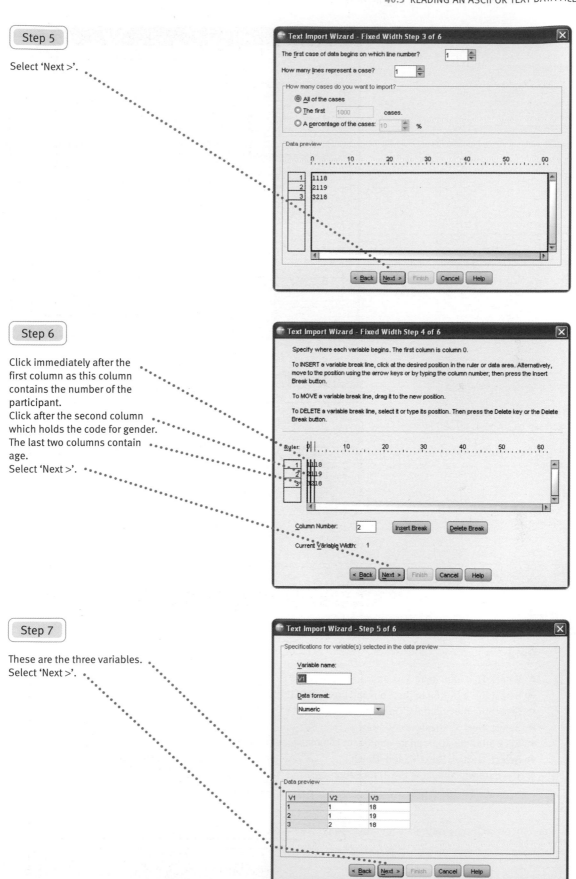

Step 5

Select 'Next >'.

Text Import Wizard - Fixed Width Step 3 of 6

The first case of data begins on which line number? 1

How many lines represent a case? 1

How many cases do you want to import?
- All of the cases
- The first 1000 cases.
- A percentage of the cases: 10 %

Data preview

1	1118
2	2119
3	3218

< Back   Next >   Finish   Cancel   Help

Step 6

Click immediately after the first column as this column contains the number of the participant.
Click after the second column which holds the code for gender.
The last two columns contain age.
Select 'Next >'.

Text Import Wizard - Fixed Width Step 4 of 6

Specify where each variable begins. The first column is column 0.

To INSERT a variable break line, click at the desired position in the ruler or data area. Alternatively, move to the position using the arrow keys or by typing the column number; then press the Insert Break button.

To MOVE a variable break line, drag it to the new position.

To DELETE a variable break line, select it or type its position. Then press the Delete key or the Delete Break button.

Ruler:

1	1118
2	2119
3	3218

Column Number: 2   Insert Break   Delete Break

Current Variable Width: 1

< Back   Next >   Finish   Cancel   Help

Step 7

These are the three variables.
Select 'Next >'.

Text Import Wizard - Step 5 of 6

Specifications for variable(s) selected in the data preview

Variable name:
V1

Data format:
Numeric

Data preview

V1	V2	V3
1	1	18
2	1	19
3	2	18

< Back   Next >   Finish   Cancel   Help

Step 8

Select 'Finish'.

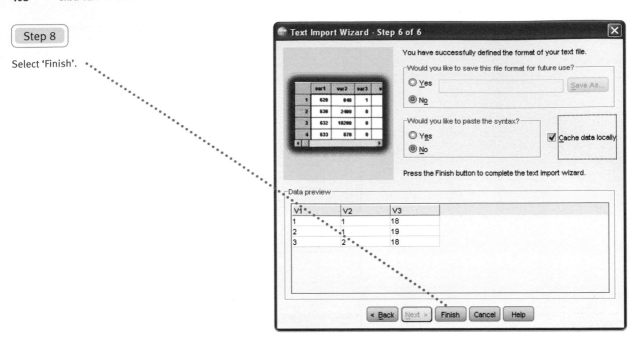

The text data have now been entered into the 'Data Editor' and can be saved as an SPSS file.

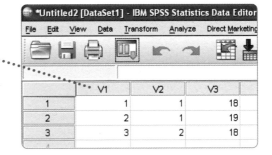

Summary of SPSS Statistics steps for inputting an ASCII or text data file

Data

- Select 'File' and 'Read Text data ...'.
- Select the ▼ button of the 'Look in:' box to find the data file.
- Select the file and then 'Open'.
- Select 'Next >', 'Next >', and 'Next >'.
- Click after the appropriate number of columns for each variable.
- Select 'Next >', 'Next >', and 'Finish'.
- The data will appear in the 'Data Editor'.

For further resources including data sets and questions, please refer to the CD accompanying this book.

CHAPTER 41

Missing values

Overview

- Sometimes in research, you may not have a complete set of data from each participant. Missing values tells the computer how to deal with such situations.

- Missing values can also be used to instruct the computer to ignore cases with a particular value(s) on particular variables.

- Typically, when coding the variables for entry into the SPSS Statistics spreadsheet, the researcher chooses a distinctive value to denote a missing value on that variable. This value needs to be outside the possible range of actual values found in the data. Typically, numbers such as 9, 99 and 999 are used for missing values.

- It is possible to have more than one missing value for any variable. For example, the researcher may wish to distinguish between circumstances in which the participant refused to give an answer to a questionnaire and cases where the interviewer omitted the question for some reason.

- One needs to be very careful when using missing values on SPSS. If a value has not been identified as a missing value for a particular variable, the computer will analyse what the researcher intended as a missing value as a real piece of data. This can seriously and misleadingly distort the analysis.

- Missing values can be used in two main ways. In listwise deletion the case is deleted from the analysis if any missing values are detected for that case. This can rapidly deplete the number of participants in a study.

- The alternative is to simply delete the case from analyses involving pairs of variables for which there is a missing value (pairwise deletion).

- It is better to specify missing values as a value rather than type nothing under an entry. This is because typos made when typing in the data are too easily confused for an actual missing value.

41.1 What are missing values?

When collecting data, information for some of the cases on some of the variables may be missing. Take, for example, the data in Table 41.1, which consists of the music and mathematics scores of 10 children with their code for gender and their age in years. There is no missing information for any of the four variables for any of the 10 cases. But suppose that the first two cases were away for the music test so that we had no scores for them. It would be a pity to discard all the data for these two cases because we have information on them for the other three variables of mathematics, gender and age. Consequently we would enter the data for these other variables. Although we could leave the music score cells empty for these two cases, what we usually do is to code missing data with a number which does not correspond to a possible value that the variable could take. Suppose the scores for the music test can vary from 0 to 10. Any number, other than 0 to 10 which are real scores, can be used to signify a missing value for the music test. We will use the number 11 as the code for a missing music score so that the values in the first two rows of the first column are 11. Additionally, it is assumed that the age for the third case is missing. The number 0 is the code for an age when the data are missing. Now we need to tell SPSS how we have coded missing data. If we do not do this, then SPSS will read these codes as real numbers.

Missing values can also be used to tell the computer to ignore certain values of a variable which you wish to exclude from your analysis. So, for example, you could use missing values in relation to chi-square to get certain categories ignored. Leaving a cell blank in the 'Data Editor' spreadsheet results in a full stop (.) being entered in the cell if it is part of the active matrix of entries. On the output these are identified as missing values but they are best regarded as omitted values. We would recommend that you do not use blank cells as a way of identifying missing values since this does not distinguish between truly missing values and keyboard errors. Normally, substantial numbers and distinctive numbers such as 99 or 999 are the best way of identifying a missing value.

SPSS offers two options for analysing data with missing values. One, listwise, ignores all of the data for an individual participant if any of the variables in the particular analysis has a missing value. This is the more extreme version. Less extreme is pairwise deletion of missing data since the data are ignored only where absolutely necessary. So if a correlation matrix is being computed, the missing values are ignored only for the correlations involving the particular variables in question. There is a limit to the number of missing values that can be included for any variable on SPSS. Figure 41.1 highlights some issues in using missing values.

Table 41.1	Scores on musical ability and mathematical ability for 10 children with their gender and age		
Music score	**Mathematics score**	**Gender**	**Age**
2	8	1	10
6	3	1	9
4	9	2	12
5	7	1	8
7	2	2	11
7	3	2	13
2	9	2	7
3	8	1	10
5	6	2	9
4	7	1	11

	Omitted data	Omitting certain types of case or individual	Different procedures for missing data

Omitted data

- Data which the participant has failed to supply may be defined as missing by the researcher.
- Certain values (e.g. 999) may be reserved for this.

Omitting certain types of case or individual

- Since the researcher defines what values he or she wishes the computer to ignore, missing values may be used to drop particular sorts of participants from the analysis.

Different procedures for missing data

- SPSS Statistics may be used to drop certain participants from the entire analysis.
- Alternatively, participants may be dropped only for those statistical analyses for which data are incomplete.

FIGURE 41.1   Why use missing values?

41.2   Entering the data

Select the file from Chapter 10 or 11 if you saved it and enter the new data as follows. Otherwise enter the data as follows.

Step 1

In 'Variable View' of the 'Data Editor' name the first four rows 'Music', 'Maths', 'Gender' and 'Age'. If using saved data, name the third and fourth rows 'Gender' and 'Age'.
Change 'Decimals' from '2' to '0'.

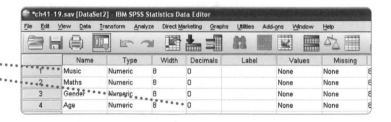

	Name	Type	Width	Decimals	Label	Values	Missing	
1	Music	Numeric	8	0		None	None	8
2	Maths	Numeric	8	0		None	None	8
3	Gender	Numeric	8	0		None	None	8
4	Age	Numeric	8	0		None	None	8

Step 2

Enter the data.
If using the saved file, change the 2 'Music' scores of cases 1 and 2 to '11' as these are now to be missing.
Change the 'Age' of case 3 to '0' as this is missing.

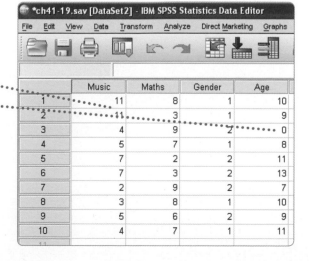

	Music	Maths	Gender	Age
1	11	8	1	10
2	11	3	1	9
3	4	9	2	0
4	5	7	1	8
5	7	2	2	11
6	7	3	2	13
7	2	9	2	7
8	3	8	1	10
9	5	6	2	9
10	4	7	1	11
11				

41.3 Defining missing values

In 'Variable View' of the 'Data Editor' select the right side of the cell for 'Music' under 'Missing'. An ellipse of three dots appears.
Select 'Discrete missing values' and type '11' in the box below.
Select 'OK'.
'11' appears in the 'Missing' cell for 'Music' (see Step 1 of Section 15.3).
Repeat this procedure for 'Age' but type in '0' instead of '11'.

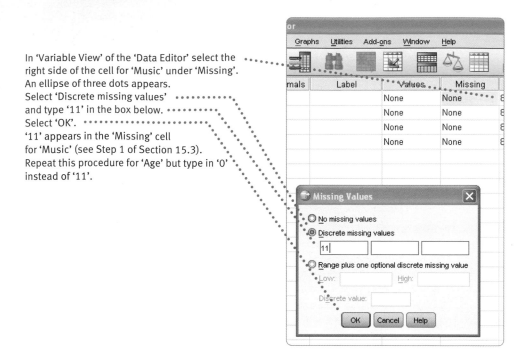

41.4 Pairwise and listwise options

We will illustrate some of the options available when you have missing data with the Correlate procedure, although similar kinds of options are available with some of the other statistical procedures.

Step 1

Select 'Analyze', 'Correlate' and 'Bivariate...'.

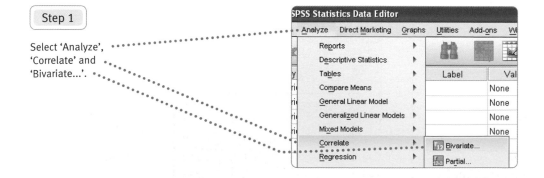

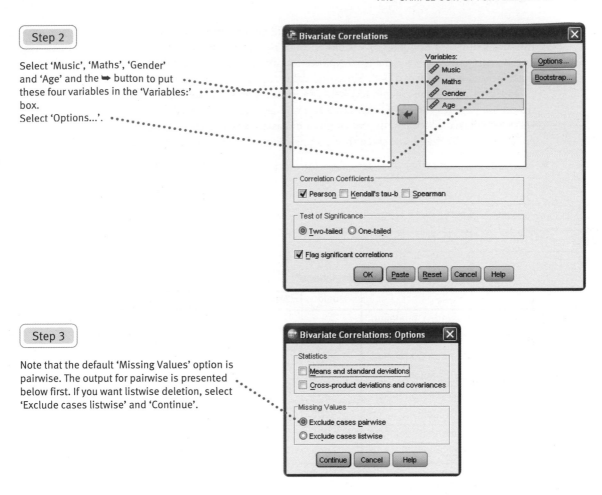

Step 2

Select 'Music', 'Maths', 'Gender' and 'Age' and the ➡ button to put these four variables in the 'Variables:' box.
Select 'Options...'.

Step 3

Note that the default 'Missing Values' option is pairwise. The output for pairwise is presented below first. If you want listwise deletion, select 'Exclude cases listwise' and 'Continue'.

41.5   Sample output for pairwise deletion

Correlations

		Music	Maths	Gender	Age	
Music	Pearson Correlation	1	-.923**	.293	.681	8 cases
	Sig. (2-tailed)		.001	.482	.092	7 cases
	N	8	8	8	7	
Maths	Pearson Correlation	-.923**	1	-.161	-.550	
	Sig. (2-tailed)	.001		.656	.125	9 cases
	N	8	10	10	9	
Gender	Pearson Correlation	.293	-.161	1	.118	10 cases
	Sig. (2-tailed)	.482	.656		.762	
	N	8	10	10	9	
Age	Pearson Correlation	.681	-.550	.118	1	
	Sig. (2-tailed)	.092	.125	.762		
	N	7	9	9	9	

**. Correlation is significant at the 0.01 level (2-tailed).

- Pairwise deletion means that a correlation will be computed for all cases which have non-missing values for any pair of variables. Since there are two missing values for the music test and no missing values for the mathematics test and gender, the number of cases on which these correlations will be based is 8. Since one value for age is missing for another case, the number of cases on which the correlation between music scores and gender is based is 7. As there are no missing values for the mathematics test and gender, the number of cases on which this correlation is based is 10. Finally, the number of cases on which the correlation between the mathematics score and age is 9 since there is one missing value for age and none for mathematics.

- *Notice that the number of cases varies for pairwise deletion of missing values.*

41.6   Sample output for listwise deletion

Correlations[a]

		Music	Maths	Gender	Age
Music	Pearson Correlation	1	-.956[**]	.354	.681
	Sig. (2-tailed)		.001	.437	.092
Maths	Pearson Correlation	-.956[**]	1	-.483	-.729
	Sig. (2-tailed)	.001		.272	.063
Gender	Pearson Correlation	.354	-.483	1	.088
	Sig. (2-tailed)	.437	.272		.852
Age	Pearson Correlation	.681	-.729	.088	1
	Sig. (2-tailed)	.092	.063	.852	

[**]. Correlation is significant at the 0.01 level (2-tailed).
a. Listwise N=7

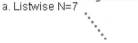

All cases are 7

- In listwise deletion, correlations are computed for all cases which have no missing values on the variables which have been selected for this procedure. In this example, the number of cases which have no missing values on any of the four variables selected is 7.

- *Notice that the number of cases does not vary in listwise deletion of missing values.*

41.7   Interpreting the output

There is little in the output which has not been discussed in other chapters. The only thing to bear in mind is that the statistics are based on a reduced number of cases.

REPORTING THE OUTPUT

Remember to report the actual sample sizes (or degrees of freedom) used in reporting each statistical analysis rather than the number of cases overall.

Summary of SPSS Statistics steps for handling missing values

Data

- In 'Variable View' of the 'Data Editor' select the right-hand side of the cell of the appropriate variable in the 'Missing' column.
- Select 'Discrete missing values' and enter in the box/es below the values used to indicate missing values.
- Select 'OK'.
- This cell can be copied and pasted in other cells to provide the same missing values.

Analysis

- In many of the analysis procedures you can select either 'Exclude cases listwise' or 'Exclude cases pairwise'.

Output

- If there is missing data and 'Exclude cases pairwise' is selected, the number of cases for the analyses carried out in a single step may differ.
- For example, in a correlation analysis involving three or more variables the number of cases for pairs of variables may differ.
- If 'Exclude cases listwise' is selected, the number of cases is the same for all pairs of variables and will be smaller.

For further resources including data sets and questions, please refer to the website accompanying this book.

CHAPTER 42

Recoding values

Overview

- From time to time researchers need to alter how certain values of a variable are recorded by the computer. Perhaps several different values need to be combined into one.

- The recoding values procedure allows you considerable flexibility to easily modify how any value has been coded numerically.

- Because SPSS Statistics can quickly recode values, it is good practice to enter as much data as possible in their original form. This leaves the researcher with the greatest freedom to recode the data. If data have been recoded by the researcher prior to entry into the SPSS data spreadsheet, it is not possible to try an alternative coding of the original data. For example, if all of the scores are entered for a measure of extraversion for each participant, SPSS can be used to recode the entries (such as reversing the scoring of an item) or to compute a total based on all or some of the extraversion items. If the researcher calculates a score on extraversion manually before entering this into SPSS, it is not possible to rescore the questionnaire in a different way without adding the original scores.

- It is usually best to keep your original data intact. So always make sure that you create a brand new variable (new column) for the recoded variable. Do not use the original variables unless you are absolutely certain that you wish to change the original data forever.

42.1 What is recoding values?

Sometimes we need to recode values for a particular variable in our data. There can be many reasons for this, including the following:

- To put together several categories of a nominal variable which otherwise has very few cases. This is commonly employed in statistics such as chi-square.

- To place score variables into ranges of scores.

- To combine items to form an overall score (see Chapter 43) where some items need to be scored in the reverse way.

Table 42.1	Scores on musical ability and mathematical ability for 10 children with their gender and age		
Music score	**Mathematics score**	**Gender**	**Age**
2	8	1	10
6	3	1	9
4	9	2	12
5	7	1	8
7	2	2	11
7	3	2	13
2	9	2	7
3	8	1	10
5	6	2	9
4	7	1	11

We may wish to categorise our sample into two or more groups according to some variable such as age or intelligence. The recoding of cases can be illustrated using the data in Table 42.1, which shows the music and mathematics scores of 10 children together with their code for gender and their age in years. These values are the same as those previously presented in Table 40.1. Suppose that we wanted to compute the correlation between the music and mathematics scores for the younger and older children. To do this, it would first be necessary to decide how many age groups we wanted. Since there are only 10 children then one might settle for two groups. Next the cut-off point in age which divide the two groups has to be decided. As it would be good to have the two groups of similar size, age 10 would make a good cutoff point with children younger than 10 falling into one group and children aged 10 or more into the other group. SPSS can be used to recode age in this way. Figure 42.1 highlights some issues in recoding values.

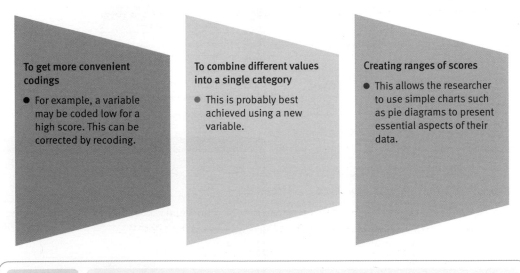

To get more convenient codings

● For example, a variable may be coded low for a high score. This can be corrected by recoding.

To combine different values into a single category

● This is probably best achieved using a new variable.

Creating ranges of scores

● This allows the researcher to use simple charts such as pie diagrams to present essential aspects of their data.

FIGURE 42.1	Why recode values?

42.2 Entering the data

Select the file from Chapter 41 or from Chapters 10 or 11 if you saved it and make the changes shown below. Otherwise enter the new data as follows.

Step 1

In 'Variable View' of the 'Data Editor' name the first four rows 'Music', 'Maths', 'Gender' and 'Age'.
If using saved data, name the third and fourth rows 'Gender' and 'Age'.
Change 'Decimals' from '2' to '0'.

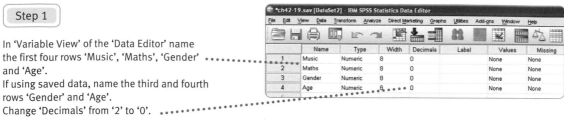

Step 2

Enter the data. If using the saved file from Chapter 41, change the two 'Music' scores of cases 1 and 2 from '11' as these are now not missing.
Change the 'Age' of case 3 from '0' as this is now not missing.

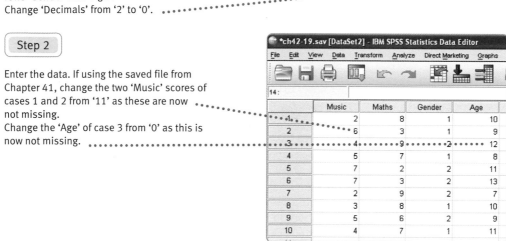

42.3 Recoding values

Step 1

Select 'Transform' and 'Recode into Different Variables...'.

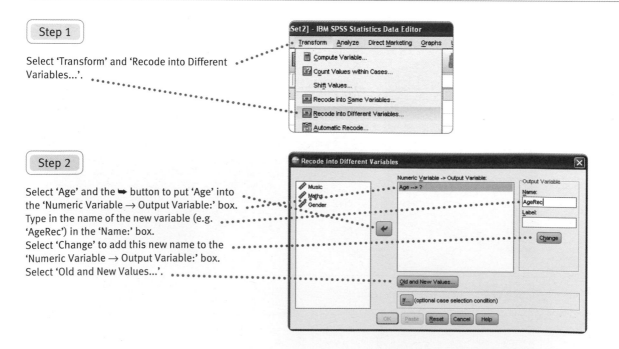

Step 2

Select 'Age' and the ➡ button to put 'Age' into the 'Numeric Variable → Output Variable:' box.
Type in the name of the new variable (e.g. 'AgeRec') in the 'Name:' box.
Select 'Change' to add this new name to the 'Numeric Variable → Output Variable:' box.
Select 'Old and New Values...'.

Step 3

Select 'Range, LOWEST through value'
and type '9' in the box below it. In the 'Value:'
box under 'New Value', type '1'.
Select 'Add' to put 'Lowest thru 9 → 1'
into the 'Old → New:' box.

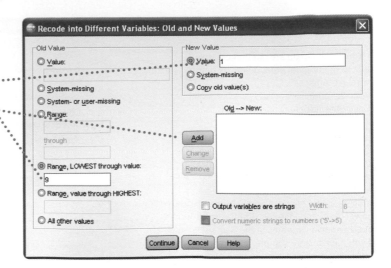

Step 4

Select 'Range, value through
HIGHEST' and type '10'
in the box below it.
In the 'Value:' box under
'New Value', type '2' in it.
Select 'Add' to put '10
thru Highest → 2' into
the 'Old → New:' box.
Select 'Continue'.
Select 'OK' (in the previous box).

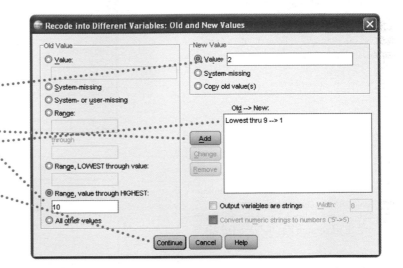

Step 5

The new variable and its values is shown
in 'Data View'. Check that these values
are what you expected.

*ch42-19.sav [DataSet2] - IBM SPSS Statistics Data Editor

File Edit View Data Transform Analyze Direct Marketing Graphs Utilities Add-ons

	Music	Maths	Gender	Age	AgeRec
1	2	8	1	10	2.00
2	6	3	1	9	1.00
3	4	9	2	12	2.00
4	5	7	1	8	1.00
5	7	2	2	11	2.00
6	7	3	2	13	2.00
7	2	9	2	7	1.00
8	3	8	1	10	2.00
9	5	6	2	9	1.00
10	4	7	1	11	2.00

Step 6

If the recoded data are not immediately next to the original data, it may be better to select 'Analyze', 'Reports' and 'Case Summaries...'.

Step 7

Select 'Age' and 'AgeRec' and the ➡ button to put these two variables in the 'Variables:' box. If you have a large sample, you could look at only, say, the first 10 rather than 100 cases by changing 100 to 10 in the 'Limit cases to first' box.
Select 'OK'.

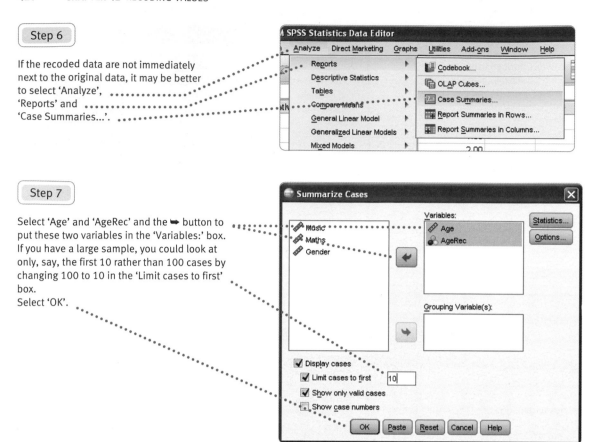

The cases are listed in the first column according to their row number.
The second column has the original values.
The third column has the recoded values.

Case Summaries[a]

		Age	AgeRec
1		10	2.00
2		9	1.00
3		12	2.00
4		8	1.00
5		11	2.00
6		13	2.00
7		7	1.00
8		10	2.00
9		9	1.00
10		11	2.00
Total	N	10	10

a. Limited to first 10 cases.

With a complex set of data it is very easy to forget precisely what you have done to your data. Recoding can radically alter the output from a computer analysis. You need to carefully check the implications of any recodes before reporting them.

42.4   Recoding missing values

Note that if there are missing values (as for 'age' in Section 42.2, Step 1) it is necessary to code these by selecting 'System- or user-missing' in the 'Old Value' section of the 'Recode into Different Variables: Old and New Values' sub-dialog box and selecting 'System-missing' in the 'New Value' section.

Always check that your recoding has worked as you intended by comparing the old and new values in the data editor for each new value for one or more cases.

42.5   Saving the Recode procedure as a syntax file

You need to keep a note of how any new variables were created. The simplest way to do this is to write it down. Another way is to save what you have done as a syntax command by using the 'Paste' option in the main box of the procedure (see Step 7 in Section 42.3). The syntax for this procedure is shown below.

A syntax command. ⋯⋯⋯⋯⋯⋯⋯⋯

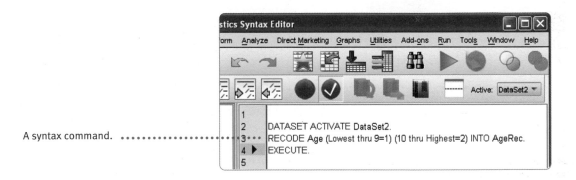

Save this command as a file. You could use this file to carry out this procedure on data subsequently added to this file or on new data with the same variables and variable names. Before Windows was developed SPSS commands were carried out with this kind of syntax command.

To check out this procedure select the column in 'Data View' containing recoded age and delete it.

Select the Syntax window, select the whole of the command in it and run it with the ▶ button on the toolbar shown above.

42.6   Adding some extra cases to Table 42.1

Add the values of these three new cases to this file. ⋯⋯⋯⋯⋯⋯⋯

11	5	4	1	8
12	6	7	2	11
13	8	4	1	9
14				

| 42.7 | Running the Recode syntax command |

Step 1

Select 'Run' and 'All' to run this syntax command.
Alternatively, highlight the command and select the ▶ button (which is hidden under the drop-down menu).

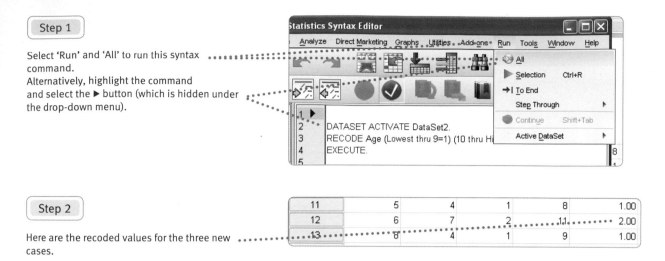

Step 2

Here are the recoded values for the three new cases.

11	5	4	1	8	1.00
12	6	7	2	11	2.00
13	8	4	1	9	1.00

Summary of SPSS Statistics steps for recoding values

Data

- Select 'Transform' and 'Recode into Different Variables …'.
- Select variable and the ▶ button to put variable into the 'Numeric Variable → Output Variable:' box.
- Type in the name of the new variable in the 'Name:' box.
- Select 'Change' to add this new name to the 'Numeric Variable → Output Variable:' box.
- Select 'Old and New Values …'.
- Type in 'Old Value' and then 'New Value'.
- Select 'Add'.
- Repeat for other values.
- When finished, select 'Continue' and 'OK'.
- Check recoding has been done correctly.
- Save procedure as a Syntax file by selecting 'Transform', 'Recode into Different Variables …' and 'Paste'.
- Label Syntax file.

For further resources including data sets and questions, please refer to the website accompanying this book.

Computing a scale score with no missing values

- Computing new variables allows you to add, subtract, etc. scores on several variables to give you a new variable. For example, you might wish to add together several questions on a questionnaire to give an overall index of what the questionnaire is measuring.

- One of the few disadvantages of SPSS Statistics is that no record is kept as to how the new variable was calculated in the first place. There are three ways of doing this: one can keep a detailed written record of the formula used to compute the new variable; if it is simple enough, the variable label could be used to describe what it is; the syntax command can be saved.

- When computing new variables, it is generally a wise precaution to do a few hand checks of cases. It is easy to inadvertently enter an incorrect formula which then gives a new variable which is not the one that you think you are creating.

43.1 What is computing a scale score?

When analysing data we may want to form a new variable out of one or more old ones. For example, when measuring psychological variables, several questions are often used to measure more or less the same thing such as the following which assess satisfaction with life:

- I generally enjoy life
- Some days things just seem to get me down
- Life often seems pretty dull
- The future looks hopeful.

Participants are asked to state how much they agree with each of these statements on the following four-point scale:

    1: Strongly agree   2: Agree   3: Disagree   4: Strongly disagree

Adding, etc. scores on several variables together

● You may wish to derive a new variable by combining several separate components together in some way. Computing totals or averages or ratios, etc. are feasible and simple on SPSS Statistics.

To make a set of scores correspond to a better distribution

● Very skewed distributions are not usually good in statistical analyses. It may be possible to get rid of some of the skew by creating a logarithmic variable from one of your variables of interest.

Other ways of combining several variables

● It is possible to combine different variable factors analytically using SPSS Statistics.
● This produces scores based on several variables (i.e. a factor) which are expressed in terms of z-scores (see Chapter 7).

FIGURE 43.1 Why compute a new scale score?

These items could be used to determine how satisfied people are with their lives by adding up each participants' responses to all four of them.

Notice a problem that frequently occurs when dealing with questionnaires: if you answer 'Strongly agree' to the first and fourth items you indicate that you enjoy life, whereas if you answer 'Strongly agree' to the second and third items you imply that you are dissatisfied with life. We want higher scores to denote greater life satisfaction. Consequently, we will reverse the scoring for the *first* and *fourth* items as follows:

1: Strongly disagree   2: Disagree   3: Agree   4: Strongly agree

The 'Recode' procedure described in Chapter 42 can be used to recode the values for the first and fourth items. Figure 43.1 outlines some issues in computing new scale score.

The data in Table 43.1 show the answers to the four statements by three individuals, and the way in which the answers to the first and fourth items have to be recoded. We will use these data to illustrate the SPSS procedure for adding together the answers to the four statements to form an index of life satisfaction.

Table 43.1	Life satisfaction scores of three respondents			
	Enjoy life (recode)	**Get me down (no recode)**	**Dull (no recode)**	**Hopeful (recode)**
Respondent 1	Agree (2 recoded as 3)	Agree (2)	St. disagree (4)	Agree (2 recoded as 3)
Respondent 2	Disagree (3 recoded as 2)	Disagree (3)	Agree (2)	St. disagree (4 recoded as 1)
Respondent 3	St. agree (1 recoded as 4)	Disagree (3)	Disagree (3)	Disagree (3 recoded as 2)

43.2   Entering the data

Step 1

In 'Variable View' of the 'Data Editor', name the four variables 'q1', 'q2', 'q3' and 'q4'. Set the number of decimal places to zero. You may wish to type in or paste the wording of each item. If you do this see Section 43.3.

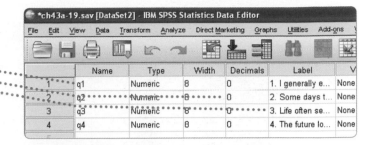

Step 2

Enter the (recoded) data, having named the variables as shown and removed the two decimal places.

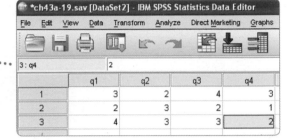

43.3   Displaying variable labels in dialog boxes

For most purposes it is sufficient to give your variables short names. However, when inputting questionnaire items it may be useful to type in or paste the complete item. This is particularly useful when seeing which items correlate or load most highly on a factor (Chapter 30) as it enables you to see at the same time the wording of the item in the output. This makes it easy to see what the factor seems to be measuring.

It is generally not useful to display a long variable label in a dialog box as only part of the label is displayed because of the limited size of the text box. Furthermore, when creating new variables as in Compute it is very difficult to see what you are doing. This is shown in the dialog box below, which is otherwise the same as the dialog box in Step 2a in Section 43.4.

The variable labels displayed.

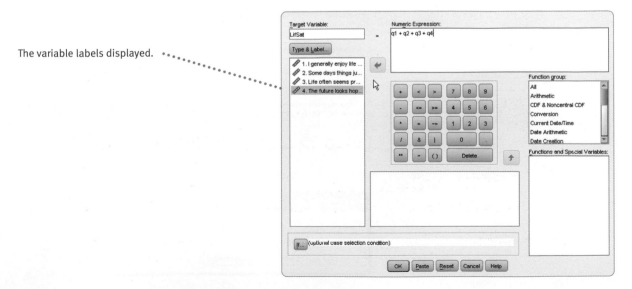

The variable labels can be either displayed or not displayed in a dialog box using the following procedure.

Step 1

Select 'Edit' and then 'Options...'.

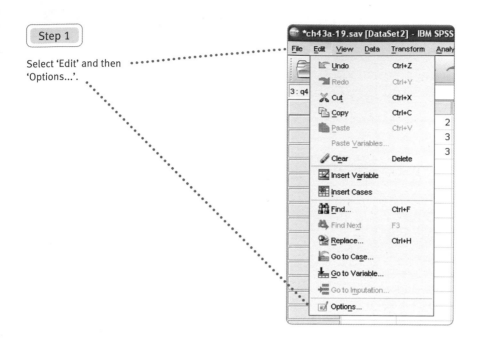

Step 2

Select 'Display labels' to display the variable labels in the dialog box.
Select 'Display names' to display only the names if there are also labels.
Select 'Apply'.

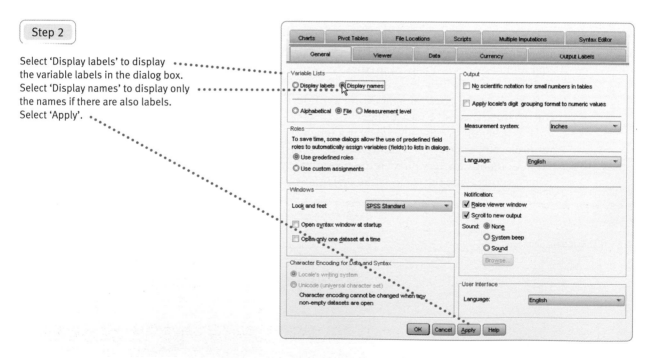

Step 3

Select 'OK'.

43.4   Computing a scale score

Step 1

Select 'Transform' and
'Compute Variable...'.

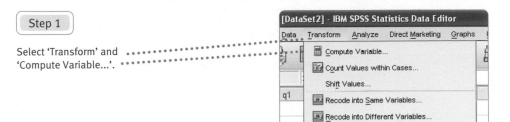

Step 2a

Type a name for the new variable in the box
under 'Target Variable:' (e.g. 'LifeSat').
Either type in or select the terms of the
expression in the 'Numeric Expression:' box.
Select 'OK'.
Select 'Paste' to save this procedure
as a syntax command.

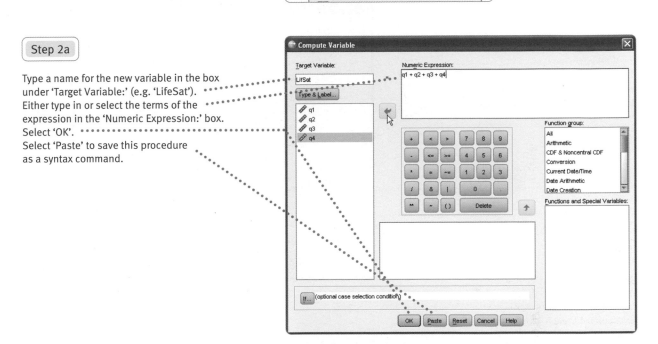

Step 2b

An alternative 'Numeric Expression' is to use
the keyword 'sum' followed by the variables in
brackets.

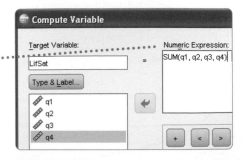

Step 3

The new variable and its values are entered into
the 'Data Editor'. Check that the values are
what they should be for a few cases.
Save the file if you want to use it again.

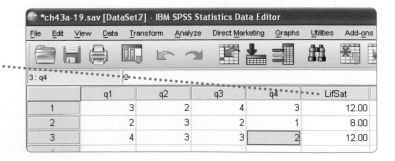

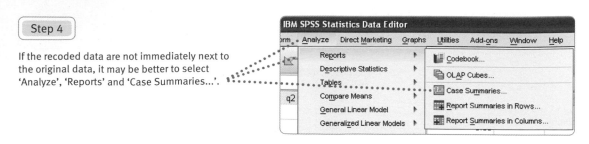

Step 4

If the recoded data are not immediately next to the original data, it may be better to select 'Analyze', 'Reports' and 'Case Summaries...'.

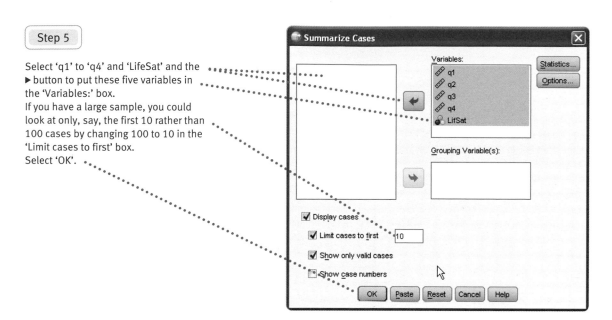

Step 5

Select 'q1' to 'q4' and 'LifeSat' and the ▶ button to put these five variables in the 'Variables:' box.
If you have a large sample, you could look at only, say, the first 10 rather than 100 cases by changing 100 to 10 in the 'Limit cases to first' box.
Select 'OK'.

Case Summaries[a]

		1. I generally enjoy life	2. Some days things just seem to get me down	3. Life often seems pretty dull	4. The future looks hopeful	LifSat
1		3	2	4	3	12.00
2		2	3	2	1	8.00
3		4	3	3	2	12.00
Total	N	3	3	3	3	3

a. Limited to first 10 cases.

The cases are listed in the first column according to their row number.
The second to fifth columns have the original values.
The sixth column has the values for the new computed variable.

43.5   Saving the Compute procedure as a syntax file

To save this procedure as a syntax file, select 'Paste' in the main box (see Step 5 in Section 43.4). This syntax command appears in the Syntax window.

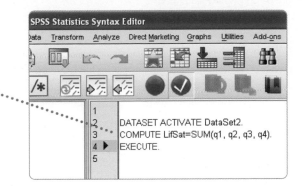

43.6   Adding some extra cases to Table 43.1

Add the values of these three new cases to this file.

3	4	3	3	2	12.00
4	1	2	1	3	.
5	2	2	3	2	.
6	3	2	3	2	.
7					

43.7   Running the Compute syntax command

Select 'Run' and 'All' to run this syntax command. Alternatively, highlight the command and select the ▶ button.

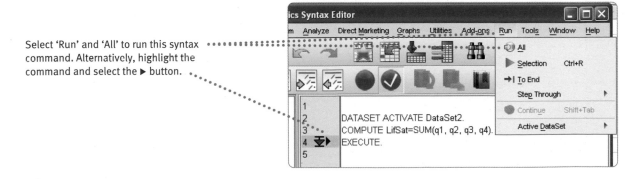

Here are the computed values for the three new cases. They are added as a new variable at the end of the data spreadsheet. Once created, the new variable may be used in the same way as other variables in SPSS.

3	4	3	3	2	12.00
4	1	2	1	3	7.00
5	2	2	3	2	9.00
6	3	2	3	2	10.00

Summary of SPSS Statistics steps for computing a scale score with no missing values

Data

- Select 'Transform' and 'Compute Variable ...'.
- Type a name for the new scale variable in the box under 'Target Variable:'.
- Either type in or select the terms of the expression in the 'Numeric Expression:' box, e.g. mean(q1, q2, etc.).
- Select 'OK'.
- Check the computation is correct.
- Save procedure as a Syntax file by selecting 'Transform', 'Compute Variable ...' and 'Paste'.
- Label Syntax file.

For further resources including data sets and questions, please refer to the webstie accompanying this book.

CHAPTER 44

Computing a scale score with some values missing

Overview

- Some of the data which go to make up a variable such as the score for an attitude or personality scale may be missing or unclear for some of your participants. For example, a few participants may have accidentally ticked or marked two of the Likert responses for the same item and not ticked an answer for the next item. In a situation like this, we would have to code the answers for these two items as missing (see Chapter 41) as we do not know what they are.

- If a large amount of data like this was missing for a participant, then it is better to discard the data for this participant. However, if only a small amount of data like this was missing it would be better not to discard these participants but to use what data we have. We could work out the mean score for these participants based on the items we have values or answers for.

- If we were working in terms of total scores rather than mean scores, then we could convert these mean scores into total scores by multiplying the mean score by the number of items for that scale.

44.1 What is computing a scale score with some values missing?

When we collect data, some items may be missing or it may not be clear what the data are. When we have a measure which is based on a number of values and some of those values are missing, then we have to decide what we are going to do in those circumstances. For example, we may be interested in measuring how quickly people respond. In order to obtain a reliable measure of their reaction time, we may obtain their reaction time to a signal on 10 separate trials and take their mean time for those 10 occasions. Now some people on some occasions may react before the signal has been given so that they have unrealistically fast reaction times. We would want to exclude these 'false starts' by coding these times as missing. If some people make many false starts, then we may feel that we have insufficient data on them to provide a reliable estimate of their reaction time and we may wish to exclude these individuals from our analysis. However, if individuals make only a few false starts, then we may be able to obtain a reasonable estimate of their reaction time. So we have to decide what our criterion is going to be for including or excluding people. We may decide that if people have more than 10 per cent of false starts, we are going to exclude them. In other words, if they have a false start on more than one trial out of the ten then we are going to exclude them. If they make a false start on only one trial, then we will include them.

Another example is where we have a questionnaire scale which consists of a number of questions or items. We may have a questionnaire which consists of only four items where people are asked to respond to these items on a six-point Likert scale where the answer 'Strongly disagree' receives a score of 1 and an answer of 'Strongly agree' a score of 6. In this case we may base our overall score for that scale if they have given clear answers to three or four items. If they have given clear answers to fewer than three items we will exclude their score on that scale. If the score for the scale is going to be based on a varying number of items, then we cannot score that scale by simply adding together or summing their scores for the items that they have answered as someone who has answered fewer items will generally obtain a lower score than someone who has answered more items. For example, if two people have answered 'Strongly agree' to all items, but one person answered all four items while the other person only answered three items, then the total score for the person answering all four items is 24 while the total score for the person only answering three items is 18. Yet both people may be similar in that they have answered 'Strongly agree' to all the items they have answered. What we have to do here is to take the mean score for the items they have answered. If we do this, then both people will have a mean score of 6, which indicates that they strongly agreed with all items. If the scale is generally interpreted in terms of a total score rather than a mean score, then we can convert this mean score into a total score by multiplying the mean score by the number of items in the scale. So, in this case, both people will have a total score of 24. Figure 44.1 highlights some issues in computing a new scale score when some values are missing.

We will illustrate how to calculate a mean score based on having answered more than a certain number of items with the data in Table 44.1 (which are the same as those in Table 31.1). There are 10 cases who have answered on a six-point Likert scale four questions that go to make up a particular index. To illustrate how this works, the first case will have missing data for the first item, the second case will have missing data for the first two items and the third case will have missing data for the first three items. Missing data are coded as zero. We would expect a mean score to be computed for the first case who has answered three items but not for the second and third case who have not.

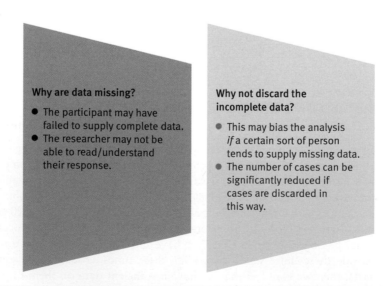

Why are data missing?

- The participant may have failed to supply complete data.
- The researcher may not be able to read/understand their response.

Why not discard the incomplete data?

- This may bias the analysis *if* a certain sort of person tends to supply missing data.
- The number of cases can be significantly reduced if cases are discarded in this way.

FIGURE 44.1   Why compute a new scale score when there are missing values?

Table 44.1	Data for 10 cases from a four-item questionnaire			
Cases	Item 1	Item 2	Item 3	Item 4
1	1	3	5	6
2	2	1	1	2
3	1	1	1	1
4	5	2	4	2
5	6	4	3	2
6	5	4	5	6
7	4	5	3	2
8	2	1	2	1
9	1	2	1	1
10	1	1	2	2

44.2   Entering the data

Retrieve the data file from Chapter 31 and enter the missing values or enter this data anew as follows.

Step 1

In 'Variable View' of the 'Data Editor' name the four variables 'Item1', 'Item2', 'Item3' and 'Item4'.
Remove the two decimal places by setting the number to 0.

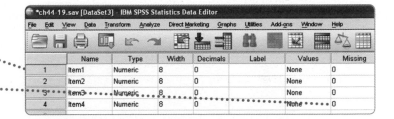

Step 2

Enter the data.
If using the saved data file from Chapter 31, enter the missing values.

Step 3

Code missing values for the first item as 0 by selecting 'Discrete missing values' and entering 0 in the box below. Select 'OK'.

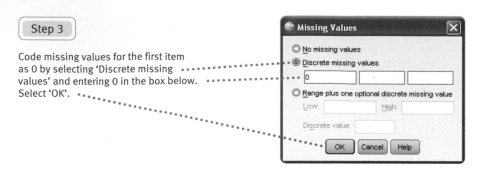

Step 4

Code the missing values of the other three variables as 0 by copying 0 in 'Missing', highlighting the three boxes below it and pasting it in them.

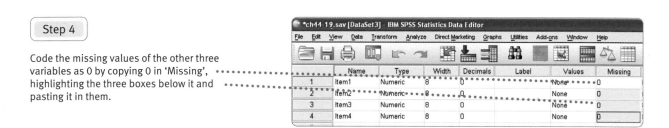

44.3   Computing a scale score with some values missing

Step 1

Select 'Transform' and 'Compute Variable...'.

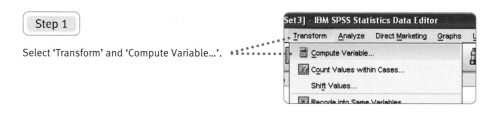

Step 2

Type a name for the new variable in the box under 'Target Variable:' (e.g. 'scale'). Either type in or select the terms of the expression in the 'Numeric Expression:' box. 'mean' is the keyword for computing a mean. '3' after the full stop indicates that this mean is based on three or more non-missing values. Select 'OK'.
Select 'Paste' to save this procedure as a syntax command.

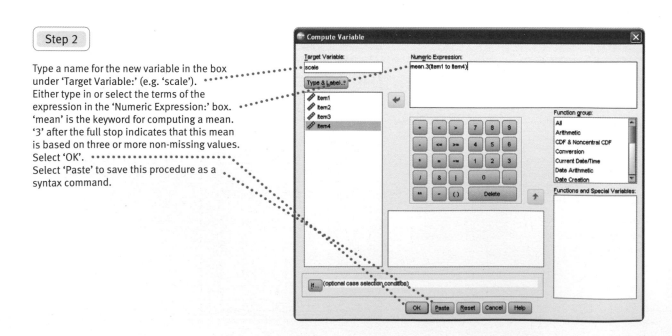

Step 3

The new variable and its values is entered into the 'Data Editor'.
Note that no mean score has been computed for the two cases that have more than one values missing.
Save the file if you want to use it again.

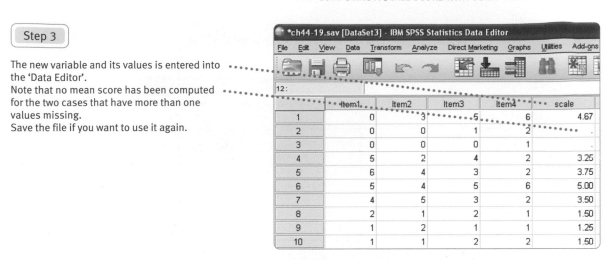

	Item1	Item2	Item3	Item4	scale
1	0	3	5	6	4.67
2	0	0	1	2	.
3	0	0	0	1	.
4	5	2	4	2	3.25
5	6	4	3	2	3.75
6	5	4	5	6	5.00
7	4	5	3	2	3.50
8	2	1	2	1	1.50
9	1	2	1	1	1.25
10	1	1	2	2	1.50

Step 4

If the new data are not immediately next to the original data, it may be better to select 'Analyze', 'Reports' and 'Case Summaries...'.

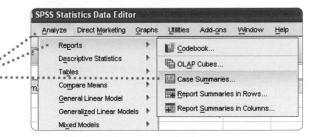

Step 5

Select 'Item1' to 'Item4' and 'scale' and the ➡ button to put these five variables in the 'Variables:' box. If you have a large sample, you could look at only, say, the first 10 rather than 100 cases by changing 100 to 10 in the 'Limit cases to first' box.
Select 'OK'.

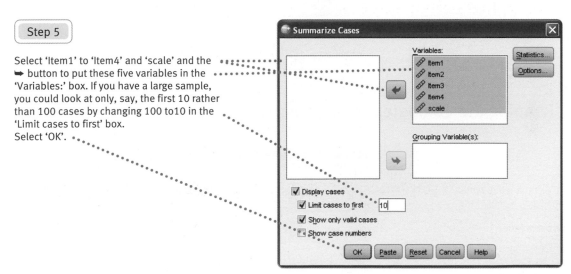

The cases are listed in the first column according to their row number.

The second to fifth columns have the original values.

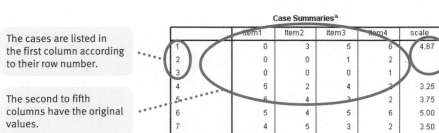

Case Summaries[a]

	Item1	Item2	Item3	Item4	scale
1	0	3	5	6	4.67
2	0	0	1	2	.
3	0	0	0	1	.
4	5	2	4	2	3.25
5	6	4	3	2	3.75
6	5	4	5	6	5.00
7	4	5	3	2	3.50
8	2	1	2	1	1.50
9	1	2	1	1	1.25
10	1	1	2	2	1.50
Total N	7	8	9	10	8

a. Limited to first 10 cases.

The sixth column has the computed mean values.

44.4 | Saving the Compute procedure as a syntax file

To save this procedure as a syntax file, select 'Paste' in the main box (see Step 5 in Section 44.3). This syntax command appears in the Syntax window.

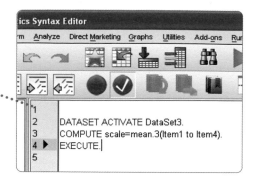

44.5 | Adding some extra cases to Table 44.1

Add the values of these three new cases to this file.

11	3	4	2	0
12	5	0	3	0
13	4	3	0	4

44.6 | Running the Compute syntax command

Select 'Run' and 'All' to run this syntax command. Alternatively, highlight the command and select the ▶ button (which is hidden behind the drop-down menu).

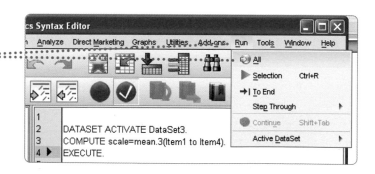

Here are the computed values for the three new cases.

11	3	4	2	0	3.00
12	5	0	3	0	
13	4	3	0	4	3.67

Summary of SPSS Statistics steps for computing a scale score with some missing values

Data

- Select 'Transform' and 'Compute Variable ...'.
- Type a name for the new scale variable in the box under 'Target Variable:'.
- Either type in or select the terms of the expression in the 'Numeric Expression:' box, e.g. mean.?(q1, q2 etc).
- Where the ? is in the numeric expression, enter the minimum number of valid items needed to form a scale.
- Select 'OK'.
- Check the computation is correct.
- Save procedure as a Syntax file by selecting 'Transform', 'Compute Variable ...' and 'Paste'.
- Label Syntax file.

For further resources including data sets and questions, please refer to the website accompanying this book.

CHAPTER 45

Computing a new group variable from existing group variables

- There are situations where you may want to compute a new group from two or more other group variables. This can be done in SPSS Statistics.

- One such situation is a significant interaction in an analysis of variance (Chapter 23) where you may wish to compare the mean of groups in one independent variable which do not fall within the same condition or level of another variable.

- An example of such a comparison is comparing never married women with married men. Marital status is one independent variable and gender is another. Coded as two separate variables it is possible to compare women and men and, say, the never married and married.

- However, it is not possible to compare never married women and married men.

- To compare these groups you would have to compute a new group variable which contained these two groups.

45.1 What is computing a new group variable from existing group variables?

There are occasions when you may want to compute or create a new group variable from existing group variables. One such occasion is where you find that the interaction effect in a factorial analysis of variance is significant. For instance, in Chapter 23 the effect of both alcohol and sleep deprivation on making mistakes in a task were analysed. There were two alcohol conditions, alcohol and no alcohol, and three sleep deprivation conditions of 4, 12 and 24 hours of sleep deprivation. Figure 45.1 shows the mean number of errors for these two independent variables.

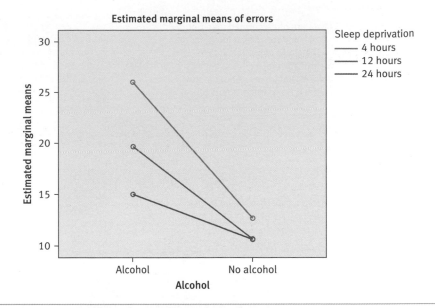

FIGURE 45.1   The effect of alcohol and sleep deprivation on errors

Although with these data there was not a significant interaction effect, an interaction effect is apparent in that the differences in mean errors for the three sleep deprivation conditions is bigger under the alcohol condition than the no alcohol condition. If the interaction had been significant, then it is possible that we would wish to determine whether the mean errors for the alcohol 4 hour sleep deprivation group or condition differed from the mean errors of the no alcohol 24 hour condition. We may also want to determine if the mean errors for the alcohol 12 hour sleep deprivation group or condition differed from the mean errors of the no alcohol 24 hour condition.

Now it is not possible to make these two comparisons with the way the two variables of alcohol and sleep deprivation are coded. With these two variables it is possible to compare the two alcohol conditions with each other and the three sleep deprivation conditions with each other but it is not possible to compare groups across these two variables. The two variables of alcohol and sleep deprivation make up the following six different groups: (1) alcohol 4 hour; (2) no alcohol 4 hour; (3) alcohol 12 hour; (4) no alcohol 12 hour; (5) alcohol 24 hour; and (6) no alcohol 24 hour. To be able to compare all six conditions with each other we could create a new variable which codes each of these six groups. For example, we could code them as numbered above. We could then analyse this new variable in a one-way analysis of variance (Chapter 21) and compare each condition with every other condition with one or more multiple comparison tests (Chapter 24).

We will illustrate how to create a new group variable with the data in Section 45.2 (which are the same as those in Table 23.1). In SPSS we need to write a syntax file involving the 'if' procedure. This procedure takes the following form:

```
if old variable one = a value & old variable two = a value compute new
variable = new value
```

We compute a new variable (e.g. group) with new values for each of the conditions in the new variable for old variables with old values. Suppose the variable name for the alcohol variable is 'Alcohol' and it has the two values of 1 and 2 representing the two alcohol conditions of alcohol and no alcohol respectively. The variable name for sleep deprivation is 'SleepDep' and it has the three values of 1, 2 and 3 denoting the three conditions of 4, 12 and 24 hours of sleep deprivation. The new variable will be called 'group'. So for the first group of alcohol and 4 hours of sleep deprivation, we would write the following syntax command:

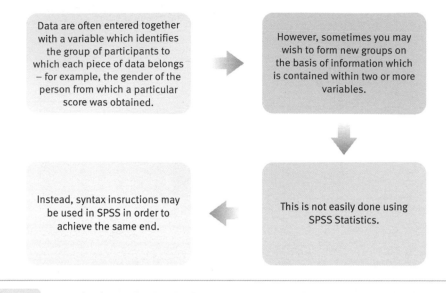

Data are often entered together with a variable which identifies the group of participants to which each piece of data belongs – for example, the gender of the person from which a particular score was obtained.	However, sometimes you may wish to form new groups on the basis of information which is contained within two or more variables.
Instead, syntax insructions may be used in SPSS in order to achieve the same end.	This is not easily done using SPSS Statistics.

FIGURE 45.2 Steps in understanding how to create new categories when original categories are specified by two different variables

```
if Alcohol = 1 & SleepDep = 1 compute group = 1
```

The value of 1 for the variable of 'group' represents participants who have had alcohol and 4 hours of sleep deprivation. We write similar syntax commands to create the five other groups.

Figure 45.2 outlines the main steps in creating a new group variable from existing group variables.

You can find out more about one-way ANOVA in Chapter 21, two-way ANOVA in Chapter 23 and multiple comparison tests in Chapter 24 of Howitt, D. and Cramer, D. (2011). *Introduction to Statistics in Psychology*, 5th edition. Harlow: Pearson.

45.2 Entering the data

Use the file from Chapter 23 if you saved it. Otherwise enter the data as follows.

Step 1

In 'Variable View' of the 'Data Editor' name the first row 'Alcohol' and name its values of '1' and '2' 'Alcohol' and 'No alcohol' respectively. Name the second row 'SleepDep' and name its values of '1', '2' and '3' '4 hrs' '12 hrs' and '24 hrs' respectively. Name the third row 'Errors'. Change 'Decimals' from '2' to '0'.

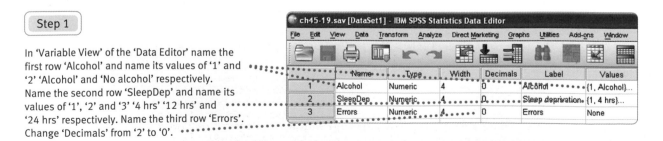

Step 2

In 'Data View' of the 'Data Editor' enter the data in the first three columns.

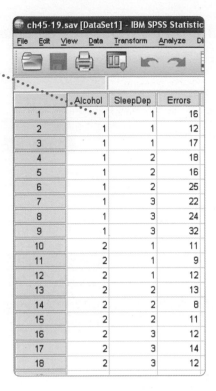

45.3 Syntax file for computing a new group variable from existing group variables

Step 1

Select 'File', 'New' and 'Syntax'.

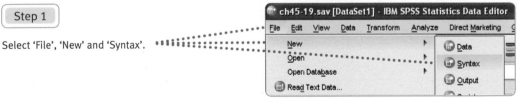

Step 2

Type in the following syntax commands.

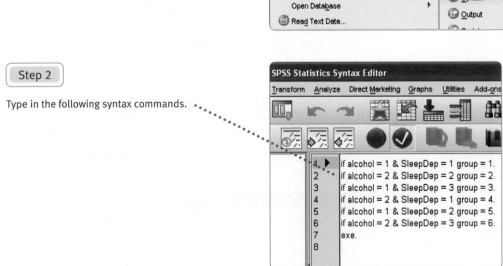

```
1.  if alcohol = 1 & SleepDep = 1 group = 1.
2   if alcohol = 2 & SleepDep = 2 group = 2.
3   if alcohol = 1 & SleepDep = 3 group = 3.
4   if alcohol = 2 & SleepDep = 1 group = 4.
5   if alcohol = 1 & SleepDep = 2 group = 5.
6   if alcohol = 2 & SleepDep = 3 group = 6.
7   exe.
8
```

45.4 Running the Compute syntax commands

Step 1

Select 'Run' and 'All' to run this syntax command.
Alternatively, highlight the command and select the ▶ button (which is hidden behind the drop-down menu).

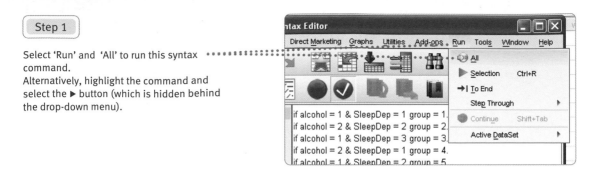

45.5 Computing a new group using menus and dialog boxes

Alternatively you can compute a new group using the 'Compute' menu option as follows.

Step 2

Here are the values for the new variable of 'group'.

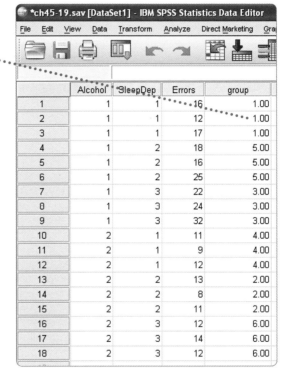

Step 1

Select 'Transform' and 'Compute Variable...'.

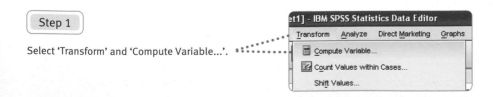

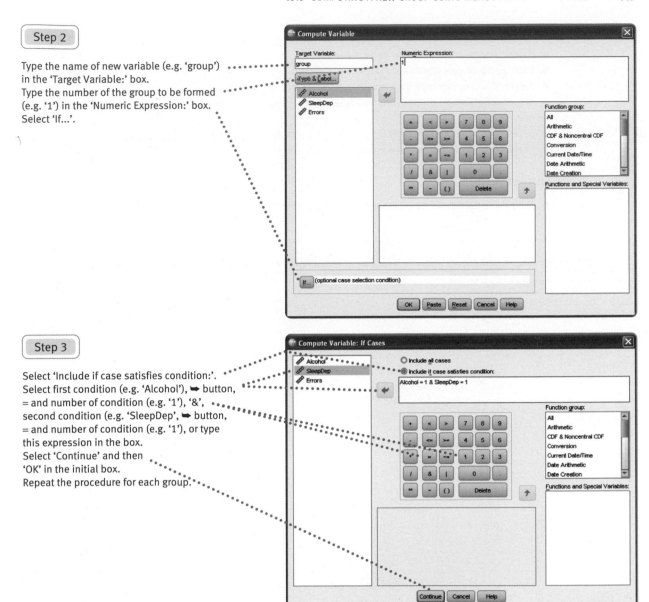

Step 2

Type the name of new variable (e.g. 'group') in the 'Target Variable:' box.
Type the number of the group to be formed (e.g. '1') in the 'Numeric Expression:' box.
Select 'If...'.

Step 3

Select 'Include if case satisfies condition:'.
Select first condition (e.g. 'Alcohol'), ➡ button, = and number of condition (e.g. '1'), '&', second condition (e.g. 'SleepDep', ➡ button, = and number of condition (e.g. '1'), or type this expression in the box.
Select 'Continue' and then 'OK' in the initial box.
Repeat the procedure for each group.

Summary of SPSS Statistics steps for computing a new group variable from existing group variables

Data

- Enter the data or retrieve data file.

Syntax file

- In the Syntax Editor type in the following kind of commands:

```
if old variable one = a value & old variable two = a value
compute new variable = new value
```

- Select 'Run' and 'All'.

Menu procedure

- Select 'Transform' and 'Compute Variable ...'.
- Enter name of new variable in 'Target Variable:' box and number of condition in 'Numeric Expression:' box.
- Select 'If ...' and 'Include if case satisfied condition:'.
- Select condition, ➡ button, = and number of condition for each condition or type in this expression in the box.
- Select 'Continue' and then 'OK'.
- Repeat the procedure for each group.

Output

- The new variable and its values are in the 'Data Editor'.

For further resources including data sets and questions, please refer to the website accompanying this book.

CHAPTER 46

Selecting cases

Overview

- This chapter explains how to select a particular subgroup from your sample. For example, you may wish to analyse the data only for young people or for women.

- It is possible to select subgroups for analysis based on multiple selection criteria such as age and gender (e.g. young women).

- Sometimes the use of subgroups leads to a much clearer understanding of the trends in the data than would be possible if, for example, one used crosstabulation by gender.

46.1   What is selecting cases?

Sometimes we may wish to carry out computations on subgroups in our sample. For example, we may want to correlate musical and mathematical ability (a) in girls and boys separately, (b) in older and younger children separately and (c) in older and younger girls and boys separately. To do this, gender needs to be coded (e.g. 1 for girls and 2 for boys) and what age is the cutoff point for determining which children fall into the younger age group and which children fall into the older age group stipulated. We will use age 10 as the cutoff point, with children aged 9 or less falling into the younger age group and children aged 10 or more falling into the older age group. Then each of the groups of interest need to be selected in turn and the appropriate statistical analyses carried out. We will illustrate the selection of cases with the data in Table 46.1, which shows the music and mathematics scores of ten children together with their code for gender and their age in years. (These values are the same as those previously presented in Table 42.1.)

Obviously the selection of cutoff points is important. You need to beware of inadvertently excluding some cases. Figure 46.1 presents some issues in selecting cases.

Music score	Mathematics score	Gender	Age
2	8	1	10
6	3	1	9
4	9	2	12
5	7	1	8
7	2	2	11
7	3	2	13
2	9	2	7
3	8	1	10
5	6	2	9
4	7	1	11

Table 46.1 — Scores on musical ability and mathematical ability for 10 children with their gender and age

Sub-groups

● Can be used to analyse subgroups of your participants separately.

Particular participants

● You can select people with a certain characteristic(s).

Removing outliers

● People with exceptionally high scores on a variable can artificially create relationships in the data which otherwise might not exist.

● It is possible to remove those potential outliers or to simply remove people with exceptionally large or exceptionally small scores.

FIGURE 46.1 — Why select cases?

46.2 Entering the data

Use the data file for the first 10 cases from Chapter 42 if you saved it. Otherwise enter the data as follows.

Step 1

In 'Variable View' of the 'Data Editor' name the first four rows 'Music', 'Maths', 'Gender' and 'Age'.
Change 'Decimals' from '2' to '0'.

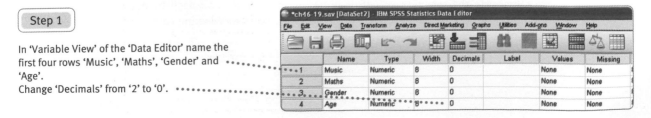

	Name	Type	Width	Decimals	Label	Values	Missing
1	Music	Numeric	8	0		None	None
2	Maths	Numeric	8	0		None	None
3	Gender	Numeric	8	0		None	None
4	Age	Numeric	8	0		None	None

Step 2

Enter the data in 'Data View' of the 'Data Editor'.

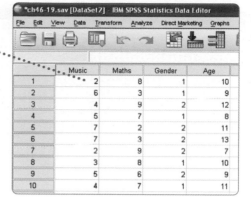

46.3 Selecting cases

Step 1

Select 'Data' and
'Select Cases...'.

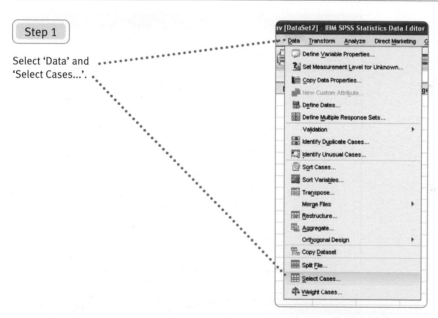

Step 2

Select 'If condition is
satisfied' and
'If...'.

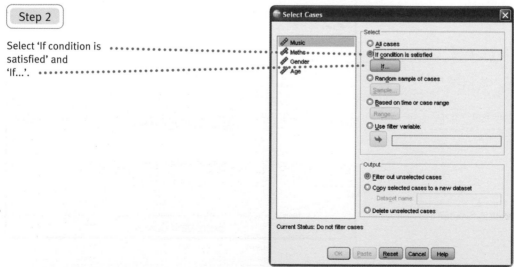

Step 3a

To select girls, select
'Gender' and the ➡ button.
Select '=' and '1'.
Alternative type in
'Gender = 1'.
Select 'Continue'.
Select 'OK' in the previous box.
Carry out the analysis you want
(e.g. correlation).

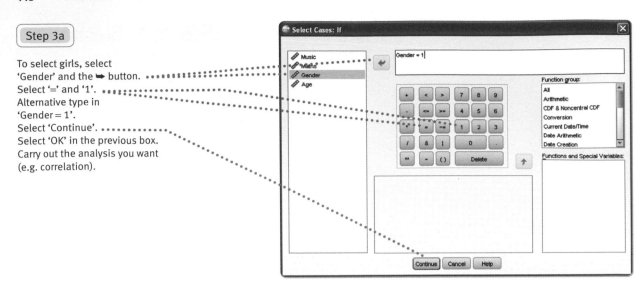

Step 3b

To select boys, select 'Data',
'Select Cases…' and 'If…'.
Replace '1' with '2'.
Select 'Continue' and 'OK'.
Conduct your analysis (e.g. correlation).

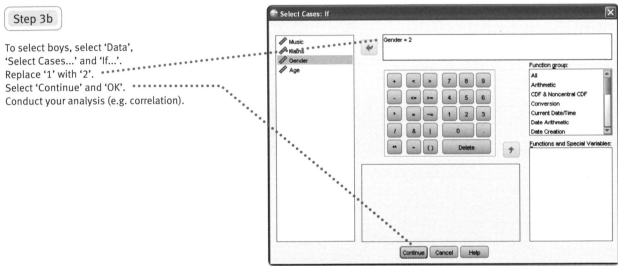

Step 3c

To select girls 9 years or
younger, select 'Data',
'Select Cases…' and
'If…'.
Replace '2' with '1'.
Select '&' (the ampersand),
'Age', the ➡ button,
'<=' (the less than or equal sign)
and '9'.
Alternatively, type this in.
Select 'Continue' and
'OK'.
Conduct your analysis
(e.g. correlation).

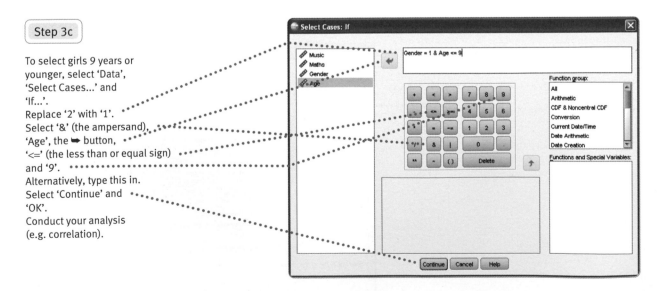

Step 3d

To select girls older than 9 years,
Select 'Data', 'Select Cases...' and 'If...'.
Replace '<=' with '>' ....
(the greater than sign).
Select 'Continue' and 'OK'. ....
Conduct your analysis
(e.g. correlation).

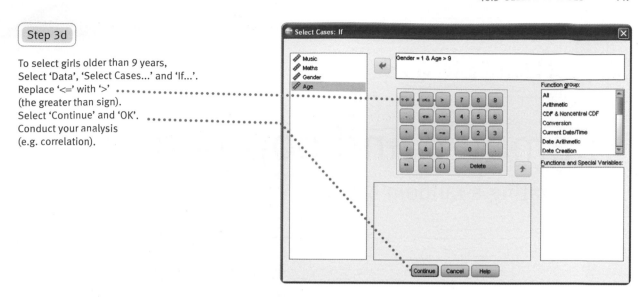

You can check that the correct sort of cases have been selected since those deselected are 'struck off' at the start of the appropriate rows of the data spreadsheet in 'Data View'.

Summary of SPSS Statistics steps for selecting cases

Data

- Select 'Data' and 'Select Cases ...'.
- Select 'If condition is satisfied' and 'If'.
- In the top right-hand box list the variables and the values you want to include or exclude.
- Select 'Continue' and then 'OK'.
- If helpful save as a syntax file.

For further resources including data sets and questions, please refer to the website accompanying this book.

CHAPTER 47

Samples and populations
Generating a random sample

Overview

- Random sampling is a key aspect of statistics. This chapter explains how random samples can be quickly generated. This is not a commonly used procedure in statistical analysis but can provide useful experience of the nature of random processes.

- One can get a better 'feel' for inferential statistics and sampling by obtaining random samples from your data to explore the variability in outcomes of further statistical analyses on these random samples. The variation in the characteristics of samples is known as sampling error and is the basis of inferential statistics.

- The procedure could be used to randomly assign participants to, say, an experimental and a control group.

- Keep an eye on the dialog boxes as you work through this chapter. You will notice options which allow you to select samples based on other criteria such as the date when the participants were interviewed.

47.1 What is generating random samples?

In this chapter, the selection of random samples from a known set of scores is illustrated. The primary aim of this is to allow those learning statistics for the first time to try random sampling in order to get an understanding of sampling distributions. This should lead to a better appreciation of estimation in statistics and the frailty that may underlie seemingly hard-nosed mathematical procedures. We will illustrate the generation of a random sample from a set of data consisting of the extraversion scores of the 50 airline pilots shown in Table 6.1.

If one wished to randomly assign participants to the experimental or control group in an experiment prior to collecting the data, then simply create a variable in the SPSS Statistics spreadsheet and enter the numbers 1 to 50 (or whatever is the expected number of participants for the study) in separate rows of the spreadsheet. By selecting a random sample of 25 individuals, you can generate a list of participants who will be in the experimental group. So if the random sample is 3, 4, 7, 9, etc. then this means that the third, fourth, seventh, ninth, etc. participant recruited will be in the experimental group. The remainder would be the control group, for example. Figure 47.1 highlights some issues in generating random samples.

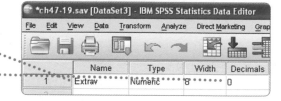

Randomisation	**Exploring randomisation**	**Random sequences**
• This involves selecting a number of participants to be in the experimental group, etc. • You need to type a simple data file with the numbers 1 to, say, 50. The random selection of half of these gives you the order number of participants to put into the experimental group.	• The concept of randomisation is very important in statistics. Using SPSS Statistics allows you to explore the practical aspects of this. • For example, you might wish to see the effects on your analysis of carrying it out on a smaller random sample of cases taken from your data set.	• SPSS Statistics does not allow you to produce a list of numbers in random order. • However, there are applets on the Web which allow you to do this. Use 'random sequence generator' or 'random number applet' to find applets which do this.

FIGURE 47.1   Why generate random samples?

You can find out more about sampling in Chapter 9 of Howitt, D. and Cramer, D. (2011). *Introduction to Statistics in Psychology*, 5th edition. Harlow: Pearson.

47.2   Entering the data

Use the data file from Chapter 6 if you saved it. Otherwise enter the data as follows.

Step 1

In 'Variable View' of the 'Data Editor' name the first variable 'Extrav'.
Remove the two decimal places.

Step 2

In 'Data View' of the 'Data Editor' enter the extraversion scores in the first column.

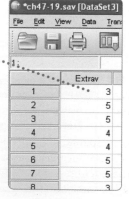

47.3   Selecting a random sample

Step 1

Select 'Data' and 'Select Cases...'.

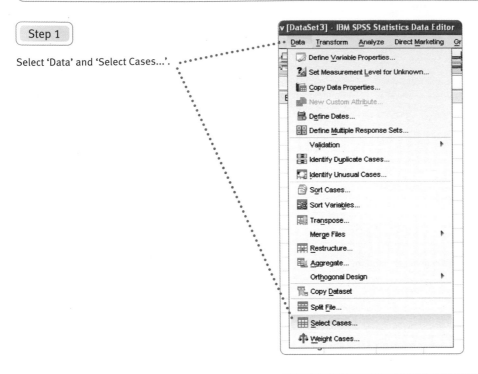

Step 2

Select 'Random sample of cases'.
Select 'Sample'.

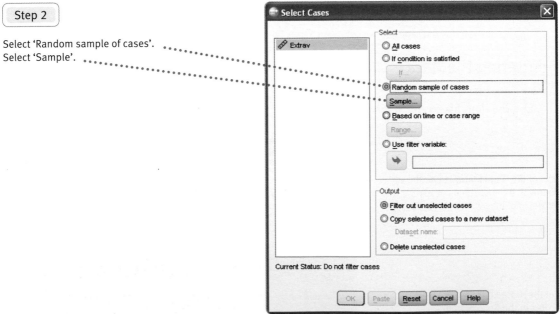

Step 3

To select about 10 per cent of all cases,
enter '10' in the box beside 'Approximately'.
To select five cases exactly, select 'Exactly',
enter '5' 'cases from the first' '50' 'cases'.
Select 'Continue'. Select 'OK' in the
previous box.

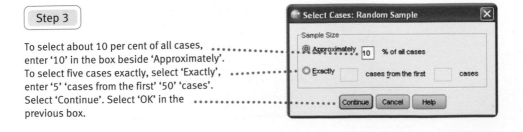

47.4   Interpreting the results

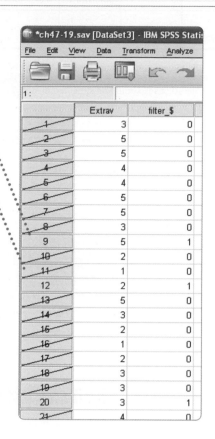

The cases that have been selected do not have a diagonal line across their case number and have a '1' in the new column created called 'filter_$'. Cases which have not been selected have a diagonal across their case number and have a '0' in the new column called 'filter_$'. Note that if you repeat this procedure, you will obtain the same sample.

47.5   Statistical analysis on a random sample

Step 1

To analyse this sample, select 'Analyze' and the analysis desired such as 'Descriptive Statistics' and 'Descriptives...'.

Step 2

Select 'Extrav' and the ➡ button to put 'Extrav' in the 'Variable(s):' box.
Select 'OK'.

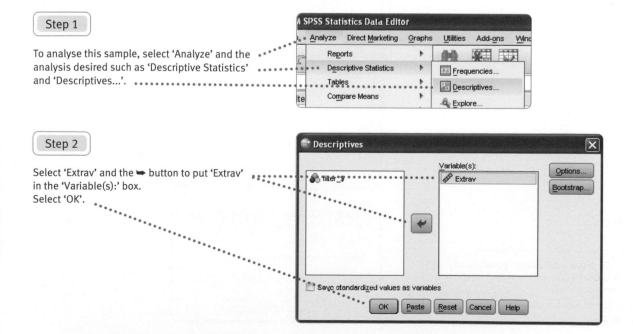

Descriptive Statistics

	N	Minimum	Maximum	Mean	Std. Deviation
Extrav	7	1	5	3.29	1.496
Valid N (listwise)	7				

Note that 7 cases were selected. If exactly 10% is required, then 5 from 50 cases should be specified in step 3 of Section 47.1.

Summary of SPSS Statistics steps for generating a random sample

Data

- Select 'Data' and 'Select Cases ...'.
- Select 'Random sample of cases' and 'Sample'.
- To select about an approximate per cent of all cases, enter that value in the box beside 'Approximately'.
- To select an exact number of cases, select 'Exactly', enter that number in the first box and the total number of cases in the second box to the right.
- Select 'Continue' and then 'OK'.

For further resources including data sets and questions, please refer to the website accompanying this book.

CHAPTER 48

Inputting a correlation matrix

Overview

- Some publications report a correlation matrix for their analysis which may also contain the means and standard deviations of the variables.

- SPSS Statistics allows you to do some further analyses on these matrices such as partial correlation, factor analysis and multiple regression.

- You need to use syntax commands to input the correlation matrix and to carry out the subsequent statistical analysis.

- The syntax commands for the statistical analysis are given first in the output when an analysis using menus and dialog boxes are run. Alternatively you can obtain these in the Syntax window by using the 'Paste' function in the main dialog box.

- We illustrate the inputting of and analysis of a correlation matrix for stepwise multiple regression with the data in Chapter 32.

- The exact way in which the matrix is referred to in the syntax command for the statistical analysis varies according to the statistical procedure. How this is done can be found by looking up the syntax command in the Help procedure.

48.1 What is inputting a correlation matrix?

If a correlation matrix has been published in a particular study, it is possible to enter this matrix into SPSS Statistics so that further analyses can be carried out. A table or matrix of correlations is sometimes presented in a publication to show the relation between the different variables. Having a matrix of correlations means that we can carry out some further statistical analyses such as partial correlation (Chapter 29), factor analysis (Chapter 30) and multiple regression (Chapters 32 to 35) which the publication may not have reported. For example, we may wish to partial out a variable which the publication has not mentioned or we may want to see what the higher order factors are in a factor analysis by carrying out a factor analysis on the correlations

Table 48.1	Data for producing a correlation matrix		
Educational achievement	**Intellectual ability**	**School motivation**	**Parental interest**
1	2	1	2
2	2	3	1
2	2	3	3
3	4	3	2
3	3	4	3
4	3	2	2

For the secondary analysis of data

- Sometimes published articles include correlation matrices of the data.
- These can potentially be analysed further using SPSS Statistics.
- For example, if you have a correlation matrix available then you may carry out a factor analysis on that data.

Where the original data have been lost or are unavailable

- If there are correlation matrices based on the data still available, then partial correlation, factor analysis, etc. may be employed.

FIGURE 48.1 Why input a correlation matrix?

between the first factors. It is possible that the publication contains means and standard deviations for the variables too. We need these to calculate unstandardised regression coefficients.

We will illustrate how to input a correlation matrix together with the means and standard deviations of the variables with the data in Table 48.1 (which are the same as those in Table 32.1). The data in this table have been weighted so as to represent 120 rather than the 6 cases presented here. This has been done so that some of the relations between the variables will be statistically significant. Small correlations become significant in large samples. Figure 48.1 outlines some issues about inputting a correlation matrix.

The means, standard deviations and correlations for these variables are presented in Table 48.2. These can be computed from the data in Table 48.1 by carrying out correlations (Chapter 10) or a multiple regression (Chapter 32). They all have to be based on the same sample of cases. In other words, the listwise procedure for missing values needs to be used.

We will carry out the same stepwise multiple regression on this matrix as was carried out in Chapter 32 so you can see that the results are essentially the same. Any differences are due to the fact that the data for this analysis have been rounded to fewer decimal places than in the original analysis. The stepwise multiple regression has to be carried out using syntax commands. As this book does not describe these commands, they can be most readily generated by simulating the statistical procedure to be used and using the Paste function to produce the relevant syntax. We do this by creating a data file which has the relevant variables in it but does not have to have any data in it.

Table 48.2	Means, standard deviations and correlations of four variables				
	Mean	**SD**	**1**	**2**	**3**
1 Achievement	2.50	.961			
2 Ability	2.67	.961	.701		
3 Motivation	2.67	.947	.369	.316	
4 Interest	2.17	.690	.127	.108	.343

$n = 120$

48.2 Syntax file for inputting a correlation matrix and running a stepwise multiple regression

Step 1

Select 'File', 'New' and 'Syntax'.

Step 2

Input this syntax in this file.
The regression syntax has been obtained by setting up a stepwise multiple regression and using Paste to enter it into its syntax into the syntax file.
The syntax 'matrix=in(*)' needs to be entered after the 'REGRESSION' keyword to tell SPSS to use the correlation matrix above.

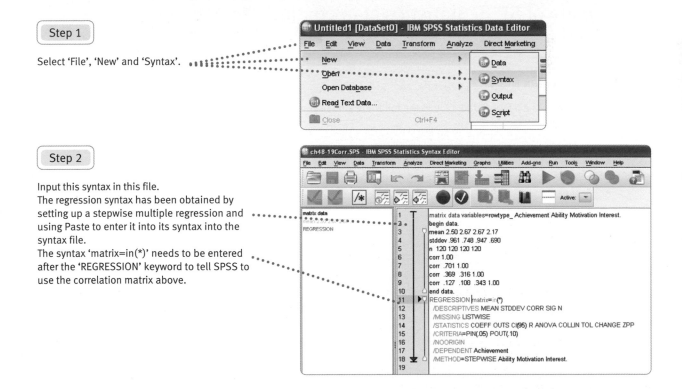

48.3 Running the syntax file

Select 'Run' and 'All' to run these syntax commands. Alternatively, highlight the commands and select the ▶ button (which is hidden behind the dropdown menu).

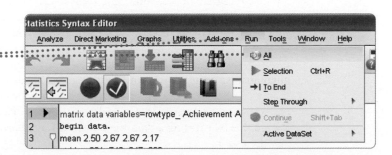

48.4 Part of the output

All the values in this table are the same as those in Chapter 32 apart for 'Std. Error of the Estimate' and the 'F Change'. This is due to rounding error.

Model Summary

Model	R	R Square	Adjusted R Square	Std. Error of the Estimate	Change Statistics				
					R Square Change	F Change	df1	df2	Sig. F Change
1	.701[a]	.491	.487	.6882459	.491	114.010	1	118	.000
2	.718[b]	.516	.507	.6745614	.024	5.836	1	117	.017

a. Predictors: (Constant), Ability
b. Predictors: (Constant), Ability, Motivation

Summary of SPSS Statistics steps for inputting a correlation matrix

Syntax file

● In the Syntax Editor type in the following kind of commands:

```
matrix data variables=rowtype_ (enter variable names).
begin data.
mean (enter means)
stddev (enter standard deviations)
n (enter number of cases for each variable
corr 1.00
corr (enter correlation) 1.00
corr (enter correlations)      1.00
corr (enter correlations)           1.00
end data.
REGRESSION matrix=in(*)
/DESCRIPTIVES MEAN STDDEV CORR SIG N
/MISSING LISTWISE
/STATISTICS COEFF OUTS R ANOVA COLLIN TOL CHANGE ZPP
/CRITERIA=PIN(.05) POUT(.10)
/NOORIGIN
/DEPENDENT (enter dependent variable)
/METHOD=(enter type of regression and then predictor variables).
```

● Select 'Run' and 'All'.

For further resources including data sets and questions, please refer to the website accompanying this book.

CHAPTER 49

Checking the accuracy of data inputting

Overview

- It is essential to enter the data accurately into the 'Data Editor'. With a small set of data it is most probably sufficient to check the inputted data against the original data. With a larger data set, this is more difficult to do.

- One strategy is to enter the same data separately into two files and to compare the value of the same cells in the two files. This kind of procedure is available as part of SPSS Statistics Data Builder but this software is expensive and not readily available.

- It is relatively easy to carry out this procedure in SPSS by combining the two files, using syntax commands to see if there is a difference between the same cells and to check and correct any differences.

- This procedure assumes that it is unlikely that the same mistake will be made twice, particularly if different people enter the data.

49.1   What is checking the accuracy of data inputting?

Mistakes can occur when data is typed into the 'Data Editor' of SPSS Statistics. When you have a small set of data, accuracy can be checked by simply comparing the 'Data Editor' with the original data. If you find this difficult to do conscientiously or if you have a large data set, then it is better to use a procedure which is more systematic and more accurate. This procedure involves entering the data twice and then comparing the two sets of data. It is better that different people are involved so that one person does not enter the data twice. Two different people are less likely to make the same mistake. The data are entered into two separate files so that data entry is independent. If there is a difference in the value of a cell between the two files, then both values cannot be correct and the original data have to be checked to see which value is correct. If there is no difference, then it is likely that the values are correct. Once all the values have been checked and corrections made as appropriate, one of the corrected files can be used for the data analysis.

Table 49.1	Correct data to be inputted		
Item 1	**Item 2**	**Item 3**	**Item 4**
1	0	1	2
2	2	3	1
2	2	3	3
3	4	0	2
3	3	4	3
4	3	2	2

Table 49.2	First data file with inputting mistakes in **red**		
Item 1	**Item 2**	**Item 3**	**Item 4**
1	0	1	2
22		3	1
2	2	3	3
3	4	0	2
3	3	4	3
4	3	2	2

Table 49.3	Second data file with inputting mistakes in **red**		
Item 1	**Item 2**	**Item 3**	**Item 4**
1	0	1	2
2	2	3	1
2	2		3
4	**3**	0	2
3	3	4	3
4	3	2	2

We will illustrate this procedure with the data files in Tables 49.1, 49.2 and 49.3. Table 49.1 shows the correct data set of the values for the responses of six people to four questionnaire items which are answered on a four-point scale where 1 represents 'Strongly agree' and 4 'Strongly disagree'. A 0 represents an item that has not been answered or where the answer is not clear. As we are just checking the accuracy of data inputting it does not matter what the numbers represent. These values have been put into a table to make them easier to read but if we were entering the data then we would be most probably reading them off a questionnaire or some other form of record. We will assume that an optical scanner is not available or is not suitable for reading this information electronically. Figure 49.1 highlights two issues about checking accuracy of data inputting.

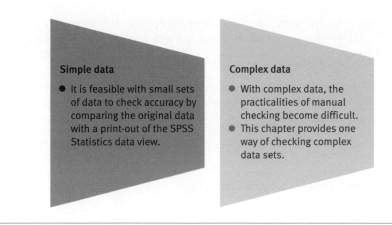

FIGURE 49.1 Why check data accuracy?

Table 49.2 shows the data file for the first time the data is entered while Table 48.3 shows the data file for the second time that the data are entered. We have introduced two mistakes in the first file in Table 49.2 and three mistakes in the second file in Table 49.3. For the second case in Table 49.2 the value for item 2 has been accidentally entered in the same cell as the value for item 1. The cell for item 2 has been left empty. For Table 49.3 the cell for item 3 for the third case has been left empty and the values of the cells for item 1 and 2 for the fourth case have been transposed.

49.2 ## Creating two data files

With a large number of variables it is better to carry out the following procedure which is designed to minimise the amount of work involved.

● Create the variable names in an SPSS file. At the end of each name add a character which cannot be easily confused with a character in the name. This character could be a hash sign (#). For example, the name of the first variable could be q1#. The reason for this will become apparent later on.

Enter the variable names in the 'Variable View'
window. It is better to identify each case with an
identification number ('idno').
Change the decimal places to '0'.

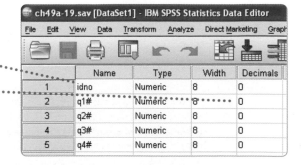

You could specify other characteristics of the variables such as the names of categories if this is helpful for you or the person entering the data. However, you should not specify any missing values when comparing the files as you will be calculating the differences between the two values. If one of the values is missing, then a difference will not be computed for that pair of values.

- Save this file with a name such as that of the person who is going to enter the data together with the character to distinguish the two forms of the name such as 'DC1#'.

- Save this file again with a different name such as 'DC2@'.

- Highlight the names of the variables in one of the files and paste them four times into a Word document and separated from one another. It is useful to do this for subsequent tasks we are going to carry out.

Paste the SPSS names of one data file into a Word file twice.

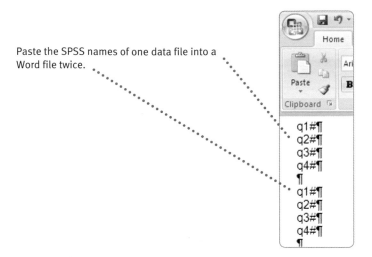

- Using the 'Replace' procedure, replace the hash sign (#) with an 'at' sign (@) for the second copy of variable names.

In 'Home' in Word 2007 select 'Replace'.
Enter '#' in 'Find what:' and '@' in 'Replace with:'.
Replace '#' in the second list of variable names.

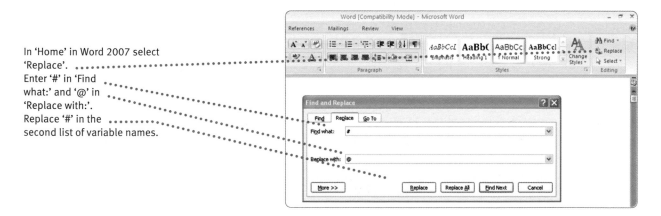

- Highlight these names ending in @ and paste them over the names in the second data file. The names in the two data files for the same variable can now be easily distinguished.

In 'Variable View' of a new data file, paste a copy of the second list of variable names ending in @.

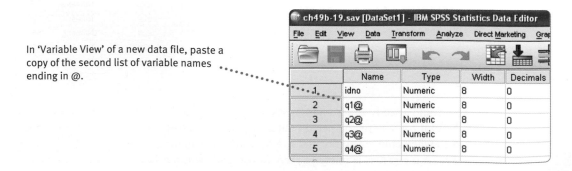

● Enter the data from Table 49.2 in the first file and the data from Table 49.3 in the second file.

Enter the data in the first file.

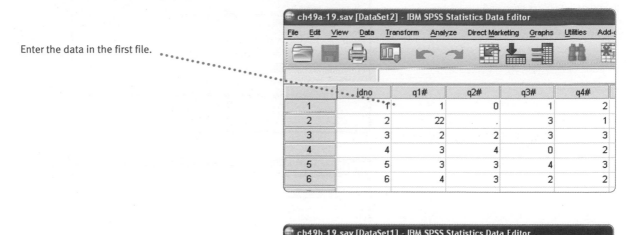

Enter the data in the second file.

49.3   Combining the two data files

With one of the data files in the Data Editor (say, DC1#), carry out the following procedure.

Step 1

Select 'Data',
'Merge Files' and
'Add Variables'.

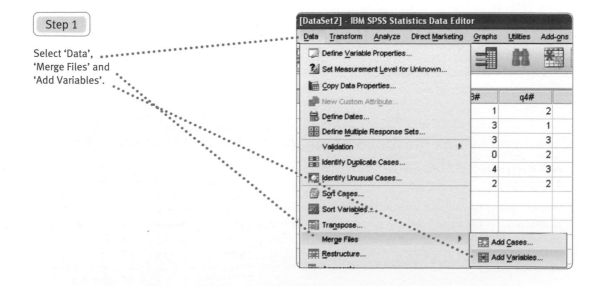

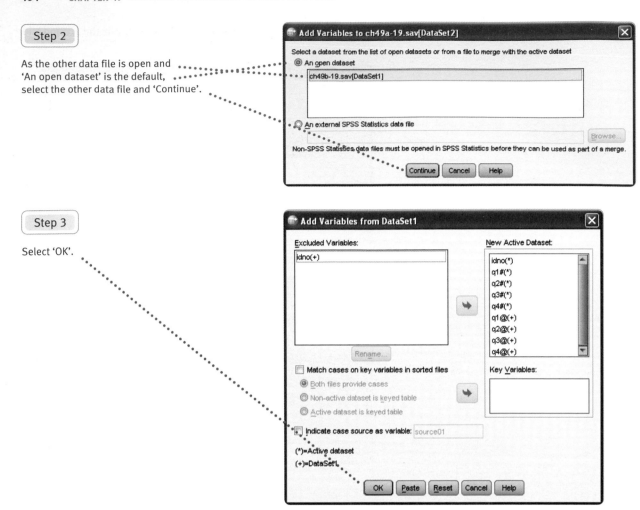

Step 2

As the other data file is open and 'An open dataset' is the default, select the other data file and 'Continue'.

Step 3

Select 'OK'.

Step 4

The variables of the second file have been put into the first file.

49.4 Creating a syntax file for computing the difference between the two entries for the same variables

We are going to subtract one entry of the variable (e.g. q1#) from the other (e.g. q1@) to create a difference score which we are going to call q1#@ using the following form of the syntax command for each of the four variables:

```
compute q1# - q1@ = q1#@.
```

If you have a large number of such commands to make, it is easier to do this in Word, copying and pasting the relevant terms into a table with seven columns and as many rows as variables. This table can then be copied and pasted into a syntax file. Add 'execute.' at the end of this file or select 'File' and 'Save' after running it to display the values of the new variables.

The first list of variable names.
The second list with '#' replaced by '@'
A third list with '#' or '@' replaced by '#@'
to be used in the syntax file.
The syntax file as a Word table.

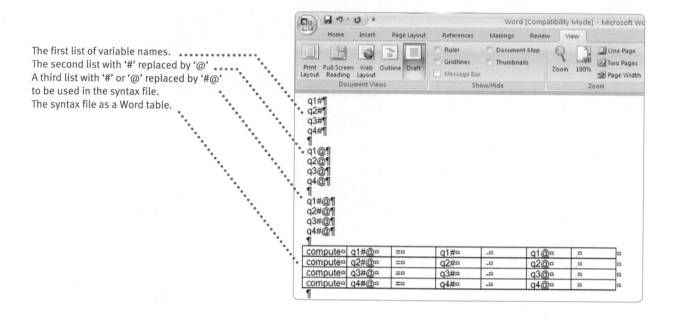

Copy this syntax file and paste it into the 'Syntax Editor'. With the first data file open, run this syntax file.

 Step 1

This is the syntax file in the 'Syntax Editor'.
Select 'Run' and 'All' to run this syntax
command. Alternatively, highlight the command
and select the ▶ button which is hidden behind
the drop-down menu.

Step 2

The difference values are presented in
'Data View'.
You need to check the original values where
there is a difference or a full stop and to make
the necessary changes. After you have done this
you need to re-run the syntax file to check that
all changes have been made.

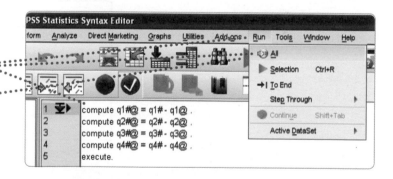

Step 3

If the difference values are not
immediately next to the original data,
it may be better to select
'Analyze',
'Reports' and
'Case Summaries...'.

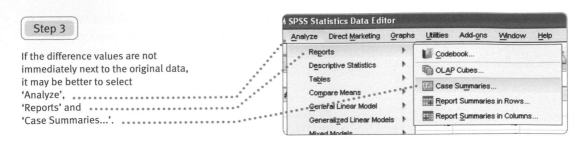

Step 4

Select 'idno', 'q1#@' to 'q4#@' and the
➡ button to put these five variables in the
'Variables:' box. If your sample is larger than
100, change the 'Limit cases to first'
to the size of your sample (or greater).
Select 'OK'.

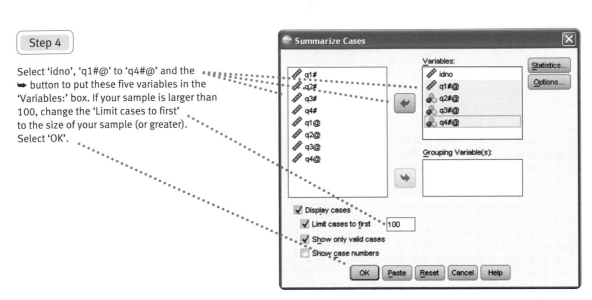

Step 5

The variables and their values are listed
in this table.

Case Summaries[a]

	idno	q1#@	q2#@	q3#@	q4#@
1	1	.00	.00	.00	.00
2	2	20.00	.	.00	.00
3	3	.00	.00	.	.00
4	4	-1.00	1.00	.00	.00
5	5	.00	.00	.00	.00
6	6	.00	.00	.00	.00
Total N	6	6	5	5	6

a. Limited to first 100 cases.

Step 6

After correcting the values, run the syntax file
again. There should be all '.00's for the all
the cells for variables ending in: '#@'.
Save this file with a new name (e.g. 'DCc.sav').

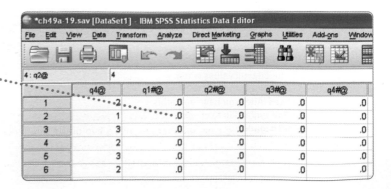

Step 7

Highlight the variables you want to delete
(i.e. 'q1@' to 'q4#@') by holding down the left
mouse key on '6', moving the cursor down to
'13', and releasing the left key.
Then delete (e.g. using the keyboard 'Del' key).
Save this file with a new name (e.g. 'DCc').

Step 8

Although this is not necessary you may like to
delete the '#' at the end of each name.
This is most easily done in the Word file by using
'Replace', copying the new variable names and
pasting them into 'Name' of 'Variable View'.
If you have later data to check, then removing '#'
is best done after all the data have been
entered and checked.

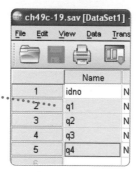

Summary of SPSS Statistics steps for checking the accuracy of data inputting

Data

- Enter the data into two files, making sure that the names for the same variables are different (e.g. age# and age@).
- In one data file, select 'Data', 'Merge Files' and 'Add Variables'.
- Select other data file, 'Continue' and then 'OK'.

Syntax

- In the Syntax editor, create a syntax file in which the two versions of the same variable is subtracted from one another to give a difference score, e.g. 'compute age#@ = age# – age@'.
- Run this syntax file.

Analysis

- Select 'Analyze', 'Reports' and 'Case Summaries ...'.
- Select the case number and up to five of the difference variables at a time and with the ▶ button to put these five variables in the 'Variables:' box.
- If your sample is larger than 100, change the 'Limit cases to first' to the size of your sample (or greater).
- Select 'OK'.

Output, Data, Syntax and Analyze

- For each difference value which is not zero, check which value has been wrongly entered (Output) and enter the correct value (Data).
- When all differences have been corrected, run the syntax file again (Syntax).
- If some differences are not zero (Analyze), check (Output) and correct these values (Data).
- When all differences are zero (Output), delete the second variable names and the difference variable names and save the file as the checked data file (Data).

For further resources including data sets and questions, please refer to the website accompanying this book.

PART 8

Linear structural relationship (LISREL) analysis

CHAPTER 50

Basics of LISREL and LISREL data entry

Overview

- This chapter introduces the main basic ideas of structural equation modelling and LISREL together with some of the basics of data entry for LISREL.

- Structural equation modelling is difficult to define but involves a range of models which a researcher puts forward and validates to indicate the pattern of (causal) relationships which lead to changes in a key variable or variables.

- LISREL stands for *Li*near *S*tructural *Rel*ationships and is one of the most widely used software programs for carrying out structural equation modelling (SEM).

- One of the advantages of the LISREL program is that measurement error in the form of alpha reliability can be controlled (Chapter 54). Variables can differ substantially in terms of their reliabilities. The reliability of a measure puts a limit on the maximum relationship it may have with another variable. So, as a consequence, relationships between variables may be distorted by variations in their reliabilities.

- A free student version of LISREL can be downloaded which can handle up to 15 variables, which is enough for many studies.

- LISREL can carry out confirmatory factor analysis in which the statistical fit of one or more theoretical models to the data can be evaluated and compared (Chapter 51).

- It can carry out path analyses where certain variables are assumed to affect or influence other variables.

- Unlike multiple regression (Chapter 34), structural equational modelling allows more than one criterion (dependent) variable to be examined at the same time, can control for measurement error and enables the statistical fit of theoretical models to be evaluated and compared.

50.1   What is LISREL and structural equation modelling?

Structural equation modelling is an advanced statistical technique which involves the researcher actively building models which account for the relationships between different variables and testing their adequacy against actual data. LISREL is an abbreviation for *Li*near *S*tructural *Re*lationships. It is the name of a particular form of software for carrying out structural equation modelling (SEM). LISREL was among the first such programs and is among the most widely used programs. A free student version of LISREL can be downloaded from the web (http://www.ssicentral.com/lisrel/student.html). This version can analyse up to 15 variables which is sufficient for the examples used in this book and for most student needs. Otherwise you will need the full version of LISREL. As SEM is a very sophisticated form of conceptual and statistical analysis and as LISREL produces a large amount of output, it is sensible to start learning this technique on a small number of variables. You can move on to using more variables as you become more familiar and skilled with the method.

It is important to understand that SEM involves a range of different statistical techniques (some of which have already been dealt with in this book) and it is not simply another statistical test. As such, it requires a developed statistical awareness on the part of its users.

Unlike multiple regression which can only analyse one criterion or dependent variable at a time (Chapter 34), SEM can analyse two or more such variables at the same time. For example, the potential effect of parents' interest in how well their children do at school can be looked at in terms of how well they perform academically as well as how interested they are in academic work. Multiple regression is not up to this analytic task for a number of reasons especially since it is able to examine only one of these two criterion variables in a single analysis.

As the name implies, SEM involves setting up a model to account for relationships among the variables in one's data. A model is simply a set or pattern of variables which predicts or explains the thing of interest. For instance, in confirmatory factor analysis, the model may specify that particular variables are expected to perfectly load on or correlate with a certain factor and not to load on or correlate with other factors. This model may be referred to as a measurement model. In terms of the example used to illustrate exploratory factor analysis in Chapter 30 we might expect that the three variables or measures of batting, darts and juggling will define the factor or construct of sensory-motor skills. The three variables or measures of crosswords, Scrabble and spelling will define the factor or construct of verbal skills. We can represent this two factor model with Table 50.1. A value of 1.00 means that the loading or correlation of that variable with that factor is expected to be perfect while a value of .00 means that no relationship is expected to exist between that variable and that factor. Furthermore, we can specify whether these two factors are expected to be unrelated (orthogonal) or related (oblique). In SEM, the variables are referred to as manifest or observed variables or as indicators. On the other hand, the factors are called latent variables because we cannot measure them directly since factors are abstractions and therefore not directly observable.

Table 50.1	Hypothetical two-factor model	
Variables	**Sensory-motor skills**	**Verbal skills**
Batting	1.00	.00
Crosswords	.00	1.00
Darts	1.00	.00
Scrabble	.00	1.00
Juggling	1.00	.00
Spelling	.00	1.00

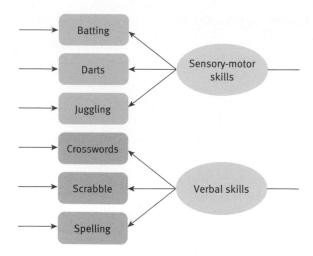

FIGURE 50.1   Path diagram of a hypothetical two-factor model similar to ones drawn by LISREL

Latent variables are assumed to affect or influence the manifest variables. This model is usually represented with a path diagram or path model like that shown in Figure 50.1:

- Manifest variables are denoted in rectangles ▭.

- Latent variables are given as circles or ovals ⬭.

- An arrow is drawn from the latent variable to the manifest variable as latent variables are thought to influence manifest latent variables →.

- Curved or straight lines with arrows at either end are drawn between variables which are expected to be related. These lines are usually referred to as two-way or double-headed arrows ↔.

So, we would draw a curved or straight two-way arrow to indicate that we expected the two factors or latent variables to be related. If we did not expect them to be related there would be no two-way arrow between them. Two-way arrows can also be drawn between manifest variables which are expected to be related. Variables may be assumed to be measured with or without error. This is indicated by drawing an arrow to the variable. If no error is expected, this can be indicated by putting .00 at the start of the arrow. The issue of the reliability of variables is important because the reliability of a variable sets a cap on the size of the correlation that it can have with another variable. This is discussed in more detail later. It is really important to understand that the models tested are ones developed by the researcher – they do not simply appear out of a statistical analysis.

Path diagrams are drawn in LISREL to specify the model that is being tested. Figure 50.1 has been drawn to illustrate the broad pattern of path diagrams in LISREL.

The relationship between variables are specified in terms of a number of structural equations which are more or less like regression equations. For example, the manifest variable of batting is a function of the product of the path coefficient between the latent variable of sensory-motor skills and the value of that latent variable together with the error (or residual) of the manifest variable:

In LISREL path coefficients are represented by different Greek letters:

● The path coefficient between a manifest and a latent variable is represented by the Greek small letter λ (lambda). For this model there would be a structural equation for each of the six manifest variables.

● The path coefficient between a predictor variable and a criterion variable is represented by the Greek small letter γ (gamma).

● The path coefficient between a predictor variable and a mediator variable is denoted by the Greek small letter β (beta).

How well the hypothesised model fits the actually observed data is assessed in SEM. Various measures of this statistical fit between model and data are available. However, there is little consensus as to which ones are the best to use. Consequently, it is common practice to report more than one measure of fit. Two measures that are frequently presented in path diagrams are: chi-square and RMSEA (Root Mean Square Error of Approximation). This chi-square is a special version of chi-square which is known as the *Normal Theory Weighted Least Squares Chi-Square*. It is called this in LISREL's detailed output.

The Normal Theory Weighted Least Squares Chi-Square compares the difference between the variance–covariance matrix of the orginal data and the variance–covariance matrix produced by the model. This is not as difficult as it sounds. Think of a correlation matrix – instead of correlations, the variance–covariance matrix contains covariances of the variables in the columns with the variables in the rows. The diagonal contains variances since the covariance of a variable with itself is what we call the variance. Like a correlation matrix, the variance–covariance matrix is symmetrical around the diagonal. If the variance–covariance matrix is the same for the data and the model then there is a perfect fit between the data and the model. However, this is not likely to be the case so the bigger the difference between the data and the model matrix, the bigger chi-square is and the more likely it is that chi-square will be statistically significant at the .05 level or less:

● A statistically significant chi-square means that the model does not fit the data well enough.

● A non-significant chi-square indicates that the model does fit the data and so confirms the model.

In other words, chi-square is used to test the hypothesis that the model and the data are different. Thus a significant value of chi-square means that the data do not support the model. A non-significant value of chi-square indicates that the model and the data are the same within the limits of sampling error.

The more values the model has to estimate the bigger chi-square is likely to be and so the more likely it is to be statistically significant. The number of values to be estimated is indicated by the degrees of freedom (*df*). So models with more degrees of freedom are more likely to yield chi-squares which are statistically significant. In other words, as a consequence, the more complex the model then the less likely it is to fit the data. Also, the bigger the sample used in data collection the bigger the chi-square value is likely to be since the chi-square measure used is multiplied by the sample size minus one. These problems make chi-square less than ideal on its own and encourage the use of other indices in addition.

Some of the other indices of fit are designed to deal with these problems:

● The RMSEA takes account of the degrees of freedom by dividing the measure of fit by the degrees of freedom. A RMSEA of about .05 or less is considered to indicate good fit while values of .10 and above denote poor and unacceptable fit.

● The Non-Normed Fit Index (NNFI) compares the hypothesised model with a null or independence model in which the variables are assumed to be unrelated. This index takes account of the degrees of freedom for each model. The NNFI is non-normed in the sense that its values can lie outside 0 and 1. A NNFI of .95 and above is thought to indicate a good or acceptable fit and values below .95 a poor and unacceptable fit. The NNFI is sometimes referred to as the Tucker–Lewis Index (TLI) which is the same as the NNFI but was independently developed earlier.

● The Comparative Fit Index (CFI) also compares the hypothesised model with a null model but is a normed index in that it values are normed to vary between 0 and 1. Values greater than 1 are changed to 1 and negative values are changed to 0. Like the NNFI, a CFI of .95 and above is thought to indicate a good or acceptable fit and values below .95 a poor and unacceptable fit.

If a model is found not to fit the data, then the model may be modified to improve the fit. LISREL provides information which shows how the model may be changed to increase its fit. However, sometimes it is not sensible to change the model in this way as the changes may be difficult to justify conceptually and theoretically.

Furthermore, unlike multiple regression, SEM can take account of the error with which variables are usually measured. It can do so in one of two ways:

● The first is based on the fact that a variable's relationship with other variables is dependent on the reliability of each of the measures, so it is possible to allow for the internal reliability of the measures used (Chapter 54) as assessed by Cronbach's alpha reliability (Chapter 31). Making such an adjustment has two related benefits:

(a) One advantage of correcting for measurement error in this way is that the path coefficients between variables will almost certainly be higher because this correction treats variables as if there was no measurement error. In terms of correlation coefficients, the highest correlation that can exist between two variables is the square root of the product of the (alpha) reliabilities of the two measures. For example, if the alpha reliability of one measure is .7 and that of the other measure is .8, then the maximum size of the correlation between those two variables is ±.75 ($\sqrt{(.7 \times .8)}$ = .75). If the alpha reliability of the two measures were a perfect 1.00, then the maximum size of the correlation would be ±1.00 ($\sqrt{(1.00 \times 1.00)}$ = 1.00).

(b) As the (alpha) reliability of measures is likely to differ, not correcting for measuring error makes it difficult to compare the relative size of coefficients. A coefficient between two variables may appear smaller than another coefficient because the alpha reliability of one or both variables may be lower than the variables for the other coefficient. Thus correcting for alpha reliability is that it enables the size of coefficients to be directly compared as they have all been standardised so that their maximum value can be ±1.00.

● Another way and one which is most commonly used to correct for measurement error is based on the relationships between the manifest variables and the latent variable which the manifest variables are believed to represent (Chapter 53). In this approach, the variance that is shared by the manifest variables is assumed to reflect the latent variable and the variance that is not shared is measurement error. Typically a latent variable may be represented by two or three manifest variables. So if one is measuring a variable with a number of questions, then those questions need to be grouped together into two or three manifest variables (Chapter 43). The way in which this is done may be rather arbitrary and doing it in another way may produce different results. For example, if we have 10 questions or items making up a scale and we want to form three groups of items, we first have to decide whether the three groups will have the same number of items and then which items are to be put together.

For the moment, that is enough of the conceptual background to SEM. We can begin to examine how LISREL, the computer program, is used. Figure 50.2 shows some of the basic steps in SEM.

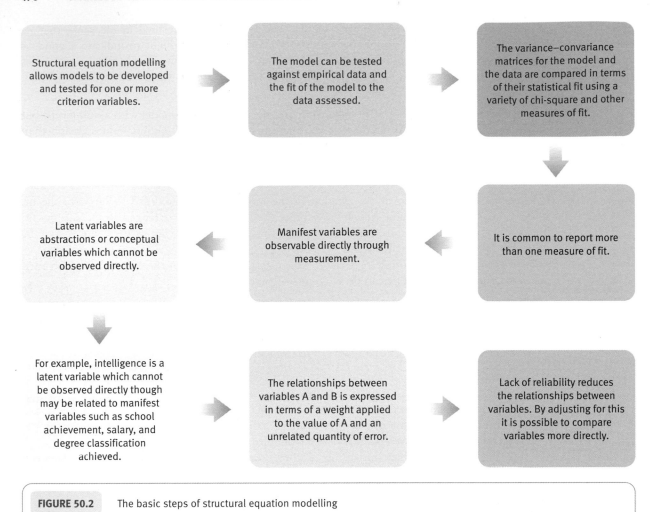

FIGURE 50.2 The basic steps of structural equation modelling

50.2 When to use structural equation modelling

SEM is a sophisticated and complicated statistical technique which should be attempted only by those who have some knowledge and experience with more basic statistical techniques such as factor analysis and multiple regression. If you are going to use this method, then it is wise to do this after you have carefully looked at your data with simpler more familiar techniques. Most undergraduate students would not be expected to use this method to analyse their data. However, it is not uncommon for doctoral students and professional researchers to adopt the modelling approach when trying to answer complex questions about the way in which things operate in the world. Some sophistication about the role of theory in research is also helpful.

Probably many of the circumstances in which students and researchers use multiple regression are ones in which SEM might be more appropriate. For example, often multiple regression is used to identify simple models which might be elaborated better with SEM.

50.3   When not to use structural equation modelling

Like correlation (Chapter 10), regression (Chapters 11 and 34) and factor analysis (Chapter 30), SEM assumes that the variables being analysed are continuous score variables (and not nominal or categorical ones) and that the relationships between variables are linear (and not curvilinear). This linear relationship is an assumption of virtually all of the statistics in this book. To the extent that these requirements are not met by your data, SEM should not be used.

50.4   Data requirements for structural equation modelling

The data for structural equation modelling take the form of a number of variables measured as scores. The closer each variable is to a continuous variable the better. This means the scores should show fine gradations for each variable; that is, there should be many different values. The assumption is that there is a linear relationship between the variables which is a standard assumption of most of the statistical techniques in this book. Avoid variables which show curvilinear relationships with each other. These may be identified from scattergrams of the relationships between pairs of variables, for example.

Nominal category or categorical variables may also be used if dichotomous or converted into dummy variables.

More than one dependent variable may be modelled at the same time unlike the case of regression, for example.

50.5   Problems in using structural equation modelling

Modelling is not part of the basic statistics that undergraduate psychology students are taught. So much statistical thinking that students learn is not very helpful when tackling SEM. For that reason it should be regarded as something for those whose understanding of psychology is well developed in terms of theory, research and statistics. SEM is fundamentally about causal relationships in a system so needs the researcher to be familiar with issues of causality.

One difficulty in using SEM is that many psychologists are unfamiliar with it and so it may be difficult to get advice and support.

You can find out more about factor analysis in Chapter 30, path analysis in Chapter 34 and alpha reliability in Chapter 31 of Howitt, D. and Cramer, D. (2011). *Introduction to Statistics in Psychology*, 5th edition. Harlow: Pearson.

50.6   The data to be analysed

To illustrate the procedure we are using the data in Table 50.2 which is the same as that in Table 30.1. We assume that you have already entered the data into an SPSS file. The data show the scores of nine individuals on six different tasks. The data are used in Chapter 51 to illustrate confirmatory factor analysis.

Table 50.2	Scores of nine individuals on six variables					
Individual	**Batting**	**Crosswords**	**Darts**	**Scrabble**	**Juggling**	**Spelling**
1	10	15	8	26	15	8
2	6	16	5	25	12	9
3	2	11	1	22	7	6
4	5	16	3	28	11	9
5	7	15	4	24	12	7
6	8	13	4	23	14	6
7	6	17	3	29	10	9
8	2	18	1	28	8	8
9	5	14	2	25	10	6

50.7 Entering data into LISREL

SPSS files can be entered into LISREL. They need to be saved as PRELIS (.psf) files. 'psf' stands for *PRELIS System File*. PRELIS is a program, like SPSS, which carries out data transformation and basic statistical analyses. There is no free student version of this software but it is not necessary to have it as data transformations such as computing scale scores (Chapter 43) can be done with SPSS. This procedure involves a number of steps and so is described separately from LISREL analyses.

Step 1

In the opening window of LISREL select 'File' and 'Import Data...'.

Step 2

Select the downward pointing arrow next to the 'Look in:' box for the directory containing the SPSS data file.

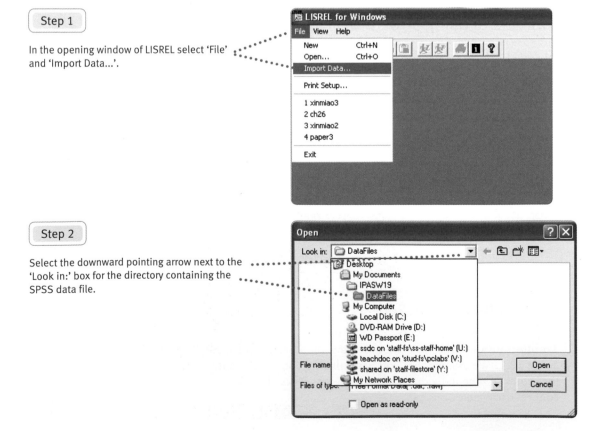

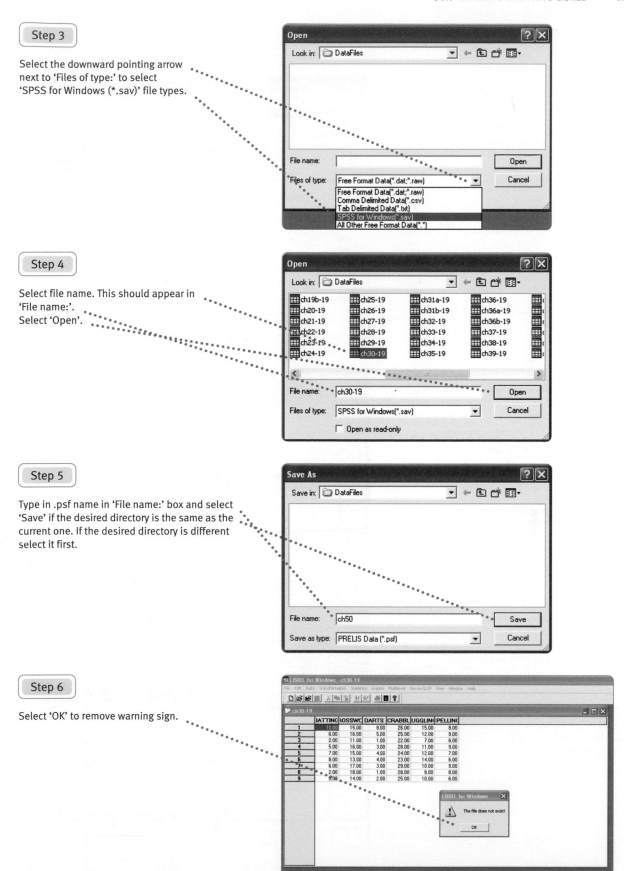

Step 3

Select the downward pointing arrow next to 'Files of type:' to select 'SPSS for Windows (*.sav)' file types.

Step 4

Select file name. This should appear in 'File name:'.
Select 'Open'.

Step 5

Type in .psf name in 'File name:' box and select 'Save' if the desired directory is the same as the current one. If the desired directory is different select it first.

Step 6

Select 'OK' to remove warning sign.

50.8 Conducting LISREL analyses

Step 1

To conduct LISREL analyses select 'File' and 'New'.
Note that there are various other procedure options such as 'Data', 'Transformation' and 'Statistics'.

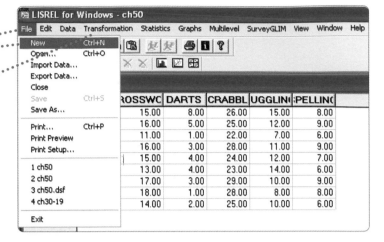

Step 2

To draw a path diagram, scroll down to 'Path Diagram' (not shown) at the bottom of the menu, select it and 'OK'.

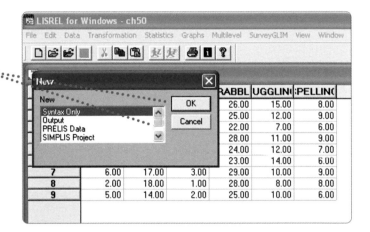

Step 3

Select the downward pointing arrow to find the directory to save the path diagram file in. Type in the 'File name:' box the name of the file for the path diagram and then select 'Save'.

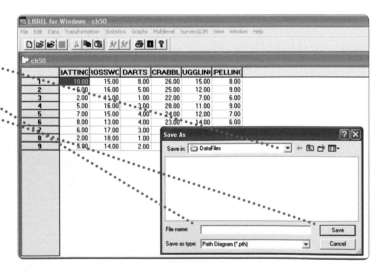

Step 4

Select 'Setup'
and 'Variables...'.

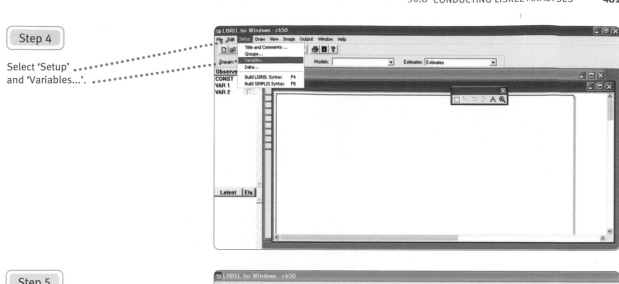

Step 5

Select
'Add/Read Variables'.

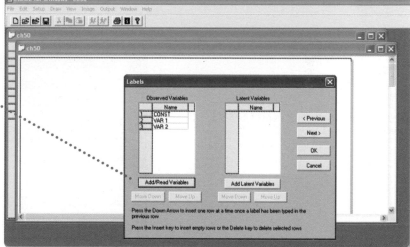

Step 6

Select downward pointing arrow
and 'PRELIS System File'.
Select 'Browse' to find the file.

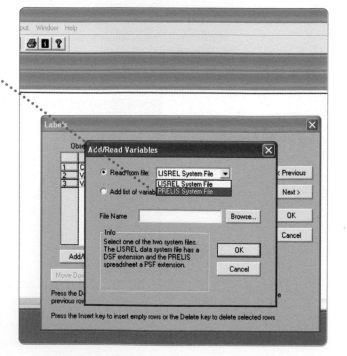

Step 7

Select desired file and then 'Open'.

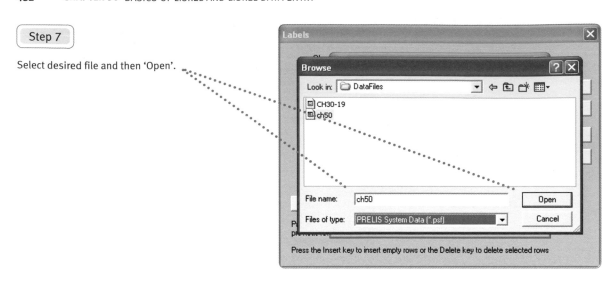

Step 8

Select 'OK'.

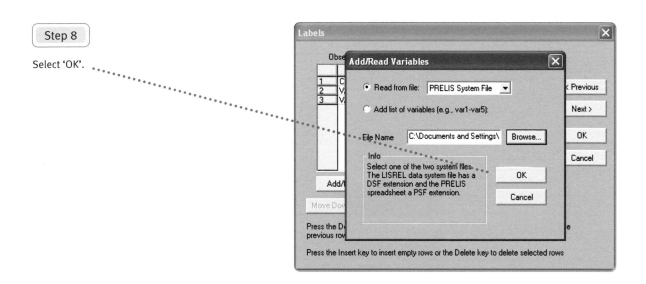

Step 9

Select 'OK'. Ignore the variable called 'CONST'.
It can be deleted if desired.

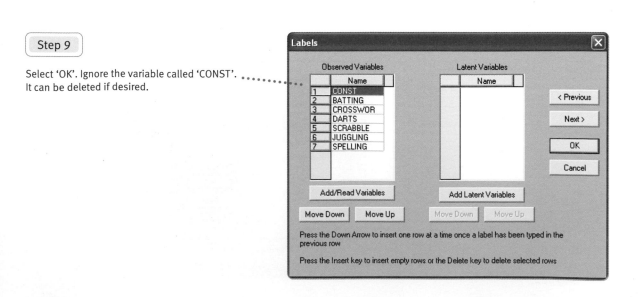

Summary of LISREL steps for data entry and analysis in LISREL

Data

- Select 'File' and 'Import Data ...'.
- Select the ▼ button of the 'Look in:' box to find the directory holding the file.
- Select the ▼ button of the 'Files of type:' box to find SPSS files.
- Select the file and then 'Open'.
- Type in .psf name in 'File name:' box and select 'Save' and then OK.

Analysis

- Select 'File' and 'New', 'Path Diagram' and then 'OK'.
- In the 'File name:' box type in name for path diagram file and then select 'Save'.
- Select 'Setup', 'Variables ...' and 'Add/Read Variables'.
- In 'Labels' box select downward pointing arrow and 'PRELIS System File'.
- Select 'Browse' to find the file and then select 'Open'.
- Select 'OK' and 'OK'.

For further resources including data sets and questions, please refer to the website accompanying this book.

CHAPTER 51

Confirmatory factor analysis with LISREL

Overview

- In confirmatory factor analysis, the researcher's expectations of the ways in which variables cluster together on factors may be tested. For example, the factor structure model may be compared with the original data and with other models.

- These models are often based on the results of factor analyses of the same variables in previous studies. These previous factor analyses can be either exploratory (see Chapter 30) or confirmatory.

- Models vary in their simplicity. The simplest model is to assume that all variables group together into a single factor. That is, all the variables are assumed to be strongly associated with one factor.

- The next simplest model is to assume that the variables group together into two factors, with one set of variables forming one factor and the other set of variables another factor.

- In this model the two factors may be assumed to be unrelated to one another so that the scores of one factor are unrelated to the scores on the other factor. This is a slightly simpler model than one assuming that the two factors are related to one another so that scores on one factor are related to scores on the other factor.

- Generally, the preference is for simpler models rather than more complex ones *if* they both fit the original data equally well.

- The statistical fit of the model to the original data can be assessed in various ways. If there is no difference or only a small difference between the model and the original data, then the model fits the data. If there is a big difference between the model and the original data, then the model is considered not to fit the data and other models may have to be tested.

- Different models for the same data can be compared to see if their statistical fits differ from one another. If there is no difference between them, the simpler model is preferred.

- The simpler model is the one which has the fewer pathways.

51.1   What is confirmatory factor analysis?

Exploratory factor analysis is dealt with in Chapter 30 of this book. It is concerned with the way in which variables can be grouped together and works on a purely empirical basis with no need to theorise about the likely structure. You will need to understand that chapter before going any further. Confirmatory factor analysis does exactly what it says – it examines whether our *expectations* about how variables will be grouped together into factors are confirmed by the data. Thus, confirmatory factor analysis tests our expectations or model of the data. Furthermore, it allows different expectations represented by different models to be compared. Our models are often based on the results of previous factor analyses of the same variables – either using exploratory or confirmatory factor analysis.

Although we may have ideas about how the variables may group together in factors in exploratory factor analysis, we do not have readily available statistical criteria for assessing whether our expectations have been met. Furthermore, we do not have statistical criteria for determining whether, say, a one factor solution is better than another factor solution. How do we decide whether allowing our factors to be correlated with one another provides a better or more accurate fit to our data than not allowing them to correlate? Confirmatory factor analysis provides us with appropriate criteria. In this sense confirmatory factor analysis can be regarded as an extension of exploratory factor analysis.

We will illustrate confirmatory factor analysis with the same example as was used to explain exploratory factor analysis (Chapter 30) and to explain some of the major ideas in structural equation modelling (Chapter 50). The data are shown in Table 51.1 and consist of the scores of nine individuals on the following six variables: (1) batting; (2) crosswords; (3) darts; (4) Scrabble; (5) juggling; and (6) spelling. As batting, darts and juggling all seem to involve coordinating perception with motor responses, we might expect the scores on these three tasks to go together. Crosswords, Scrabble and spelling, on the other hand, seem to reflect an ability with words and so we might expect the scores on these three tasks to group together. It seems likely that how well people perform on the sensory-motor tasks will be unrelated to how well they perform on the verbal tasks in which case we would expect these two factors to be unrelated to one another.

Table 51.1	Scores of nine individuals on six variables				
Individual	**Batting**	**Crosswords**	**Darts**	**Scrabble**	**Juggling**	**Spelling**
1	10	15	8	26	15	8
2	6	16	5	25	12	9
3	2	11	1	22	7	6
4	5	16	3	28	11	9
5	7	15	4	24	12	7
6	8	13	4	23	14	6
7	6	17	3	29	10	9
8	2	18	1	28	8	8
9	5	14	2	25	10	6

Table 51.2	Hypothetical unrelated two-factor model	
Variables	**Sensory-motor skills**	**Verbal skills**
Batting	1.00	.00
Crosswords	.00	1.00
Darts	1.00	.00
Scrabble	.00	1.00
Juggling	1.00	.00
Spelling	.00	1.00

In confirmatory factor analysis we represent these expectations in terms of the relationship or loading of each variable with the two hypothesised factors as shown in Table 51.2. In other words, Table 51.2 presents our model of the relationship between various skills and the factors sensory-motor skills and verbal skills. Thus since we believe that the first factor reflects the sensory-motor tasks and the second factor the verbal tasks, the three sensory-motor scores will have a loading or coefficient of 1.00 with the first factor and a loading or coefficient of .00 with the second factor. The three verbal scores will have a loading of .00 on the first factor and a loading of 1.00 on the second factor. A loading of 1.00 indicates a perfect relationship between a variable and a factor while a loading of .00 denotes no relationship between a variable and a factor. In the table there is no simple way of showing whether the factors themselves are expected to be related or unrelated. We could indicate the expected relationship between the factors in the title of the table or in a note to the table. In our title for Table 51.2, we have mentioned that it is an unrelated two-factor model, thus informing the reader that the two factors should be uncorrelated.

An alternative way of representing this model is in terms of a path diagram as shown in Figure 51.1. The six variables are manifest or observed variables and are depicted as rectangles. The two factors are latent variables and are shown as ovals or ellipses. A relationship between an observed variable and a factor is portrayed with an arrow going from the factor to an observed

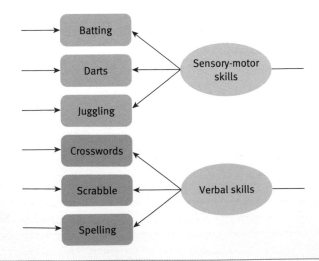

FIGURE 51.1   Path diagram of a hypothetical two-factor model similar to that drawn by LISREL

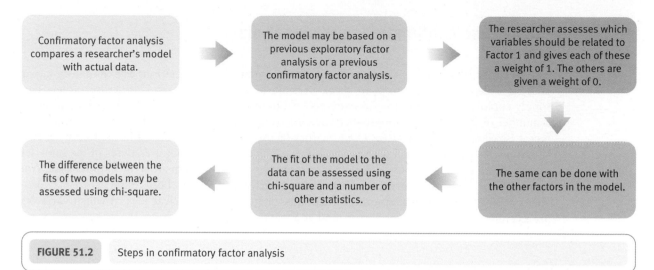

FIGURE 51.2   Steps in confirmatory factor analysis

variable as it is assumed that the factor affects the performance or score on that variable. The absence of an arrow between a factor and an observed variable means that there is no relationship between the factor and the observed variable. We would indicate the expectation that variables were related by drawing a straight or slightly curved two-way or double-headed arrow between them. The absence of such an arrow means that there is no relationship between them.

The extent to which this particular model provides an adequate fit to the data is given by various indices which were outlined in Chapter 50:

- If the Normal Theory Weighted Least Squares Chi-Square is not significant, this means that the model gives a good fit to the data. Conversely, if it is significant, then the model does not fit the data.

- A Root Mean Square Error of Approximation (RMSEA) of about .05 or less means that the model provides a good fit to the data while values above this indicate a poor or inadequate fit.

- A Non-Normed Fit Index (NNFI) and a Comparative Fit Index (CFI) of .95 and above are thought to indicate a good or acceptable fit and values below .95 a poor and unacceptable fit.

The fit for any two models to the data may be compared with the Normal Theory Weighted Least Squares Chi-Square. For example, the fit of a two-factor unrelated factors model may be compared to the fit of a two-factor related factors model. For our data, as will be calculated later in this chapter, the chi-square fit is 5.50 with nine degrees of freedom for the unrelated two-factor model (based on an orthogonal exploratory factor analysis) and 5.50 with eight degrees of freedom for the related two-factor model (based on an oblique factor analysis). The model with the smaller chi-square is substracted from the model with the bigger chi-square. The model with the bigger chi-square also tends to be the model with the more degrees of freedom. The statistical significance of the difference in chi-square is looked up in a significance table for chi-square for the difference in degrees of freedom. The difference in degrees of freedom is 1 (9 – 8 = 1) in this case. With one degree of freedom chi-square has to be 3.84 or larger to be significant at the .05 level. As the difference in chi-square is 0, chi-square is not statistically significant. In this case, the simpler unrelated two-factor model may be preferred to the related two-factor model. Tables of the significance of chi-square are found in some statistics textbooks such as *ISP* where an explanation may be found of how to use them. Alternatively, the Web has many sites containing tables of chi-square such as http://home.comcast.net/~sharov/PopEcol/tables/chisq.html and http://people.richland.edu/james/lecture/m170/tbl-chi.html. Figure 51.2 shows some of the key steps in confirmatory factor analysis.

51.2 When to use confirmatory factor analysis

Confirmatory factor analysis is generally used in the same kind of situations as exploratory factor analysis except that the aim is to check or confirm a particular factor structure rather than simply explore what the factor structure may be. In other words, it is used to determine whether a number of variables may be grouped together to form a factor and to determine whether that factor is related to other factors. Confirmatory factor analysis may be used in the following two situations. One is where you wish to check the results of previous factor analyses of the same set of variables. These previous factor analyses may have been confirmatory or exploratory. You may wish to see whether an alternative factor structure which has not been tested before may provide a better fit to the data. Another situation is where you have a clear idea of which variables are expected to load on which factors and you want to see whether this is the case.

51.3 When not to use confirmatory factor analysis

If you have no clear idea as to which variables are likely to load or be most strongly associated with which factors, it is better to use exploratory rather than confirmatory factor analysis.

51.4 Data requirements for confirmatory factor analysis

The data for confirmatory factor analysis may be dichotomous variables such as 'Yes' or 'No' responses to statements and should be at least ordinal in the sense that higher numbers indicate more of the quality being measured. There needs to be variation in the data for a variable. If there is no variation, then we are dealing with a constant rather than a variable. There is some controversy about the number of participants required for a study using factor analysis. The suggestions vary widely according to different sources. Generally speaking, you would not carry out a confirmatory factor analysis with fewer than about 50 participants. Of course, for student work then a smaller sample than this would be common and in this context probably acceptable as part of learning and assessment. However, one also finds suggestions that for a factor analysis to be stable then one should have a minimum of about 10 participants per variable. Sometimes the figure given is greater than this. Basically, the more participants the better, though confirmatory factor analysis may be illuminating even if one is at the lower end of acceptable numbers of participants.

51.5 Problems in the use of confirmatory factor analysis

A problem in confirmatory factor analysis can be that the analysis stops before it is completed. It is not always easy to know the reason for this. One possibility is that two or more variables may be very highly correlated with each other. If this is the case then the correlation matrix between the variables will reveal it. In these circumstances, only one of the highly correlated variables should be included in the analysis. In the event that it is not possible to sort out the problem, it is usually better to use exploratory factor analysis instead.

You can find out more about factor analysis in Chapter 30 of Howitt, D. and Cramer, D. (2011). *Introduction to Statistics in Psychology*, 5th edition. Harlow: Pearson.

51.6   The data to be analysed

The computation of a confirmatory factor analysis is illustrated with the data in Table 51.1. This consists of scores on six variables for nine individuals. The small number of cases is for illustrative purposes only. Normally, you would not carry out a confirmatory factor analysis on such a small number of cases.

51.7   Entering the data

Carry out the steps in Sections 50.6 and 50.7 to have the appropriate data file.

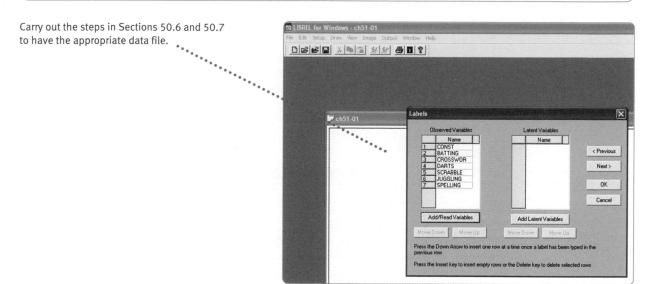

51.8   Confirmatory factor analysis

Step 1

Select 'Add Latent Variables', type in name of latent variable in 'Add Variables' box and select 'OK'.
Repeat for second latent variable.
Select 'OK'.

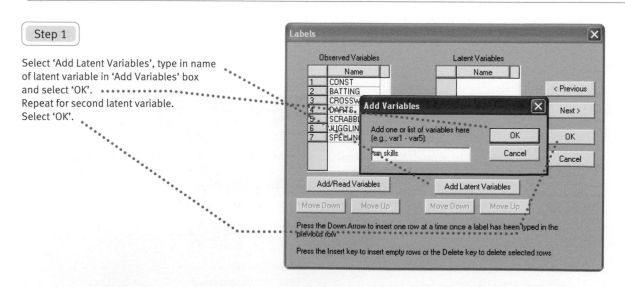

Step 2

Start to draw the path diagram by selecting the first observed variable and moving it over to the expanded diagram window.

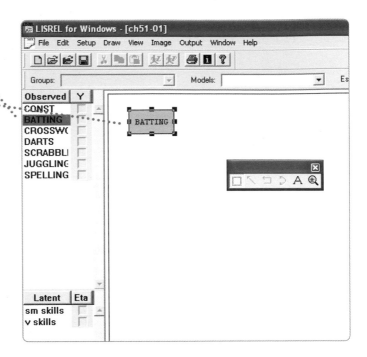

Step 3

Select the arrow in the draw box and draw arrows between latent and manifest variables. Ensure start of arrow is well within the oval of the latent variable.

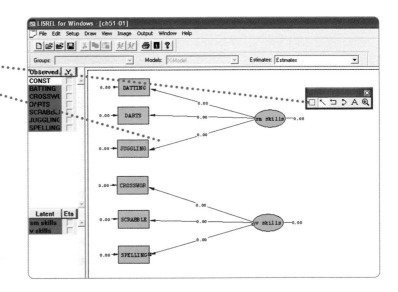

Step 4

Select 'Setup' and 'Build LISREL Syntax'.

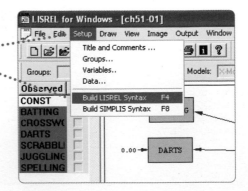

Step 5

These are the syntax commands for the path diagram in Step 3.

```
TI
DA NI=6 NO=0 MA=CM
RA FI='C:\Documents and Settings\ssdc\My Documents\IPASW19\DataFiles\ch50.psf'
SE
 1 3 5 2 4 6 /
MO NX=6 NK=2 TD=SY
LK
'sm skills' 'v skills'
FR LX(1,1) LX(2,1) LX(3,1) LX(4,2) LX(5,2) LX(6,2)
PD
OU
```

Step 6

Select 'File' and 'Run' to run the syntax commands.

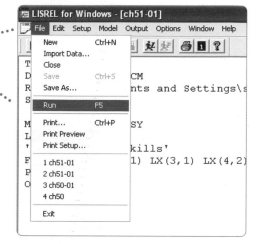

Step 7

These are the results for the path diagram. The unstandardised 'Estimates' are shown.
For other statistics select the downward pointing arrow in the 'Estimates:' box.
Note there is a two-headed arrow between the latent variables which means that this is a related two-factor model.

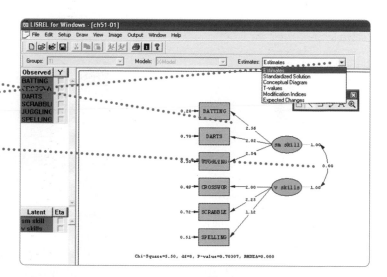

Step 8

To display more detailed output, select 'Window' and the file with the '.out' suffix. Although this file may be more difficult to understand, it may provide a more accessible record of the results than the path diagram. With some models where the paths are close together, the results may be clearer in the output than in the path diagram.

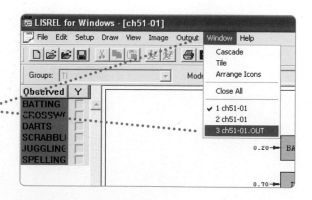

Step 9

This is the start of the output file.

Step 10

To delete the two-headed arrow between the two latent variables, select it and then 'Edit' and 'Delete'.

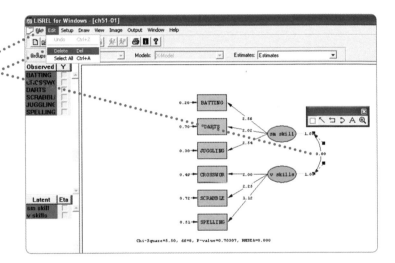

Step 11

To save the syntax for this new path model, select 'Setup' and 'Build LISREL Syntax'. Note the two-headed arrow between the two latent variables has gone.

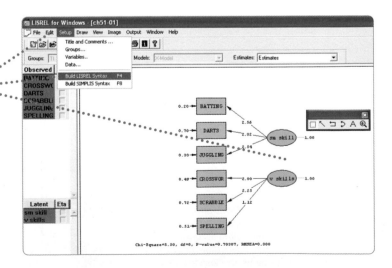

Step 12

This is the syntax for the unrelated two-factor model. Note the extra command which defines the difference between the two models.

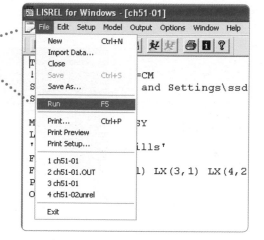

```
TI
!DA NI=6 NO=9 MA=CM
SY='C:\Documents and Settings\ssdc\My Documents\IPASW19\LISREL\ch51-01.DSF'
SE
 1 3 5 2 4 6 /
MO NX=6 NK=2 TD=SY
LK
'sm skill' 'v skills'
FI PH(2,1)
FR LX(1,1) LX(2,1) LX(3,1) LX(4,2) LX(5,2) LX(6,2)
PD
OU
```

Step 13

To run the syntax file select 'File' and 'Run'.

LISREL for Windows - [ch51-01]

File Edit Setup Model Output Options Window Help

New	Ctrl+N
Import Data...	
Close	
Save	Ctrl+S
Save As...	
Run	**F5**
Print...	Ctrl+P
Print Preview	
Print Setup...	
1 ch51-01	
2 ch51-01.OUT	
3 ch51-01	
4 ch51-02unrel	
Exit	

Step 14

These are the results for the unrelated two-factor model.
The unstandardised estImates are shown.

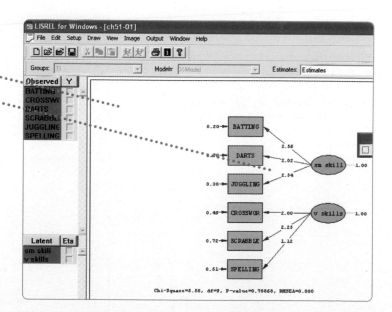

Step 15

For more detailed output select 'Window' and
the .OUT file in the OUT window.

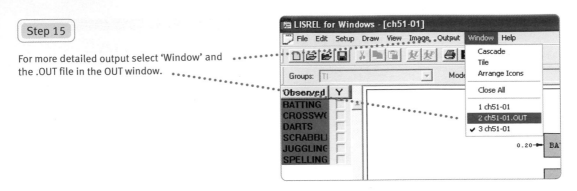

Step 16

For details about the fit indices only select
'Output' and 'Fit Indices'.

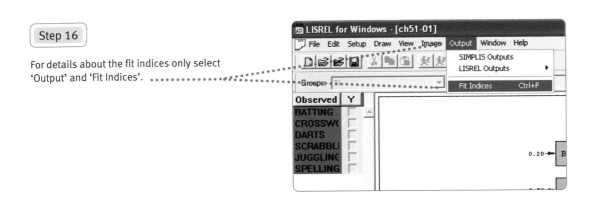

Step 17

These are the four main fit indices mentioned in
this book.

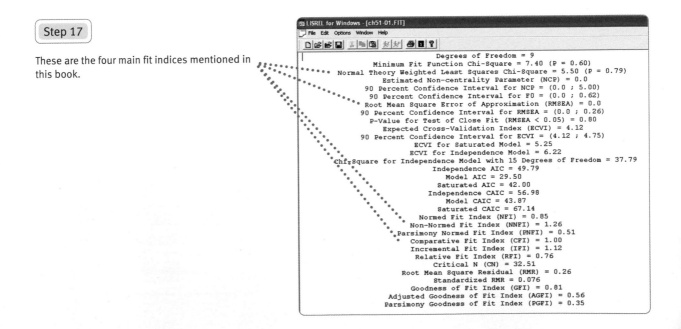

Step 18

To obtain other statistics in the OUT window,
in the Syntax window select 'Output' and
'Selections...'.

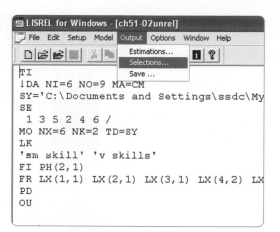

Step 19

This box allows other details to be selected
such as 'Completely Standardized Solution'
for the standardized coefficients.
Select desired options and then 'OK'.

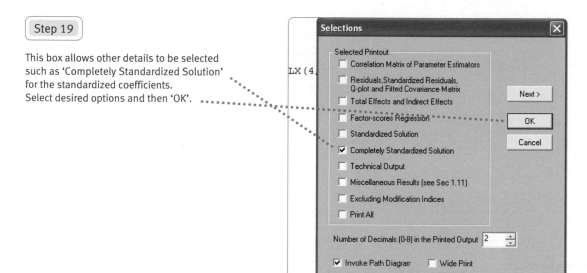

Step 20

Note 'SC' has been added to
the last syntax command.
To run this syntax file,
select 'File' and 'Run'.

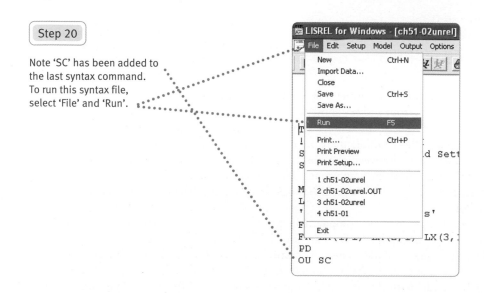

Step 21

To view output, select 'Window' and output file.

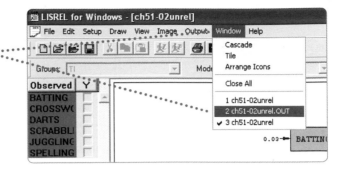

Step 22

Scroll to the bottom of the output file to see the values for the completely standardised solution.

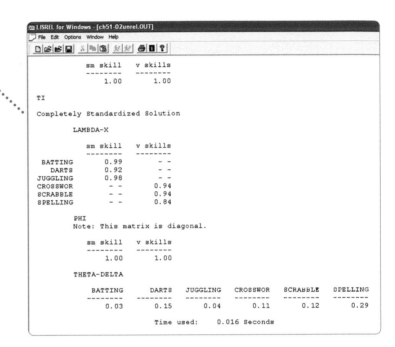

51.9   Interpreting the output

Assuming the path diagram is the correct one, the first step is to check the fit indices. For the unrelated two-factor model, all four indices show that the model fits the data. If the model did not fit the data, then one could look at the modification indices to see which paths need to be changed to produce a better fit. In this example there are no modification indices. If the model fits the data, the standardised coefficients between the observed and latent variables should be examined to check the size and direction of the coefficients. The coefficients should be relatively large. In this case the coefficients are large with the smallest one being .84 and being between spelling and the verbal skills factor.

The fit for two models may be compared in terms of the Normal Theory Weighted Least Squares Chi-Square. For example, the fit of an unrelated two-factor model may be compared with the fit of a related two-factor model. This chi-square is 5.50 with nine degrees of freedom for the unrelated two-factor model and 5.50 with eight degrees of freedom for the related two-factor model. Usually the model with the smaller chi-square is substracted from the model with the bigger chi-square. The model with the bigger chi-square tends to be the model with the more degrees of freedom. The statistical significance of the difference in chi-square is looked up in a significance table for chi-square for the difference in degrees of freedom. The difference in degrees of freedom is 1 (9 − 8 = 1) in this case. With one degree of freedom chi-square has to be 3.841 or larger to be significant at the .05 level or less. As the difference in chi-square is 0, chi-square is not statistically significant. In this case, the simpler unrelated two-factor model may be preferred to the related two-factor model.

REPORTING THE OUTPUT

● The exact way of reporting the results of a confirmatory factor analysis will depend on the purpose of the analysis. One way of describing the results would be as follows:

The statistical fit of the unrelated and the related two-factor models were compared using the maximum likelihood estimation method of confirmatory factor analysis with LISREL 8.80 Student. The fit of each model was assessed in terms of the four indices of the Normal Theory Weighted Least Squares Chi-Square (chi-square), Root Mean Square Error of Approximation (RMSEA), the Non-Normed Fit Index (NNFI) and the Comparative Fit Index (CFI). For the unrelated two factor model chi-square was 5.50 $p = .79$, RMSEA 0.00, NNFI 1.26 and CFI 1.00. For the related two factor model chi-square was 5.50 $p = .70$, RMSEA 0.00, NNFI 1.21 and CFI 1.00. Both models provided a satisfactory fit to the data as chi-square was not significant, RMSEA was less than .05 and NNFI and CFI were greater than .95. As the difference in chi-square between the two models was not significant, chi square difference (1) = 0.00, *p ns*, the unrelated two-factor model was seen as providing the more parsimonious model. The two factors were interpreted as assessing sensory-motor and verbal skills respectively as batting, darts and juggling were significantly related to the sensory-motor factor and crosswords, Scrabble and spelling to the verbal factor.

The standardised coefficients for the accepted model may be presented in a path diagram such as Step 14 or in a table such as Table 51.3. It is better to draw the path diagram yourself rather than use the one provided by LISREL as you will be able to provide full variable names.

Table 51.3	Standardised lambda coefficients of the unrelated two-factor model	
Variable	**Factor 1**	**Factor 2**
Skill at batting	.99***	.00
Skill at crosswords	.92**	.00
Skill at darts	.98***	.00
Skill at Scrabble	.00	.94***
Skill at juggling	.00	.94***
Skill at spelling	.00	.84**

**p < .01; ***p < .001 (two-tailed).

Summary of LISREL steps for confirmatory factor analysis

Analysis

● Select 'File' and 'New', 'Path Diagram' and then 'OK'.
● In the 'File name:' box type in name for path diagram file and then select 'Save'.
● In 'Labels' box select downward pointing arrow and 'PRELIS System File'.
● Select 'Browse' to find the file and then select 'Open'.
● Select 'OK' and 'OK'.

- Select 'Setup', 'Variables ...' and 'Add Latent Variables'.
- Type in name of latent variable in 'Add Variables' box and select 'OK'.
- Repeat for other latent variables.
- Draw path diagram by moving variables into the path diagram window and then draw arrows between them.
- Select 'Setup' and 'Build LISREL Syntax'.
- Select 'File' and 'Run' to run the syntax commands.

Output

- Output is shown in the path diagram window as well as the output window.

For further resources including data sets and questions, please refer to the website accompanying this book.

CHAPTER 52

Simple path analysis with measurement error uncorrected

Overview

- A useful way to begin structural equation modelling is to examine simple path models. A path model is a pattern of relationships among variables in which causal directions are identified.

- For our purposes, we can use a basic model which involves a third variable which may mediate the effect between two other variables (Chapter 34).

- In the analysis, the three variables in the structural equation model are assumed to be measured without error. This is an assumption made in other statistics such as multiple regression.

- In structural equation models we refer to the 'unstandardised maximum likelihood coefficients'. These are similar to regression coefficients in multiple regression.

- Two measures of fit of the model to the data are used: (1) the Normal Theory Weighted Least Squares Chi-Square and the (2) the Root Mean Square Error of Approximation (RMSEA), both of which were discussed in some detail in Chapter 50.

- A model may be termed saturated (or alternatively just-identified) because it accounts for all of the variation in the data. It happens because the number of covariances equals the number of pathways.

- However, if a pathway is removed from the model, the remaining model is described as being over-identified. This is because there is more covariance than the model accounts for. In other words, an over-identified model fails to account completely for the data.

- If there is no significant difference between the fits of the two models with the data, the simpler or more parsimonious model is preferred.

- A model needs to be interpreted in terms of the size, direction and statistical significance of the path coefficients.

52.1 What is simple path analysis with measurement error uncorrected?

A simple mediational analysis involves the statistics of correlations (Chapter 10) and multiple regression (Chapter 34). Both simple and multiple regression are used to explain one variable (the criterion) in terms of one or more other variables. There are various other terms to refer to the variable being explained (the criterion) including the consequent, dependent, effect or outcome variable. They all refer to the same thing. The terms employed to refer to the explanatory variables also vary, including terms such as causal, independent or predictor variables. We will generally use the term 'criterion variable' to refer to the variable being explained and the term 'predictor variable' to refer to the variables being used to explain the criterion variable in the model. The term 'explanation' is used in specific ways in modelling. It often refers to the size and the direction of the association between the predictor and the criterion variables. Alternatively, it is the percentage of the variance in the criterion variable that is explained or shared with the predictor variables. These two measures refer to much the same thing since an association and the percentage of variance explained or shared are closely related. The stronger the association is between two variables, the greater the proportion or percentage of their variance they share.

A mediator or intervening variable is one that is thought to wholly or partially explain the relation between a predictor and a criterion variable. As the mediator variable is also thought to explain the criterion variable, it is also a predictor variable but it is one whose effect on the criterion variable is assumed to come later on in the sequence of variables assumed to affect the criterion variable. Suppose, for example, we think that more intelligent people are likely to do better at their schoolwork because of their intelligence. Now being intelligent does not necessarily mean that you will do better at school because you may be more interested in and spend more time doing activities other than schoolwork, in which case your schoolwork is likely to suffer. However, being more intelligent may mean that you are more interested in and spend more time doing schoolwork because you find it challenging, you think it relatively easy and you do well in it. So part or all of the relationship between intellectual ability and educational achievement may be due to how motivated you are to do well at school. The more motivated you are, the better you do at school. So school motivation may wholly or partly mediate the assocation between intelligence and educational achievement. In other words, intelligence may lead directly to educational achievement but intelligence may result in motivational factors which lead to educational achievement. Thus there are two paths between intelligence and educational achievement.

It may be useful to draw out the relation between these three variables in terms of a simple path diagram as shown in Figure 52.1 where the predictor variable of intellectual ability is listed first and to the left on the page and the criterion variable of educational achievement is listed last

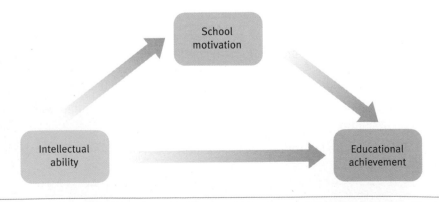

FIGURE 52.1   Partial mediation of school motivation on educational achievement

FIGURE 52.2   Total mediation of school motivation on educational achievement

and to the furthest right on the page. This variable is called an **exogenous** variable as its explanation lies outside the model. The other two variables are known as **endogenous** variables as they are partly explained within the model. The mediator or intervening predictor variable of school motivation is placed between these other two variables and above them. The direction of the relation between the three variables is indicated by arrows pointing to the right. So there is a right pointing arrow between intellectual ability and educational achievement, between intellectual ability and school motivation and between school motivation and educational achievement. If we thought that the relation between intellectual ability and educational achievement was totally mediated by school motivation, then we would omit the arrow between intellectual ability and educational achievement as shown in Figure 52.2.

The question is the extent to which the model which is proposed is supported by the empirically derived relationships between the variables – i.e. the data. This involves calculating the regression weights between, in this case, our three variables. For example, if intellectual ability is not predictive of school motivation or if school motivation is not predictive of educational achievement, then the model or part of the model which has school motivation as a mediating variable is not supported. However, if intellectual ability is predictive of educational achievement through school motivation then the model is supported. Nevertheless, it remains necessary to see the extent to which the simpler model of intellectual ability directly affecting educational achievement is also true.

We can see this more concretely if we add standardised regression weights to the previous figures. This has been done in Figure 52.3. Before we get on to the complexities of the LISREL analysis, it is necessary to do some clear thinking about our model so we have added some

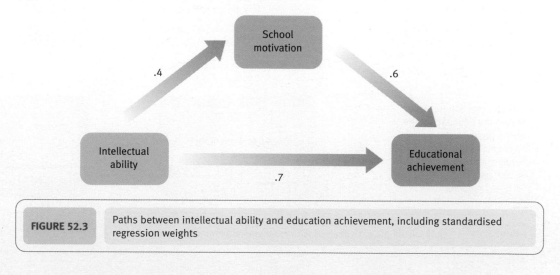

FIGURE 52.3   Paths between intellectual ability and education achievement, including standardised regression weights

somewhat arbitrary regression weights, although LISREL will calculate proper ones for you. We can see that intellectual ability is actually quite strongly predictive of educational achievement since the standardised regression weight is .7. This is larger but not greatly so than the relationship (.6) between our possible mediating variable school motivation and educational achievement. Thus (assuming that these standardised regression weights are statistically significant) both (a) the simple direct model of intellectual ability leading directly to educational achievement and (b) the more complex indirect model which has school motivation mediating between intellectual ability and educational achievement are supported.

The question, though, is the extent to which the complex model adds anything over and above what the simple model does. Remember that the basic principle is that the simpler model is preferred to the more complex model unless it can be shown that the complex model contributes to a significantly greater fit between the model and the data. So we need to calculate the standardised regression between intellectual ability and educational achievement controlling for (or partialling) school motivation. If partialling in this way takes away the statistical significance of the simpler model then we would prefer the more complex model. However, if partialling in this way made no difference then we would prefer the simpler model. But there is a third possibility which is that the size of the relationship changes somewhat but what remains is still statistically significant. In this case, a model combining the simple and the complex models is to be preferred. Obviously things can get somewhat more complex than this. The important thing to remember is that the process is of suggesting and then testing models.

In path analyis the terms just-identified and over-identified are often used. The just-identified model accounts for all of the variation in the data and is sometimes also known as a saturated model. The model in Figure 52.1 is an example of this. It should be fairly obvious that this model must completely explain the data collected since there are as many pathways as there are different correlations or path coefficients. However, if we take the model in Figure 52.2 then this fails to take into account all of the different correlations or path coefficients simply because everything that is available is not included in the model – after all, it does not include one of the pathways. This sort of 'incomplete' model is referred to as being over-identified. This distinction is of theoretical more than practical significance as it is unlikely that you will wish to develop models which have a large number of pathways rather than something simpler and more parsimonious. Figure 52.4 shows the key steps in a simple path analysis with measurement error uncorrected.

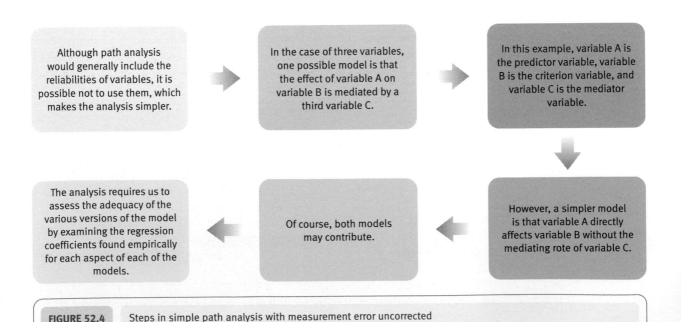

FIGURE 52.4   Steps in simple path analysis with measurement error uncorrected

52.2 When to use simple path analysis with measurement error uncorrected

A simple mediational analysis should be used when the following three conditions are met. Firstly, the mediator variable should provide a plausible explanation for the relation between the predictor and the criterion variable. For example, age should not be used as a mediator variable in this example since intellectual ability cannot affect age. Intellectual ability does not cause people to be older. Secondly and thirdly, there is a significant or substantial correlation between the predictor and the mediator variable and between the mediator and the criterion variable as this suggests that these variables are related. It is not necessary that there should also be a significant or substantial correlation between the predictor and the criterion variables as this relation may be suppressed by the mediator variable.

52.3 When not to use simple path analysis with measurement error uncorrected

A simple mediational analysis should not be carried out under the following three conditions:

- If no plausible explanation can be provided as to why the mediator variable may explain the relation between the predictor and the criterion variable.

- If there is little or no relation between the predictor and the mediator variable and/or between the mediator and the criterion variable. Since the variables are unrelated they cannot explain the relationship between the predictor and criterion variables.

- If you have generally very high correlations between the variables in your analysis. This may make some standardised regression coefficients greater than ±1.00, which is not really possible since a perfect regression coefficient of ±1.00 exists only when a variable is effectively being correlated with itself. Different variables should not lead to such a situation.

52.4 Data requirements for a simple path analysis with measurement error uncorrected

A basic data requirement for any form of simple or multiple regression is that of homoscedasticity. This means that if one examines the scatterplot of one variable against another variable, the variance around the line of best fit should be the same irrespective of, say, whether one examines the top end, middle or bottom end of the line of best fit. This can be roughly assessed by plotting the line of best fit in a scatterplot and seeing if this assumption seems to have been met for each of the predictor variables. If you look at Figure 11.1 (Chapter 11), then it seems that the requirement of homoscedacity is met since the variance around the line of best fit seems similar all the way along the line of best fit, though there is not a lot of data points on which to base this judgement in this case.

It is important the criterion variable should be a continuous score which is normally distributed. If the criterion is a binomial variable, then it is far better to use binomial logistic regression (Chapter 39). If it is a multinomial variable, then multinomial logistic regression should be used (Chapter 38).

52.5 Problems in the use of simple path analysis with measurement error uncorrected

Care needs to be taken in interpreting the results of a simple path analysis with no measurement error. If some of the variables are very highly correlated, then two potential problems may occur. One problem is the standardised regression coefficients may be greater than ±1.00, which indicates that the two variables are more than perfectly correlated. Two variables are only perfectly correlated if they are, in effect, the same variable, which should not be the case when the variables are different. Another problem is the sign of the partial regression coefficient may be reversed.

You can find out more about simple mediational analysis in Chapter 34 of Howitt, D. and Cramer, D. (2011). *Introduction to Statistics in Psychology*, 5th edition. Harlow: Pearson.

52.6 The data to be analysed

We will illustrate the computation of a simple path analysis without measurement error with the data shown in Table 52.1. This data consist of scores for six individuals on the three variables of educational achievement, intellectual ability and school motivation respectively. This is the same data that we used in Chapter 32 for the stepwise multiple regression except that we will ignore the fourth variable of parental interest for the sake of simplicity.

The correlations between these three variables is shown in Table 52.2 (see Chapter 10 for information about correlation). We can see that the correlation between the causal variable (intellectual ability) and the moderator variable (school motivation) is .32, which is acceptable and is significant for a sample of this size.

Because this is for illustrative purposes and to save space, we are going to enter the data 20 times to give us a respectable amount of data to work with. Obviously you would *not* do this if

Table 52.1	Data for a simple mediational analysis	
Educational achievement	**Intellectual ability**	**School motivation**
1	2	1
2	2	3
2	2	3
3	4	3
3	3	4
4	3	2

Table 52.2	A correlation matrix for the three variables	
	1 Achievement	**2 Ability**
2 Ability	.70	
3 Motivation	.37	.32

your data were real. It is important to use quite a lot of research participants or cases for path analysis. Ten or 15 times your number of variables would be reasonably generous. Of course, you can use less for data exploration purposes.

52.7 Entering the data

Copy and paste 29 times the initial six cases in the SPSS data file for Chapter 34 so you have 180 cases. Carry out the steps in Sections 50.6 and 50.7 so you have the appropriate data.

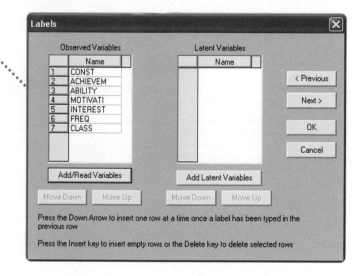

52.8 Simple path analysis with measurement error uncorrected

Step 1

Click on the box beside 'ACHIEVEM' and 'MOTIVATI' to indicate these are the endogenous variables as opposed to the exogenous ones.

Step 2

Click on 'ACHIEVEM' and move it to right pane.

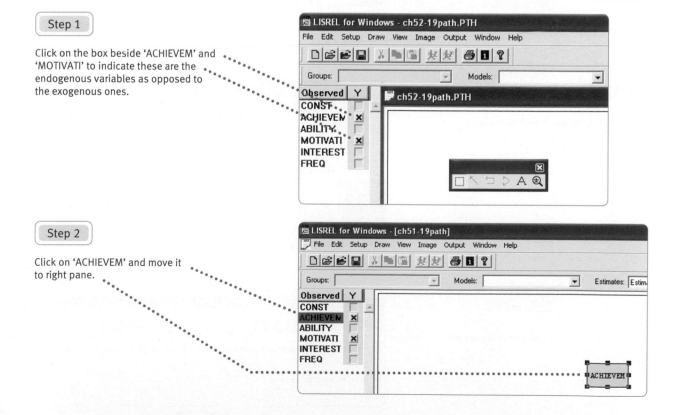

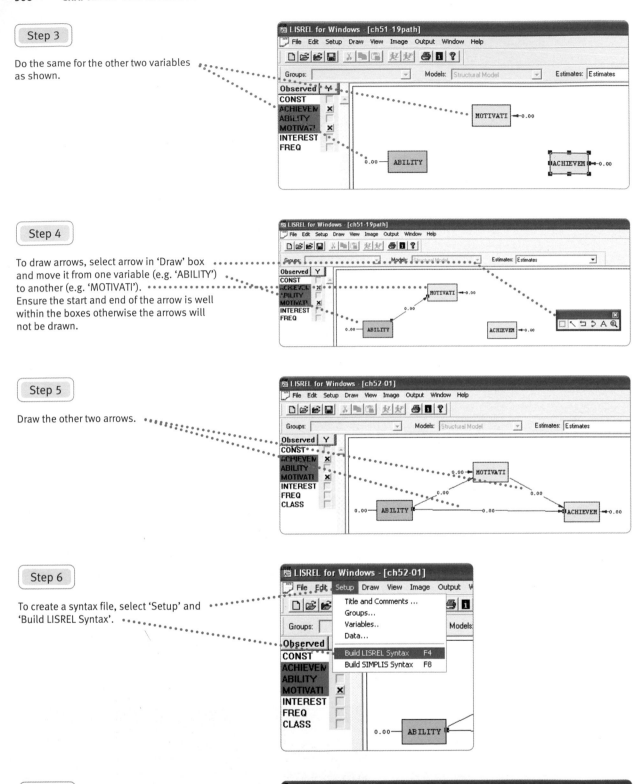

Step 3

Do the same for the other two variables as shown.

Step 4

To draw arrows, select arrow in 'Draw' box and move it from one variable (e.g. 'ABILITY') to another (e.g. 'MOTIVATI').
Ensure the start and end of the arrow is well within the boxes otherwise the arrows will not be drawn.

Step 5

Draw the other two arrows.

Step 6

To create a syntax file, select 'Setup' and 'Build LISREL Syntax'.

Step 7

These are the syntax commands for the path diagram in Step 1.

```
TI
DA NI=6 NO=0 MA=CM
RA FI='C:\Documents and Settings\ssdc\My Documents\IPASW19\DataFiles\ch52.psf'
SE
 1 3 2 /
MO NX=1 NY=2 BE=FU GA=FI PS=SY
FR BE(1,2)  GA(1,2)  PS(2,1)
PD
OU |
```

Step 8

Select 'File' and 'Run' to run
the syntax commands.

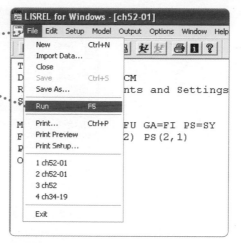

Step 9

These are the results for the path diagram.
The unstandardised 'Estimates' are shown.
For other statistics select the downward
pointing arrow in the 'Estimates:' box.

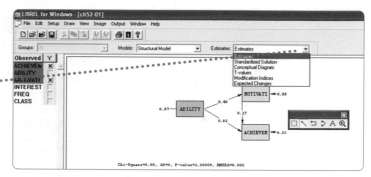

Step 10

To show the standardised solution in the OUT
window, in the Syntax window select 'Output'
and 'Selections...'.

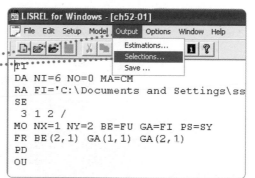

Step 11

Select 'Total Effects and Indirect Effects',
'Standardized Solution' and 'OK'.
Note that this step can occur after
Step 3 as done in Chapters 52 and 53
to save running the syntax file twice.

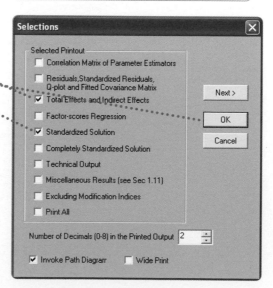

Step 12

Note 'EF' and 'SC' have been added
to the last syntax command.
To run this syntax file,
select 'File' and 'Run'.

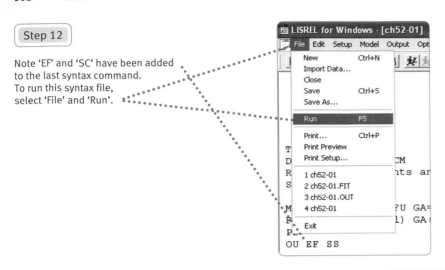

Step 13

To view output, select 'Window' and output file.
Note for this screenshot, the standardised
solution has been selected for the path
diagram.

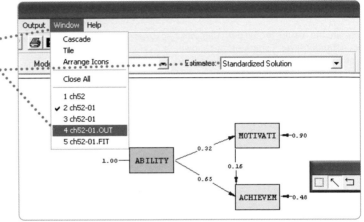

Step 14

These are the standardised coefficients.

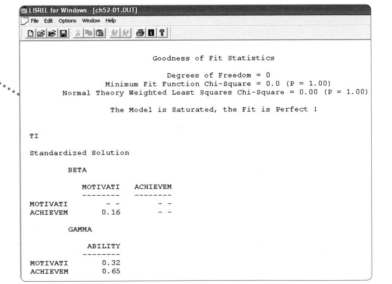

Step 15

These are the standardised effects.

```
Standardized Total and Indirect Effects

          Standardized Total Effects of X on Y

                    ABILITY
                    --------
MOTIVATI            0.32
ACHIEVEM            0.70

          Standardized Indirect Effects of X on Y

                    ABILITY
                    --------
MOTIVATI             - -
ACHIEVEM            0.05

          Standardized Total Effects of Y on Y

                  MOTIVATI    ACHIEVEM
                  --------    --------
MOTIVATI            - -         - -
ACHIEVEM           0.16         - -
```

Step 16

The T-value is the third value.
The unstandardised maximum likelihood
coefficient is the first value. Its standard
error is the second value. The T-value is the
unstandardised coefficient divided by its
standard error. Any difference is due to
rounding error (.17/.06 = 2.83). In the
'Selections' box in Step 7, decimal places
can be increased to 8. In this case five decimal
places gives 2.98 (.16667/.058867 = 2.98).

```
LISREL Estimates (Maximum Likelihood)

        BETA

              MOTIVATI    ACHIEVEM
              --------    --------
MOTIVATI        - -         - -

ACHIEVEM        0.17        - -
               (0.06)
                2.98


        GAMMA

              ABILITY
              --------
MOTIVATI        0.40
               (0.09)
                4.45

ACHIEVEM        0.83
               (0.07)
               11.79
```

Step 17

To delete a path, select the draw toolbox
and the path to be deleted.
Select 'Edit' and 'Delete'.

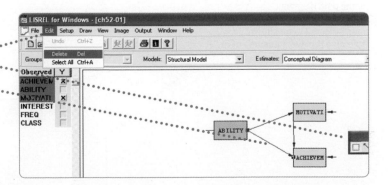

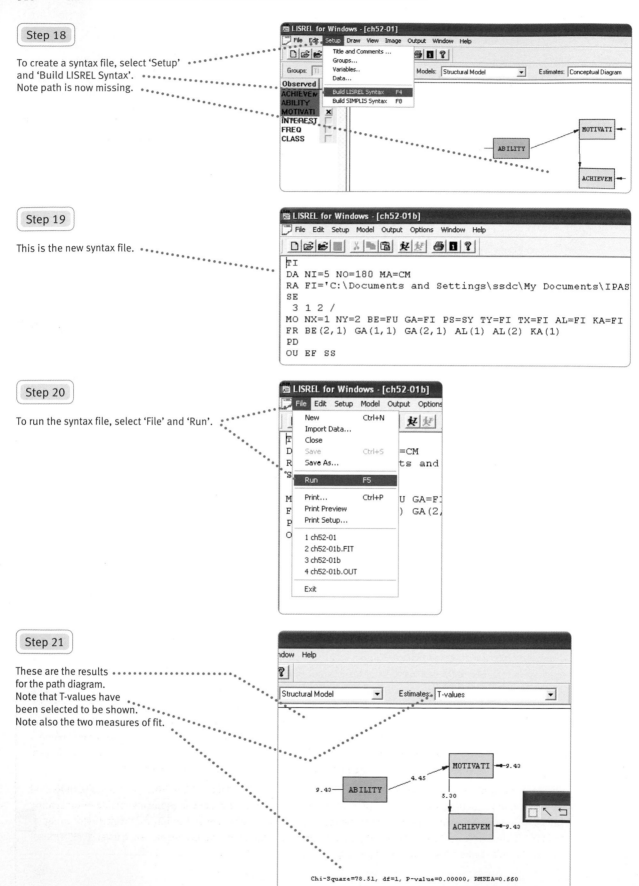

Step 18

To create a syntax file, select 'Setup' and 'Build LISREL Syntax'. Note path is now missing.

Step 19

This is the new syntax file.

```
TI
DA NI=5 NO=180 MA=CM
RA FI='C:\Documents and Settings\ssdc\My Documents\IPAS
SE
 3 1 2 /
MO NX=1 NY=2 BE=FU GA=FI PS=SY TY=FI TX=FI AL=FI KA=FI
FR BE(2,1) GA(1,1) GA(2,1) AL(1) AL(2) KA(1)
PD
OU EF SS
```

Step 20

To run the syntax file, select 'File' and 'Run'.

Step 21

These are the results for the path diagram. Note that T-values have been selected to be shown. Note also the two measures of fit.

Chi-Square=78.51, df=1, P-value=0.00000, RMSEA=0.650

52.9   Interpreting the output

The statistical fit of the saturated or just-identified model is given just below the path diagram and in the output. As this model is saturated or just-identified, the fit is perfect. The next step is to examine the standardised path coefficients and their statistical significance. The standardised path coefficient between intellectual ability and educational achievement is .65 and statistically significant at the .001 level or less. The standardised path coefficient between intellectual ability and school motivation is .32 and statistically significant at the .001 level or less. The standardised path coefficient between school motivation and educational achievement is .16 and statistically significant at the .05 level or less. Note that these results are the same as those for the multiple regression analysis on page 338 in Chapter 34.

These results indicate that educational achievement is more strongly associated with intellectual ability than school motivation. Both intellectual ability and school motivation are positively associated with educational achievement so that greater intellectual ability and school motivation is related to greater educational achievement. Intellectual ability is also positively associated with school motivation so that greater intellectual ability is related to greater school motivation. As the relation between intellectual ability and school motivation and the relation between school motivation and educational achievement are significant, this means that the relation between intellectual ability and educational achievement is partly mediated by school motivation. The mediation is partial because there is little reduction in the size of the relation between intellectual ability and educational achievement when school motivation is controlled or partialled out.

We can go on to check this by determining the fit of a model in which there is no relation between intellectual ability and educational achievement. This model assumes that the relation between these two variables is totally mediated by school motivation. As the chi-square for this model is statistically significant, the model does not provide a satisfactory fit to the data. A model which assumes that the relation between intellectual ability and educational achievement is totally mediated by school motivation is inadequate. In other words, we need a model which has a path between intellectual ability and school motivation. In this case, there is no need to compare the fit of the two models as the simpler or nested model does not fit the data. If this model had fitted the data, we would compare the fit of the two models to see whether the chi-square difference was significant. If it was not significant, we would go for the simpler model. If it was significant, we would go for the model with the better fit.

REPORTING THE OUTPUT

- One of the simplest ways of reporting the results of this simple path analysis with no measurement error is as follows:

  A model assuming that educational achievement was directly affected by intellectual ability and indirectly affected by school motivation was tested with maximum likelihood structural equation model using LISREL 8.80 Student. All variables were manifest variables. As this model was just-identified it provided a perfect fit to the data. Educational achievement was significantly positively associated with both intellectual ability, $\beta = .65$, two-tailed $p < .001$, and school motivation, $\beta = .16$, two-tailed $p < .05$. School motivation was also significantly positively associated with intellectual ability, $\beta = .32$, two-tailed $p < .001$. A model postulating that the relation between intellectual ability and educational achievement was totally mediated by school motivation did not fit the data, the Normal Theory Weighted Least Squares Chi-Square(1) = 78.51, $p < .001$ and RMSEA = 0.66. Educational achievement was largely explained in terms of intellectual ability with school motivation having a small mediating effect.

- It may be useful to illustrate these results with the path diagram shown in Figure 52.5. When this is done in journal articles, the correlation coefficients and the statistical symbols are not usually shown. When inspecting or presenting your results it may be useful to include them so that it is easier to see the effect of controlling for the mediator variable. Asterisks indicate the coefficients are statistically significant. One asterisk indicates a significance level of .05 or less, two asterisks .01 or less and three asterisks .001 or less.

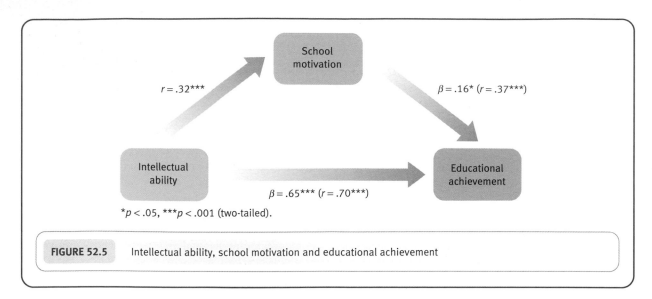

FIGURE 52.5   Intellectual ability, school motivation and educational achievement

Summary of LISREL steps for simple path analysis with measurement error uncorrected

Analysis

- Select 'File' and 'New', 'Path Diagram' and then 'OK'.
- In the 'File name:' box type in name for path diagram file and then select 'Save'.
- Select 'Setup', 'Variables …' and 'Add/Read Variables'.
- In 'Labels' box select downward pointing arrow and 'PRELIS System File'.
- Select 'Browse' to find the file and then select 'Open'.
- Select 'OK'.
- Click box for endogenous variables.
- Draw path diagram by moving variables into the path diagram window and then draw arrows between them.
- Select 'Setup' and 'Build LISREL Syntax'.
- Select 'File' and 'Run' to run the syntax commands.

Output

- Output is shown in the path diagram window as well as the output window.

For further resources including data sets and questions, please refer to the website accompanying this book.

Some other statistics in SPSS Statistics

Some other statistical methods provided by SPSS Statistics but not described in this book are shown below in terms of their options on the 'Analyze' menu, submenu and dialog box options.

Analyze menu	Analyze submenu	Dialog box
Descriptive	Crosstabs . . .	Lambda
Statistics		Uncertainty coefficient
		Gamma
		Somers' d
		Kendall's tau-b
		Kendall's tau-c
		Risk
		Eta
Correlate	Bivariate . . .	Kendall's tau-b
Regression	Curve Estimation . . .	
	Ordinal . . .	
	Probit . . .	
	Nonlinear . . .	
	Weight Estimation . . .	
	2-Stage Least Squares . . .	
	Optimal Scaling . . .	
Loglinear	Logit . . .	
Classify	TwoStep Cluster . . .	
	K-Means Cluster . . .	
	Hierarchical Cluster . . .	
	Discriminant . . .	

Analyze menu	Analyze submenu	Dialog box
Dimension Reduction	Correspondence Analysis . . .	
	Optimal Scaling . . .	
Scale	Multidimensional Scaling . . .	
	Multidimensional Scaling [PROXSCAL] . . .	
	Multidimensional Scaling [ALSCAL]	
Nonparametric	Binomial . . .	
Tests	Runs . . .	
	1-Sample K-S . . .	
	(Kolmogorov–Smirnov)	
	2 Independent Samples . . .	Kolmogorov–Smirnov Z
		Wald–Wolfowitz runs
		Moses extreme reactions
	K Independent Samples . . .	Jonckheere–Terpstra
		Median
	2 Related Samples . . .	Marginal Homogeneity
	K Related Samples . . .	Kendall's W
		Cochran's Q
Survival	Life Tables . . .	
	Kaplan–Meier . . .	
	Cox Regression . . .	
	Cox w/ Time-Dep Cov . . .	

GLOSSARY

A priori test: A test of the difference between two groups of scores when this comparison has been planned ignorant of the actual data. This contrasts with a *post hoc* test which is carried out after the data have been collected and which has no particularly strong expectations about the outcome.

Adjusted mean: A mean score when the influence of one or more covariates has been removed especially in analysis of covariance.

Alpha level: The level of risk that the researcher is prepared to mistakenly accept the hypothesis on the basis of the available data. Typically this is set at a maximum of 5% or .05 and is, of course, otherwise referred to as the level of significance.

Analysis of covariance (ANCOVA): A variant of the analysis of variance (ANOVA) in which scores on the dependent variable are adjusted to take into account (control) a covariate(s). For example, differences between conditions of an experiment at pre-test can be controlled for.

Analysis of variance (ANOVA): An extensive group of tests of significance which compare means on a dependent variable. There may be one or more independent (grouping) variables or factors. ANOVA is essential in the analysis of most laboratory experiments.

Association: A relationship between two variables.

Bar chart: A picture in which frequencies are represented by the height of a set of bars. It should be the areas of a set of bars but SPSS ignores this and settles for height.

Bartlett's test of sphericity: A test used in MANOVA of whether the correlations between the variables differ significantly from zero.

Beta or Type II level: The risk that we are prepared to accept of rejecting the null hypothesis when it is in fact true.

Beta weight: The standardised regression weight in multiple regression. It corresponds to the correlation coefficient in simple regression.

Between-groups design: A design where different participants are allocated to different groups or conditions.

Between-subjects design: See Between-groups design.

Bimodal: A frequency distribution with two modes.

Bivariate: Involving two variables as opposed to univariate which involves just one variable.

Bivariate correlation: A correlation between two variables.

Block: A subset of variables which will be analysed together in a sequence of blocks.

Bonferroni adjustment: A method of adjusting significance levels for the fact that many statistical analyses have been carried out on the data.

Boxplot: A diagram indicating the distribution of scores on a variable. It gives the median in a box, the upper and lower sides of which are the upper and lower values of the interquartile range. Lines at each side of the box identify the largest and smallest scores.

Box's M: A statistical test which partly establishes whether the data meet the requirements for a MANOVA analysis. It examines the extent to which the covariances of the dependent variables are similar for each of the groups in the analysis. Ideally, then, Box's M should not be significant. The test is used in MANOVA though its interpretation is complex.

Case: The basic unit of analysis on which data are collected such as individuals or organisations.

Categorical variable: A nominal or category variable.

Category variable: A variable which consists of categories rather than numerical scores. The categories have no particular quantitative order. However, usually on SPSS they will be coded as numbers.

Cell: The intersection of one category of a variable with another category of one or more other variables. So if a variable has categories A, B and C and the other variable has categories X, Y and Z, then the cells are A with X, A with Y, A with Z, B with X, B with Y, etc. It is a term frequently used in ANOVA as well as with chi-square tables (i.e. cross-tabulation and contingency tables).

Chart: A graphical or pictorial representation of the characteristics of the data.

Chart Editor window: In SPSS it is a Window which can be opened to refine a chart.

Chi-square distribution: A set of theoretical probability distributions which vary according to the degrees of freedom and which are used to determine the statistical significance of a chi-square test.

Chi-square test, Pearson's: A test of goodness-of-fit or association for frequency data. It compares the observed data with the estimated (or actual) population distribution (this is usually based on combining two or more samples).

Cluster analysis: A variety of techniques which identify the patterns of variables or cases which tend to be similar to each other. No cluster analysis techniques are dealt with in this book as they are uncommon in psychology. Factor analysis often does a similar job.

Cochran's Q test: A test of whether the frequencies of a dichotomous variable differ significantly for more than two related samples or groups.

Coefficient of determination: The square of Pearson's correlation coefficient. So a correlation of .4 has a coefficient of determination of .16. It is useful especially since it gives a numerically more accurate representation of the relative importance of different correlation coefficients than the correlation coefficients themselves do.

Common variance: The variance that two or more variables share.

Communality: The variance that a particular variable in an analysis shares with other variables. It is distinct from error variance and specific variance (which is confined to a particular variable). It mainly appears in factor analysis.

Component matrix: A table showing the correlations between components and variables in factor analysis.

Compute: In SPSS this procedure allows the researcher to derive new variables from the original variables. For example, it would be possible to sum the scores for each participant on several variables.

Condition: One of the groups in ANOVA or the *t*-test.

Confidence interval: A more realistic way of presenting the outcomes of statistical analysis than, for example, the mean or the standard deviation would be. It gives the range within which 95 per cent or 99 per cent of the most common means, standard deviations, etc. would lie. Thus instead of saying that the mean is 6.7 we would say that the 95 per cent confidence interval for the mean is 5.2 to 8.2.

Confirmatory factor analysis: A test of whether a particular model or factor structure fits a set of data satisfactorily.

Confounding variable: Any variable which clouds the interpretation of a correlation or any other statistical relationship. Once the effects of the confounding variable are removed, the remaining relationship presents a truer picture of what is going on in reality.

Contingency table: A frequency table giving the frequencies in all of the categories of two or more nominal (category) variables tabulated together.

Correlation coefficient: An index which gives the extent and the direction of the linear association between two variables.

Correlation matrix: A matrix of the correlations of pairs of variables.

Count: The number of times (frequency) a particular observation (score or category, for example) occurs.

Counterbalancing: If some participants take part in condition A of a study first followed by condition B later, then to counterbalance any time or sequence effects other participants should take part in condition B first followed by condition A second.

Covariance: The variance which two or more score variables have in common (i.e. share). It is basically calculated like variance but instead of squaring each score's deviation from the mean, the deviation of variable X from its mean is multiplied by the deviation of variable Y from its mean.

Covariate: A variable which correlates with the variables that are the researcher's main focus of interest. In the analysis of covariance it is the undesired influence of the covariate which is controlled for.

Cox and Snell's R^2: The amount of variance in the criterion variable accounted for by the predictor variables. It is used in logistic regression.

Cramer's V: Also known as Cramer's phi, this correlation coefficient is usually applied to a contingency or cross-tabulation table greater than 2 rows × 2 columns.

Critical value: Used when calculating statistical significance with statistical tables. It is the minimum value of the statistical calculation which is statistically significant (i.e. which rejects the null hypothesis).

Cronbach's alpha: A measure of the extent to which cases respond in a similar or consistent way on all the variables that go to make up a scale.

Data Editor Window: The data spreadsheet in which data are entered in SPSS.

Data handling: The various techniques to deal with data from a study excluding its statistical analysis. It would include data entry into the spreadsheet, the search for errors in data entry, recoding variables into new values, computing new variables and so forth.

Data View: The window in SPSS which allows you to see the data spreadsheet.

Degrees of freedom: The number of components of the data that can vary while still yielding a given population value for characteristics such as mean scores. All other things being equal, the larger the degrees of freedom the more likely it is that the research findings will be statistically significant.

Dependent variable: A variable which potentially may be affected or predicted by other variables in the analysis. It is sometimes known as the criterion or outcome variable.

Descriptive statistics: Indices which describe the major characteristics of variables or the relationships between variables. They include measures of central tendency (mean, median and mode for example) and measures of spread (range, variance, etc.).

Deviation: Usually the difference between a score and the mean of the set of scores.

Dialog box: A rectangular picture in SPSS which allows the user to select various procedures.

Dichotomous: A nominal (category) variable with just two categories. Gender (male/female) is an obvious example.

Direct oblimin: A rotation procedure for making factors in a factor analysis more meaningful or interpretable. Its essential characteristic is that the factors are not required to be uncorrelated (independent) of each other.

Discriminant (function) analysis: A statistical technique for score variables which maximises the difference(s) between two or more groups of participants on a set of variables. It generates a set of 'weights' which are applied to these variables.

Discriminant function: Found mainly in discriminant (function) analysis. A derived variable based on combining a set of variables in such a way that groups are as different as possible on the discriminant function. More than one discriminant function may emerge but each discriminant function is uncorrelated with the others.

Discriminant score: An individual's score on a discriminant function.

Dummy coding: Used when analysing nominal (category) data to allow such variables to be used analogously to scores. Each category of the nominal (category) variable is made into a separate dummy variable. If the nominal (category) variable has three categories A, B and C then two new variables, say A versus not A and B versus not B are created. The categories may be coded with the value 1 and 0. It would not be used where a variable has only two different categories.

Dummy variable: A variable created by dummy coding.

Effect size: A measure of the strength of the relationship between two variables. Most commonly used in meta-analysis. The Pearson correlation coefficient is a very familiar measure of effect size. Also commonly used is Cohen's *d*. The correlation coefficient is recommended as the most user-friendly measure of effect size as it is very familiar to most of us and easily understood.

Eigenvalue: The variance accounted for by a factor. It is simply the sum of the squared factor loadings. The concept is also used for discriminant functions.

Endogenous variable: Any variable in path analysis that can be explained on the basis of one or more variables in that analysis.

Eta: A measure of association for non-linear (curved) relationships.

Exact significance: The precise significance level at and beyond which a result is statistically significant.

Exogenous variable: A variable in path analysis which is not accounted for by any other variable in that analysis.

Exploratory factor analysis: The common form of factor analysis which finds the major dimensions of a correlation matrix using weighted combinations of the variables in the study. It identifies combinations of variables which can be described as one or more superordinate variable or factor.

Exponent or power: A number with an exponent or power superscript is multiplied by itself by that number of times. Thus 3^2 means 3×3 whereas 4^3 means $4 \times 4 \times 4$.

Extraction: The process of obtaining factors in factor analysis.

F ratio: The ratio of two variances. It can be used to test whether these two variances differ significantly using the *F*-distribution. It can be used on its own but is also part of the *t*-test and ANOVA.

Factor, in analysis of variance: An independent or subject variable but is best regarded as a variable on which groups of participants are formed. The variances of these groups are then compared using ANOVA. A factor should consist of a nominal (category) variable with a small number of categories.

Factor, in factor analysis: A variable derived by combining other variables in a weighted combination. A factor seeks to synthesise the variance shared by variables into a more general variable to which the variables relate.

Factor matrix: A table showing the correlations between factors and the variables.

Factor scores: Standardised scores for a factor. They provide a way of calculating an individual's score on a factor which precisely reflects that factor.

Factorial ANOVA: An analysis of variance with two or more independent or subject variables.

Family error rate: The probability or significance level for a finding when a family or number of tests or comparisons are being made on the same data.

Fisher test: Tests of significance (or association) for 2×2 and 2×3 contingency tables.

Frequency: The number of times a particular category occurs.

Frequency distribution: A table or diagram giving the frequencies of values of a variable.

Friedman's test: A non-parametric test for determining whether the mean ranks of three or more related samples or groups differ significantly.

Goodness-of-fit index: A measure of the extent to which a particular model (or pattern of variables) designed to describe a set of data actually matches the data.

Graph: A diagram for illustrating the values of one or more variables.

Grouping variable: A variable which forms the groups or conditions which are to be compared.

Harmonic mean: The number of scores, divided by the sum of the reciprocal ($1/x$) of each score.

Help: A facility in software with a graphical interface such as SPSS which provides information about its features.

Hierarchical agglomerative clustering: A form of cluster analysis, at each step of which a variable or cluster is paired with the most similar variable or cluster until one cluster remains.

Hierarchical or sequential entry: A variant of regression in which the order in which the independent (predictor) variables are entered into the analysis is decided by the analyst rather than mathematical criteria.

Hierarchical regression: *see* Hierarchical or sequential entry.

Histogram: A chart which represents the frequency of particular scores or ranges of scores in terms of a set of bars. The height of the bar represents the frequency of this score or range of scores in the data.

Homogeneity of regression slope: The similarity of the regression slope of the covariate on the criterion variable in the different groups of the predictor variable.

Homogeneity of variance: The similarity of the variance of the scores in the groups of the predictor variable.

Homoscedasticity: The similarity of the scatter or spread of the data points around the regression line of best fit in different parts of that line.

Hypothesis: A statement expressing the expected or predicted relationship between two or more variables.

Icicle plot: A graphical representation of the results of a cluster analysis in which x's are used to indicate which variables or clusters are paired at which stage.

Identification: The extent to which the parameters of a structural equation model can be estimated from the original data.

Independence: Events or variables being unrelated to each other.

Independent groups design: A design in which different cases are assigned to different conditions or groups.

Independent *t*-test: A parametric test for determining whether the means of two unrelated or independent groups differ significantly.

Independent variable: A variable which may affect (predict) the values of another variable(s). It is a variable used to form the groups in experimental designs. But it is also used in regression for the variables used to predict the dependent variable.

Inferential statistics: Statistical techniques which help predict the population characteristics from the sample characteristics.

Interaction: This describes outcomes in research which cannot be accounted for on the basis of the separate influences of two or more variables. So, for example, an interaction occurs when two variables have a significant influence when combined.

Interaction graph: A graph showing the relationship of the means of three or more variables.

Interquartile range: The range of the middle 50 per cent of a distribution. By ignoring the extreme quarter in each direction from the mean, the interquartile range is less affected by extreme scores.

Interval data: Data making up a scale in which the distance or interval between adjacent points is assumed to be the same or equal but where there is no meaningful zero point.

Just-identified model: A structural equation model in which the data are just sufficient to estimate its parameters.

Kaiser or Kaiser–Guttman criterion: A statistical criterion in factor analysis for determining the number of factors or components for consideration and possible rotation in which factors or components with eigenvalues of one or less are ignored.

Kendall's tau (τ): An index of the linear association between two ordinal variables. A correlation coefficient for non-parametric data in other words.

Kolmogorov–Smirnov test for two samples: A non-parametric test for determining whether the distributions of scores on an ordinal variable differ significantly for two unrelated samples.

Kruskal–Wallis test: A non-parametric test for determining whether the mean ranked scores for three or more unrelated samples differ significantly.

Kurtosis: The extent to which the shape of a bell-shaped curve is flatter or more peaked than a normal distribution.

Latent variable: An unobserved variable that is measured by one or more manifest variables or indicators.

Level: Used in analysis of variance to describe the different conditions of an independent variable (or factor). The term has its origins in agricultural research where levels of treatment would correspond to, say, different amounts of fertiliser being applied to crops.

Levels of measurement: A four-fold hierarchical distinction proposed for measures comprising nominal, ordinal, equal interval and ratio.

Levene's test: An analysis of variance on absolute differences to determine whether the variances of two or more unrelated groups differ significantly.

Likelihood ratio chi-square test: A form of chi-square which involves natural logarithms. It is primarily associated with log-linear analysis.

Line graph: A diagram in which lines are used to indicate the frequency of a variable.

Linear association or relationship: This occurs when there is a straight line relationship between two sets of scores. The scattergram for these data will be represented best by a straight line rather than a curved line.

Linear model: A model which assumes a linear relationship between the variables.

LISREL: The name of a particular software designed to carry out linear structural relationship analysis also known as structural equation modelling.

Loading: An index of the size and direction of the association of a variable with a factor or discriminant function of which it is part. A loading is simply the correlation between a variable and the factor or discriminant function.

Log likelihood: An index based on the difference between the frequencies for a category variable(s) and what is predicted on the basis of the predictors (i.e. the modelled data). The bigger the log likelihood the poorer the fit of the model to the data.

−2 log likelihood (ratio) test: Used in logistic regression, it is a form of chi-square test which compares the goodness of fit of two models where one model is a part of (i.e. nested or a subset of) the other model. The chi-square is the difference in the −2 log likelihood values for the two models.

Logarithm: The amount to which a given base number (e.g. 10) has to be multiplied by itself to obtain a particular number. So in the expression 3^2, 2 would be the logarithm for the base 3 which makes 9. Sometimes it is recommended that scores are converted to their logarithms if this results in the data fitting the requirements of the statistical procedure better.

Logistic or logit regression: A version of multiple regression in which the dependent, criterion or outcome variable takes the form of a nominal (category) variable. Any mixture of scores and nominal (category) variables can act as predictors. The procedure uses dummy variables extensively.

Log-linear analysis: A statistical technique for nominal (category) data which is essentially an extension of chi-square where there are three or more independent variables.

Main effect: The effect of an independent or predictor variable on a dependent or criterion variable.

Manifest variable: A variable which directly reflects the measure used to assess it.

Mann–Whitney test: A non-parametric test for seeing whether the number of times scores from one sample are ranked significantly higher than scores from another unrelated sample.

Marginal totals: The marginal totals are the row and column total frequencies in cross-tabulation and contingency tables.

Matched-subjects design: A related design in which participants are matched in pairs on a covariate or where participants serve as their own control. In other words, a repeated or related measures design.

Matrix: A rectangular array of rows and columns of data.

Mauchly's test: A test for determining whether the assumption that the variance–covariance matrix in a repeated measures analysis of variance is spherical or circular.

Maximum likelihood method: A method for finding estimates of the population parameters of a model which are most likely to give rise to the pattern of observations in the sample data.

McNemar test: A test for assessing whether there has been a significant change in the frequencies of two categories on two occasions in the same or similar cases.

Mean: The everyday numerical average score. Thus the mean of 2 and 3 is 2.5.

Mean square: A term for variance estimate used in analysis of variance.

Measure of dispersion: A measure of the variation in the scores such as the variance, range, interquartile range, and standard error.

Median: The score which is halfway in the scores ordered from smallest to largest.

Mixed ANOVA: An ANOVA in which at least one independent variable consists of related scores and at least one other variable consists of uncorrelated scores.

Mixed design: *See* Mixed ANOVA.

Mode: The most commonly occurring score or category.

Moderating or moderator effect: A relationship between two variables which differs according to a third variable. For example, the correlation between age and income may be moderated by a variable such as gender. In other words, the correlation for men and the correlation for women between age and income is different.

Multicollinearity: Two or more independent or predictor variables which are highly correlated.

Multimodal: A frequency distribution having three or more modes.

Multiple correlation or R: A form of correlation coefficient which correlates a single score (A) with two or more other scores (B + C) in combination. Used particularly in multiple regression to denote the correlation of a set of predictor variables with the dependent (or outcome) variable.

Multiple regression: A parametric test to determine what pattern of two or more predictor (independent) variables is associated with scores on the dependent variable. It takes into account the associations (correlations) between the predictor variables. If desired, interactions between predictor variables may be included.

Multivariate: Involving more than two variables.

Multivariate analysis of variance (MANOVA): A variant of analysis of variance in which there are two or more *dependent* variables combined. MANOVA identifies differences between groups in terms of the combined dependent variable.

Nagelkerke's R^2: The amount of variance in the criterion variable accounted for by the predictor variables.

Natural, Naperian or Napierian logarithm: The logarithms calculated using 2.718 as the base number.

Nested model: A model which is a simpler subset of another model and which can be derived from that model.

Non-parametric test: A statistical test of significance which requires fewer assumptions about the distribution of values in a sample than a parametric test.

Normal distribution: A mathematical distribution with very important characteristics. However, it is easier to regard it as a bell-shaped frequency curve. The tails of the curve should stretch to infinity in both directions but this, in the end, is of little practical importance.

Numeric variables: Variables for which the data are collected in the form of scores which indicate quantity.

Oblique factors: In factor analysis, oblique factors are ones which, during rotation, are allowed to correlate with each other. This may be more realistic than orthogonal rotations. One way of looking at this is to consider height and weight. These are distinct variables but they correlate to some degree. Oblique factors are distinct but they can correlate.

Odds: Obtained by dividing the probability of something occurring by the probability of it not occurring.

Odds ratio: The number by which the odds of something occurring must be multiplied for a one unit change in a predictor variable.

One-tailed test: A version of significance testing in which a strong prediction is made as to the direction of the relationship. This should be theoretically and empirically well founded on previous research. The prediction should be made prior to examination of the data.

Ordinal data: Numbers for which little can be said other than the numbers give the rank order of cases on the variable from smallest to largest.

Orthogonal: Essentially means at right angles.

Orthogonal factors: In factor analysis, orthogonal factors are factors which do not correlate with each other.

Outcome variable: A word used especially in medical statistics to denote the dependent variable. It is also the criterion variable. It is the variable which is expected to vary with variation in the independent variable(s).

Outlier: A score or data point which differs substantially from the other scores or data points. It is an extremely unusual or infrequent score or data point.

Output window: The window of computer software which displays the results of an analysis.

Over-identified model: A structural equation model in which the number of data points is greater than the number of parameters to be estimated, enabling the fit of the model to the data to be determined.

Paired comparisons: The process of comparing each variable mean with every (or most) other variable mean in pairs.

Parameter: A characteristic such as the mean or standard deviation which is based on the population of scores. In contrast, a statistic is a characteristic which is based on a sample of scores.

Parametric: To do with the characteristics of the population.

Parametric test: A statistical test which assumes that the scores used come from a population of scores which is normally distributed.

Part or semi-partial correlation: The correlation between a criterion and a predictor when the predictor's correlation with other predictors is partialled out.

Partial correlation: The correlation between a criterion and a predictor when the criterion's and the predictor's correlation with other predictors have been partialled out.

Participant: Someone who takes part in research. A more appropriate term than the archaic and misleading 'subject'.

PASW Statistics: The name for SPSS in 2008–9. PASW stands for Predictive Analytic Software.

Path diagram: A diagram in which the relationships (actual or hypothetical) between variables are presented.

Pathway: A line in a path diagram depicting a relationship between two variables.

Phi: A measure of association between two binomial or dichotomous variables.

Pivot table: A table in SPSS which can be edited.

Planned comparisons: Testing whether a difference between two groups is significant when there are strong grounds for expecting such a difference.

Point biserial correlation: A correlation between a score variable and a binomial (dichotomous) variable – i.e. one with two categories.

Population: All of the scores from which a sample is taken. It is erroneous in statistics to think of the population as people since it is the population of scores on a variable.

Post hoc **test:** A test to see whether two groups differ significantly when the researcher has no strong grounds for predicting or expecting that they will. Essentially they are unplanned tests which were not stipulated prior to the collection of data.

Power: In statistics the ability of a test to reject the null hypothesis when it is false.

Principal component analysis: Primarily a form of factor analysis in which the variance of each variable is set at the maximum value of 1 as no adjustment has been made for communalities. Probably best reserved for instances in which the correlation matrix tends to have high values which is not common in psychological research.

Probability distribution: The distribution of outcomes expected by chance.

Promax: A method of oblique rotation in factor analysis.

Quantitative research: Research which at the very least involves counting the frequency of categories in the main variable of interest.

Quartimax: A method of orthogonal rotation in factor analysis.

Randomisation: The assignment of cases to conditions using some method of assigning by chance.

Range: The difference between the largest and smallest score of a variable.

Ratio data: A measure for which it is possible to say that a score is a multiple of another score such as 20 being twice 10. Also there should be a zero point on the measure. This is a holy grail of statistical theory which psychologists will never find unless variables such as time and distance are considered.

Recode: Giving a value or set of values another value such as recoding age into ranges of age.

Regression coefficient: The weight which is applied to a predictor variable to give the value of the dependent variable.

Related design: A design in which participants provide data in more than one condition of the experiment. This is where participants serve as their own controls. More rarely, if samples are matched on a pairwise basis to be as similar as possible on a matching variable then this also constitutes a related design if the matching variable correlates with the dependent variable.

Related factorial design: A design in which there are two or more independent or predictor variables which have the same or matched cases in them.

Reliability: Internal reliability is the extent to which items which make up a scale or measure are internally consistent. It is usually calculated either using a form of split-half reliability in which the score for half the items is correlated with the score for the other half of the items (with an adjustment for the shortened length of the scale) or using Cronbach's alpha (which is the average of all possible split-half reliabilities). A distinct form of reliability is test–retest reliability which measures consistency over time.

Repeated-measures ANOVA: An analysis of variance which is based on one or more related factors having the same or similar cases in them.

Repeated-measures design: A design in which the groups of the independent variables have the same or matched cases in them.

Residual: The difference between an observed and expected score.

Residual sum of squares: The sum of squares that are left over after other sources of variance have been removed.

Rotation: *see* Rotation of factors.

Rotation of factors: This adjusts the factors (axes) of a factor analysis in order to make the factors more interpretable. To do so, the number of high and low factors loadings are maximised whereas the number of middle-sized factor loadings are made minimal. Originally it involved plotting the axes (factors) on graph paper and rotating them physically on the page, leaving the factor loadings in the same points on the graph paper. As a consequence, the factor loadings change since these have not moved but the axes have.

Sample: A selection or subset of scores on a variable. Samples cannot be guaranteed to be representative of the population but if they are selected at random then there will be no systematic difference between the samples and the population.

Sampling distribution: The theoretical distribution of a particular size of sample which would result if samples of that size were repeatedly taken from that population.

Saturated model: A model (set of variables) which fully accounts for the data. It is a concept used in log-linear analysis and structural equation modelling.

Scattergram: *see* Scatterplot.

Scatterplot: A diagram or chart which shows the relationship between two score variables. It consists of a horizontal and a vertical axis which are used to plot the scores of each individual on both variables.

Scheffé test: A *post hoc* test used in analysis of variance to test whether two group means differ significantly from each other.

Score statistic: A measure of association in logistic regression.

Scree test: A graph of the eigenvalues of successive factors in a factor analysis. It is used to help determine the 'significant' number of factors prior to rotation. The point at which the curve becomes flat and 'straight' determines the number of 'significant' factors.

Select cases: The name of an SPSS procedure for selecting subsamples of cases based on one or more criteria such as the gender of participants.

Sign test: A non-parametric test which determines whether the number of positive and negative differences between the scores in two conditions with the same or similar cases differ significantly.

Significance level: The probability level at and below which an outcome is assumed to be unlikely to be due to chance.

Simple regression: A test for describing the size and direction of the association between a predictor variable and a criterion variable.

Skew: A description given to a frequency distribution in which the scores tend to be in one tail of the distribution. In other words, it is a lop-sided frequency distribution compared with a normal (bell-shaped) curve.

Sort cases: The name of an SPSS procedure for ordering cases in the data file according to the values of one or more variables.

Spearman's correlation coefficient: A measure of the size and direction of the association between two variables rank ordered in size.

Sphericity: Similarity of the correlations between the dependent variable in the different conditions.

Split-half reliability: The correlation between the two halves of a scale adjusted for the number of variables in each scale.

SPSS: A computer program, now called IBM SPSS Statistics, which performs many important data analyses and statistical procedures. It can be regarded as the standard program worldwide.

Squared Euclidean distance: The sum of the squared differences between the scores on two variables for the sample.

Standard deviation: Conceptually the average amount by which the scores differ from the mean.

Standard error: Conceptually, the average amount by which the means of samples differ from the mean of the population.

Standard or direct entry: A form of multiple regression in which all of the predictor variables are entered into the analysis at the same time.

Standardised coefficients or weights: The coefficients or weights of the predictors in an equation are expressed in terms of their standardised scores.

Stepwise entry: A form of multiple regression in which variables are entered into the analysis one step at a time. In this way, the most predictive predictor is chosen first, then the second most predictive predictor is chosen second, having dealt with the variance due to the first predictor, and so forth.

Sum of squares: The total obtained by adding up the squared differences between each score and the mean of that set of scores. The 'average' of this is the variance.

Syntax: Statements or commands for carrying out various procedures in computer software.

Test–retest reliability: The correlation of a measure taken at one point in time with the same (or very similar) measure taken at a different point in time.

Transformation: Ways of adjusting the data to meet the requirements for the data for a particular statistical technique. For example, the data could be changed by taking the square root of each score, turning each score into a logarithm, and so forth. Trial and error may be required to find an appropriate transformation.

Two-tailed test: A test which assesses the statistical significance of a relationship or difference in either direction.

Type I error: Accepting the hypothesis when it is actually false.

Type II error: Rejecting the hypothesis when it is actually true.

Under-identified model: A structural equation model in which there are not enough data points to estimate its parameters.

Unique variance: Variance of a variable which is not shared with other variables in the analysis.

Univariate: Involving one variable.

Unplanned comparisons: Comparisons between groups which were not stipulated before the data were collected but after its collection.

Unstandardised coefficients or weights: The coefficients or weights which are applied to scores (as opposed to standardised scores).

Value label: The name or label given to the value of a variable such as 'Female' for '1'.

Variable label: The name or label given to a variable.

Variable name: The name of a variable.

Variable View: The window in SPSS 'Data Editor' which shows the names of variables and their specification.

Variance: The mean of the sum of the squared difference between each score and the mean of the set of scores. It constitutes a measure of the variability or dispersion of scores on a quantitative variable.

Variance ratio: The ratio between two variances, commonly referred to in ANOVA (analysis of variance).

Variance–covariance matrix: A matrix containing the variance of the variables and the covariance between pairs of variables.

Varimax: In factor analysis, a procedure for rotating the factors to simplify understanding of the factors which maintains the zero correlation between all of the factors.

Wald statistic: The ratio of the beta coefficient to its standard error. Used in logistic regression.

Weights: An adjustment made to reflect the size of a variable or sample.

Wilcoxon signed-rank test: A non-parametric test for assessing whether the scores from two samples that come from the same or similar cases differ significantly.

Wilks' lambda: A measure, involving the ratio of the within-groups to the total sum of squares, used to determine if the means of variables differ significantly across groups.

Within-subjects design: A correlated or repeated-measures design.

Yates's continuity correction: An outmoded adjustment to a 2×2 chi-square test held to improve the fit of the test to the chi-square distribution.

z-score: A score expressed as the number of standard deviations a score is from the mean of the set of scores.

INDEX

Note: Entries in **bold** refer to terms described in the Glossary. Page numbers in **red** refer to Chapters 53–56 which are available on the website at www.pearsoned.co.uk/howitt